# Encompassing a Fractal World

# Encompassing a Fractal World

## The Energetic Female Core in Myth and Everyday Life— A Few Lessons Drawn from the Nepalese Himalaya

Gil Daryn

LEXINGTON BOOKS

A division of
ROWMAN & LITTLEFIELD PUBLISHERS, INC.
*Lanham • Boulder • New York • Toronto • Oxford*

LEXINGTON BOOKS

A division of Rowman & Littlefield Publishers, Inc.
A wholly owned subsidiary of The Rowman & Littlefield Publishing Group, Inc.
4501 Forbes Boulevard, Suite 200
Lanham, MD 20706

PO Box 317
Oxford
OX2 9RU, UK



British Library Cataloguing in Publication Information Available

**Library of Congress Cataloging-in-Publication Data**

Daryn, Gil, 1963–
Encompassing a fractal world : the energetic female core in myth and everyday life : a few lessons drawn from the Nepalese Himalaya / Gil Daryn.
p. cm.
Includes bibliographical references and index.
ISBN-13: 978-0-7391-1173-4 (cloth : alk. paper)
ISBN-10: 0-7391-1173-6 (cloth : alk. paper)
1. Hinduism—Nepal—Thamghar. 2. Brahmans—Nepal—Thamghar—Social life and customs. 3. Hindu cosmology. 4. Rice planting rites—Nepal—Thamghar. I. Title.
BL1156.8.T53D37 2006
306.6'945095496—dc22 2006002538

Printed in the United States of America

∞™ The paper used in this publication meets the minimum requirements of American National Standard for Information Sciences—Permanence of Paper for Printed Library Materials, ANSI/NISO Z39.48-1992.

Dedicated to the memory of my parents, Nechama and Dr. Ely Daryn, and to the three women in my life, Noa, Adi and Shira, for their encompassing maya (ahava).

# Contents

List of Figures xi

List of Photographs xiii

A Number of Preliminary Notes xvii

Acknowledgments xix

**1** An Overview 1
- Introduction 1
- From Geometrical Monstrosities to Hindu Cosmology 2
- General Outline of the Book 8
- A High-Caste Village in Central Nepal 12
- Unpredictability and Fieldwork in Thamghar 16
- Notes 21

**Part I: Man: Understanding Mistrust**

**2** Himalayan Uncertainties 27
- Introduction 27
- The Consequence of Misappropriation 30
- House, Rice, and Self 40
- A Number of Concluding Remarks 46
- Notes 49

**3** Encompassing the Ambivalent Female Core 53
Introduction 53
Women in Flux 56
Multiple Perspectives and Perceptual Resolution 67
Matrimonial Encompassment 71
Endnote: From the Dialectics of Images to a Vicious Cycle 77
Notes 82

**Part II: Cosmos: From Image to Cosmology**

**4** On Hindu Divine Sexuality and Marriage 91
Introduction 91
Ascetic Divine Sex 92
Procreating Immortality 96
Birth into Agony or "The Nemesis of Reproduction" 102
Conclusion 106
Notes 106

**5** The *Jagya*—Ethnography 111
Introduction 111
*Jagya Lāunu*—The Construction of the *Jagya* 117
*Hom* 128
Final Rites 132
Notes 135

**6** The *Jagya*—an Analysis 141
Introduction 141
Initial Performances 147
Orchestrating a Micro-Macrocosmic Matrimony 151
Incorporating Divine *Śakti* 166
On Ritual Efficacy 170
A Concluding Remark 177
Notes 178

**Part III: Rice: From Must to Trust—an Ontogeny of Rice**

**7** An Agricultural *Jagya* 185
Introduction 185
Fashioning a *Jagya* of Rice 194
The *Kunyā*: A Fireless *Hom* in Bhūme's Procreative Temple of Rice 201

Another Encompassment 209
From the Vedic Fire Altar to the Hindu Temple via Vegetal Connections 213
Endnote 216
Notes 217

**8** The "World Upside-Down" or a Himalayan Inversion of Hierarchy and Trust 221
Introduction 221
*Dāī Hālnu* 222
Analysis 228
Playing with Images: Illusion of Demons and Trust 233
Conclusion 235
Notes 236

**9** Conclusion 237
Notes 243

Appendix A: A Personal Note 245
Finding a Home in the Himalayas 245
Note 249

Appendix B: Glossary 251

Bibliography 263

Index 283

About the Author 297

# List of Figures


| | | |
|---|---|---|
| Figure 1.1 | Mandelbrot's fractal images | 3 |
| Figure 5.1 | A simple *jagya* | 119 |
| Figure 5.2 | An elaborated *jagya* | 120 |
| Figure 5.3 | A *rekhi* | 126 |
| Figure 6.1 | Sorcery's reversible causal nexus of agency according to Gell | 174 |
| Figure 6.2 | The *jagya*'s causal nexus | 176 |
| Figure 7.1 | Annual cycle of main village crops | 187 |
| Figure 8.1 | Hierarchy of wealth | 231 |

# List of Photographs


1.1 Mahaboudha temple in Patan, the Katmandu valley. 7
1.2 Thamghar—general view of the village's upper part. 13
1.3 Sharmananda Subedi's house. 18
5.1 A *purān*. 114
5.2 An eighty-four-year-old *kartā*, his wife, and other family members offer grains to the *kalas* (representing Viṣṇu), which is situated on the *bedi* at the center of a (simple) *jagya*, during a *purān* performed before the *chaurāsī godan*. The white mud pedestal to the right, where the *tulasi* (here missing) is normally planted, forms one of the *jagya*'s corners. 115
5.3 A simple *jagya* viewed from the east. Early in the morning, the officiating priest (left) explains the procedures of the *jagya* to the *kartā* (right). At the center of the *jagya*, behind one of the *liṅgas*, is the *bedi*. In front of each *liṅga* a special *rekhi* for its *kalas* is already prepared. 119
5.4 A *kartā* and his wife (touching his leg) during *saṅkalpa*. 130
5.5 A *rekhi* with the offering placed at its edges and the *samidhā* piled in the center, forming a triangular shape at the top. As the *hom* is performed within the house for Bastu *pūjā*, no *bedi* is constructed and the *samidhā* are piled upon some soil taken from the *khāldo*. 131
5.6 *Snān* at the end of a *jagya*. 133
5.7 Eknath Lamichane's two younger sisters with the large Dasaī *achetā ṭikā* on their foreheads. An identical but smaller *ṭikā* is given in ordinary *pūjās*. 134

6.1 The *khāldo* in the center of Bastu *pūjā*'s inner *jagya*. A *kalas* can be seen under each *liṅga*. 160
6.2 Bringing Bastu *liṅga* from the forest just prior to the performance of *parsinu*. 161
6.3 The *kartā*, helped by his priest (in white), placing Bastu *liṅga* in the *khāldo* at the center of the inner *jagya*, alongside one of the house's (black) beams. The colorful cloth seen above the *kartā*'s head is the *canduwā*, which is tied between the *jagya*'s *liṅgas*. Various *naibbede* are usually placed on the *canduwā* and distributed as *prasād* at the end of the *jagya*. 162
7.1 Ripened rice. 186
7.2 A Bhūme tree-temple situated just above a rice field. 190
7.3 Amrit Bhandari plows the flooded terraces prior to transplantation. 193
7.4 *Byāṛ*—rice seedling nursery, three weeks after *biu rākhnu*. 196
7.5 A young *buhāri* during transplantation. The picture was taken on a rainy day, hence her head is covered with a plastic sheet. 197
7.6 Flooded rice-fields during transplantation. Closer, three green terraces of *byāṛs* can be seen among others in various stages of preparations prior to transplantation. 198
7.7 *Bhakāri bādhnu*—Muktinath Adhikari with a mud *ṭikā* and seedlings (flowers) behind his ears. 199
7.8 Rice fields a few weeks prior to the harvest. 202
7.9 The *buṛi biṭo*. 204
7.10 During the construction of a *kunyā*, the *buṛi biṭo* is almost totally covered by the surrounding rice-sheaves. 205
7.11 A field-owner's wife prepares the *juro* and adorns it with red leaves and flowers. On the right is the wall of the *kunyā* and a sack full of threshed rice grains that will be used as the following year's seeds. 206
7.12 Threshing an impure type of rice and the seeds for the next season before the *kunyā* is completed. 207
7.13 Sharmananda Subedi prepares *nāg pūjā* near the *kunyā* prior to *dāī hālnu*. Both the *nāg* and *nāgini* (made of dough) are placed on a banana leaf with various offerings in front of them. 208
7.14 A *kunyā* in the morning fog during early November. 209

8.1 *Dāī hālnu*—the men stand in a circle and thresh the rice by beating the sheaves on the ground. Behind them, on the left (with a man standing on top) is the half-threshed *kunyā*. 223
8.2 Children spur the oxen walking around the *miyo* (right). 224
8.3 Retiring to sleep in the straw around the pile of rice. 225
8.4 The morning after *dāī hālnu*: carrying the straw to the village in the morning mist. 226
8.5 A pile of rice (with the *miyo* behind it) at the center of the *khalo garo*. 227
8.6 Pouring the rice into a *bhakāri* on the first floor. 228

# A Number of Preliminary Notes

## PSEUDONYMS

Thamghar is a pseudonym and the names of all the individuals referred to in this book have been changed.

## TRANSLITERATION

All foreign words have been italicized. In transliterating Nepali or Sanskrit words I followed the standard system of R. L. Turner's *Comparative and Etymological Dictionary of the Nepali Language* (1996), adding local pronunciations in brackets. For words not mentioned by Turner, I have given my own approximation according to local pronunciation. Personal names are given without diacritics and written as they would be in ordinary English in a phonetic manner, such as Shiva and not Śiva.

Well-known Sanskrit terms used in Nepali are given according to the standard form, such as, *pūjā* and not *puja*, and I have not italicized terms, particularly names that are extremely common in the relevant literature or in works on high-caste Nepalese. Nepali words are pluralized as in English, by adding an "s". All terms are explained in the text when they first appear, and if mentioned more than once are also in the glossary, but for the benefit of the non-Indologist, I have added more explanations and translations when this seems appropriate.

# Acknowledgments

I have accumulated many debts in the course of the research for and writing of this book. My prolonged fieldwork would not have been possible without the assistance and kindness of many people, of whom I can mention only a few. Above all, I would like to thank the people of the village of Thamghar, who literally made my research possible, and to whom I remain forever grateful. Their hospitality, goodwill, extraordinary generosity, and willingness to open their hearts and share with me their rich knowledge and the intricacies of their life are still vivid in my mind. In particular, I would like to thank my two wonderful host families who adopted me as one of their own and went out of their way to make my stay with them a memorable experience, assisting me in every possible manner. One of these was the family of my excellent research assistant whose help, unfailing good humor, and personal friendship were crucial to the success of my fieldwork.

Warm thanks must also go to Mrs. Sudha Joshi and her wonderful extended family who provided me with a home away from home while in Katmandu. I would also like to pay tribute to Mr. Mahesh Shrestha, Mr. Madhu Chitrakar, Mrs. Prabha Kaini, and their families for their generous hospitality, and to Dr. Tika Pokharil for his assistance. In particular, I would like to mention the late Professor Gopal Singh Nepali who generously gave of his time and thoughtful advice during fieldwork. I would also like to thank Mr. Kilnath Bastakoti and Mr. Ramesh Pandey (the Royal Nepalese priest) for the knowledge they so kindly shared with me.

Thanks must go to Tribhuvan University's Centre of Nepalese and Asian Studies (CNAS) and the Department of Social Anthropology, to which I was affiliated during much of the research, for their generous hospitality. I am deeply indebted to the staff of the Israeli Embassy in Katmandu for their extraordinary cooperation and assistance, and in particular I would like to mention the Ambassador Mrs. Esther Smilag-Efrat, whose kindness and goodwill surpassed all my expectations and whose help at times was indeed crucial.

A number of people have read and made particularly valuable comments on earlier drafts of this book. I wish to thank Professor Johnny Parry and Dr. James Laidlaw for their careful comments and thoughtful criticism. For their important comments on various parts of this text, I would also like to thank Professor Ted Riccardi, Professor Lawrence A. Babb, Professor Dame Marilyn Strathern, Professor Bill Watson, Professor McKim Marriott, Dr. Marta Levitt, Dr. Susan Bayly and Dr. Jaro Stacul. Special thanks are due to Professor Oz Almog who introduced me to the fractal concept. Many thanks also go to Mrs. Marilyn Lehrer for her encouragement, constructive advice, and friendship, to Dr. Ada Gansach for crucial practical suggestions during fieldwork, to Mr. Jeremy Vine for his computer support, to Mr. Ithamar Theodor for his assistance with Sanskrit, and to Mr. Udaya Neopane for his excellent assistance with Nepali.

With special gratitude I would like to mention the financial support of the following bodies and individuals, whose generous help was vital for the completion of my research: Trinity Hall Research Bursary (Cambridge), Fredrick Williamson Memorial Fund (Cambridge), Worts Traveling Scholars Fund (Cambridge), Richards Fund (Cambridge), The Wyse Fund (Cambridge), H. M. Chadwick Fund (Cambridge), The Friends of the Hebrew University in Jerusalem (London), Mrs. B. Curty, Mr. C. Morris, and Mr. I. Karten. This manuscript was completed during my British Academy Postdoctoral Fellowship, at the Department of Anthropology, the School of Oriental and African Studies, University of London. I would like to thank SOAS for its generous assistance toward publication of this book.

Throughout, my work was profoundly inspired by my late father, Dr. Ely Daryn and by Mr. Yigal Tepper, whose inquisitive minds, courageous research attitude, and lasting fondness for challenging conventional wisdom and accepted "axioms" greatly fashioned my own intellectual journey.

Finally, this book may have never seen the light of day without the constant love, support, and assistance of my wife, who meticulously read, commented, and edited the text numerous times. Yet it would probably never have even gone as far as being registered on a computer file were it not for the birth of our first daughter, Adi, who brought with her rays of happiness, which were able to illuminate much of what, at the time, seemed a rather incomprehensible ethnographic picture.

# 1

# An Overview

## INTRODUCTION

Benoit Mandelbrot's fractal concept, first introduced in the 1970s,[1] has gained considerable momentum over the years and is today considered to be the general modality in which chaos "organizes" itself in nature (Bak 1997). Although Mandelbrot himself implied and demonstrated the applicability of the fractal concept to various other realms of life[2] it has so far made, and unjustly so, only a relatively modest impression on the humanities and social sciences and has only recently begun gaining ground in these fields of knowledge.[3]

This book interweaves the fractal concept with the anthropology of Hinduism and Indology. More specifically, the fractal notion is invoked in order to explore the life of present-day high-caste Nepalese Hindus while shedding new light on a number of cardinal issues and rather perplexing themes within the broader study of ancient and contemporary Hindu culture and society. This book attempts to accomplish this through a gradual examination and analysis of the meaning, significance, and far-reaching personal and social implications of the particular fractalic marriage imagery found among Nepalese Brāhmaṇs while drawing upon an extensive set of comparative material. The latter ranges from Vedic cosmogonies and sacrifice, through Puranic mythology, to contemporary ethnographic accounts from Nepal and India. In a nutshell, the Nepalese Brāhmaṇs studied herein view the conjugal bond as an ideal state where the female is encompassed within her husband's body—what I prefer to term "matrimonial encompassment." The book probes the deep involvement of this imagery in the manner in which interpersonal relationships, the kinship milieu, ritual worlds, and rice cultivation are drawn and orchestrated together, as well as the paths of its manipulation, in various forms, for achieving a range of goals and eliciting particular effects. It thus opens up fruitful paths for investigating questions of immense import both ethnographically and theoretically and, among others, it examines the origins, significance, and operation of similar imagery within the larger historical body of Hindu thought and religious praxis.

This book's main thrust is not that the fractal concept may neatly bring together much of what has already been written about Hindu culture, which may seem obvious and is further elaborated upon below, but instead it argues the case for the aforementioned *gendered* fractal dimension, as implied by a glimpse into the life of the Brāhmaṇs living in the remote Nepalese village of Thamghar, as well as ancient Hindu myth, ritual, and temple architecture. In effect, demonstrating how this gendered fractal dimension implicitly permeates Hindu modes of life and imagination, the book illustrates the way in which it should be employed as the embodiment of a powerful analytical tool—a *key notion* for understanding some of the manners by which many past and contemporary Hindus alike organize their basic perceptions of cosmos and self. This interdisciplinary study, offering a comparative examination of Hindu life through the new and important angle of the gendered fractal concept it introduces, attempts to "join up the dots" and move beyond isolated local village-based studies in order to bridge the gulf between anthropology and Hindu studies.

As may already be apparent and as will become evident as the chapters unfold, the fractal concept embodies not only the most general framework in which I wish to locate the present study but also serves as a central metaphor, which is able to closely convey the complexities inherent in the local perceptions of the high-caste Nepalese studied, as well as ancient and contemporary Hindus elsewhere more generally. I will thus begin with a brief note about this concept, its wide-ranging cultural applicability, its relevance to Hindu studies, and the manner in which it will be employed herein.

## FROM GEOMETRICAL MONSTROSITIES TO HINDU COSMOLOGY

The term fractal[4] is primarily used to denote a family of seemingly complex self-similar shapes possessing some highly unusual properties, formed by the reiteration of very simple rules, in which the whole and its parts are identical in all but scale. One may regard *self-similarity* across scale as the most salient characteristic of a fractal reality, object, or image. Self-similarity may be simply defined as symmetry at every scale and implies the recursion of pattern inside pattern (Gleick 1987: 103). At times it may be rather abstract and counterintuitive as in the degree of irregularity found on each scale of manifestation

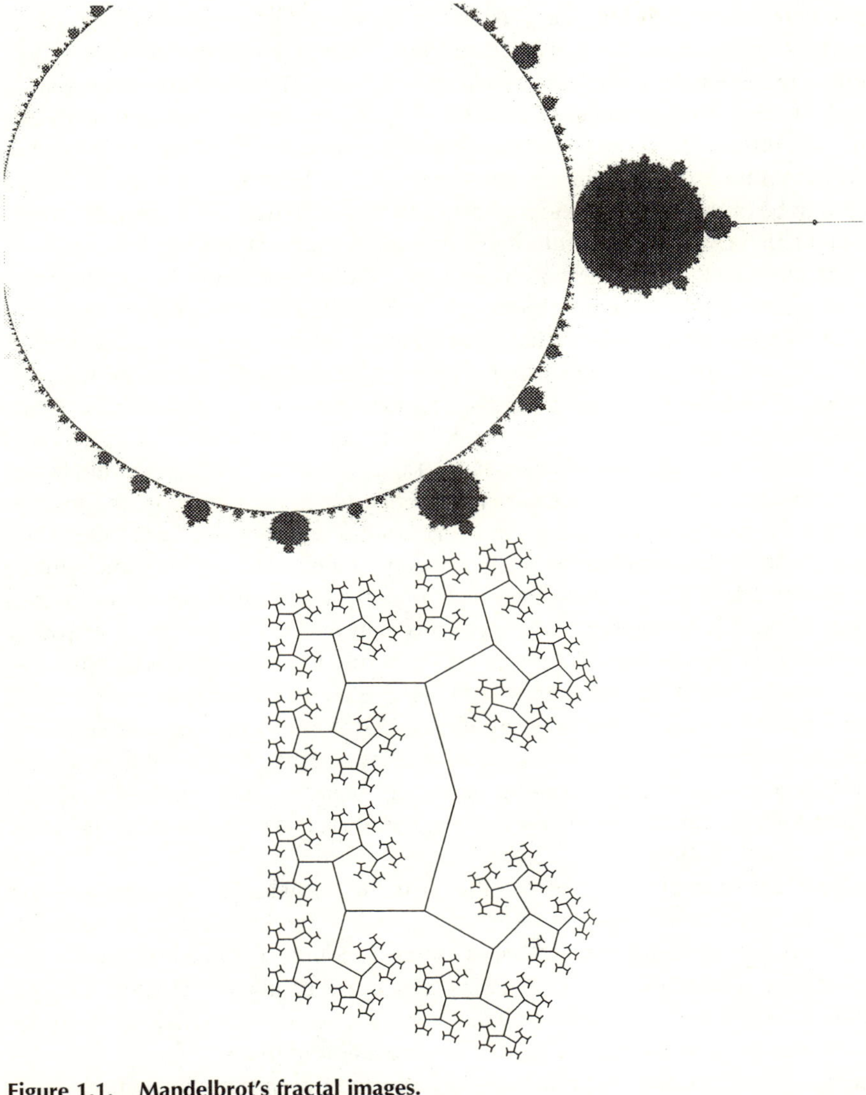

**Figure 1.1. Mandelbrot's fractal images.**
Courtesy of Benoit B. Mandelbrot, *The Fractal Geometry of Nature* (San Francisco, CA: W. H. Freeman and Company, 1982), 155, 189.

of the coast of Britain, to use Mandelbrot's well-known example; regardless of scale, be it a single pebble on the beach, a large rock, an aerial photo of part of the coastline or an atlas map—all will exhibit the same degree of inherent roughness.[5] Often, however, self-similarity is easily recognizable and visually apparent in a salient manner, as in Mandelbrot's famous complex fractal geometrical contrivances (see figure 1.1),[6] where the whole and each of its parts, situated on a set of ever decreasing scales, possess an identical shape, creating increasing complexity. Mandelbrot's (1982) basic argument is that in order to capture the essence of fractalic shapes, ordinary Euclidean dimensions do not suffice. As a result, he introduces the concept of a fractal dimension, which unlike ordinary dimensions such as 0, 1, 2, or 3 is *fractional*, that is, cannot be expressed in whole numbers.[7] In this way, fractal shapes may be defined and the degree of regularity of irregular objects is made calculable.

Mathematics aside, the notion of a fractal dimension will be employed at present for defining and marking various fractals. This will be done rather intuitively and will mainly be based on visually apparent or imagined similarity. For example, the fractal dimension of a Russian doll would be the self-similar shape being reiterated on the various scales of its figurines nested one within the other. Likewise, the fractal dimension of the carved fractal god that Gell (1998: 138) provides is the shape of the god itself. One further and rather intuitive property of fractals needs to be mentioned here, namely, their rather subjective and relative aspect; for a single observer, the details of a fractal object (or its "effective dimensions")[8] depend on the relation of the object to the observer, that is, the scale of the object observed and the distance of observation. Hence, for example, the degree of detail exhibited by a map obviously depends on the scale to which it is drawn, while the manner in which the distance of observation determines both the details and apparent dimensions of an object can be demonstrated by imagining a ball of thread observed from different distances as follows. From far away, the ball will appear as a mere dot of zero dimensions. Coming closer, an observer will easily identify its three dimensions but, from very close, one thread will come into view, and thus the ball will appear unidimensional.[9]

Fractals are perhaps "good to think with" to paraphrase Levi-Strauss—not only for modern scientists but for many people elsewhere and have been that way from time immemorial. This is clearly evident in studies of various ancient and contemporary societies, such as those of ancient Greece, Europe, Polynesia, Melanesia, Africa, native America, and India.[10] The idea of a fractal seems to lie at the root of the notion, prevalent in many "premodern" societies, of a microcosm that represents, is affected by, and provides the medium for operation on the macrocosm.[11] Thus in effect, all imaginings of the human body, the house, temple, ritual arena, village, or city as analogous microcosms appear to involve the implicit assumption of a fractal cosmology.[12] In fact, as recently demonstrated by Jackson (2004), fractals seem to be almost omnipresent the world over, in ancient and contemporary cultures alike, and in almost every field of human endeavor, and have clearly been central to humanity long before being recognized in mathematics and science.

As will become apparent below, the fractal notion is particularly prevalent and highly relevant to Hindu thought, myth, and practice,[13] and fractality is implied in and is often fundamental to the way many Hindus view their world. As highlighted by the following examples, Hindus not only seem to imagine fractals but also tend to literally construct them, and although not using the term itself, they employ this concept to organize their basic perceptions of cosmos and self. A case in point is the pan-Indian *Upaniṣadic* notion of the identity of the Brāhmaṇ, the all-encompassing, transcendent, cosmic "soul" with its fragmentary parts, each of which is immanent in every living entity in the form of the personal *ātman* (soul). Put differently, the Brāhmaṇ and its individual fragments are self-similar, thus evoking a "spiritual" fractal. This inherent fractality seems in turn to embody the cornerstone of the Hindu theory of transmigration as well as the concomitant quest for "liberation."[14]

The fractal notion is also implicit in ancient Hindu texts; according to Daniélou (1964: 43), the *Ṛg Veda* declares that "man is the fragmentary-universe . . . or microcosm" and thus they "appear as two parallel beings similar to one another":[15]

> Two birds, beautiful of wing, inseparable friends, dwell together on the same tree (the universe). One of them (the individual being) eats the fruit [of action], the other (the individual Being) looks on but does not eat (original notes).[16]

A number of other, frequently discussed, prominent Hindu notions that imply fractality merit a brief mention. These include the ideas of a kingdom as a microcosm of the world and of human bodies as temples,[17] of the human being as microcosm seen as the "city of Brahman" which holds the entire cosmic space in its hidden heart,[18] the notion that "all divinities have their abode in man," and the perception of the human body as an embodiment of all the cosmic creative principles and potencies.[19] Likewise, the view of temples as bodies[20] or as the kingdom in microcosm,[21] and the perception of temples and cities as small-scale manifestations of the kingdom or universe[22] all suggest a perception of a fractal cosmology. In fact, all apparent paradoxes such as the unity of body and cosmos,[23] the relativity of "inside" and "outside"[24] or the simultaneity of the Hindu gods' transcendence and immanence[25] evoke and may also be resolved by the notion of a fractalic reality. In a similar vein, White (2000) mentions a number of prominent patterns of fractalic imagery within the corpus of ideas that comes under the title of Tantra.

Likewise, though I seriously doubt whether Dumont (1980) would have embraced the fractal concept, since it does not seem to concur with his notion of hierarchy as the encompassment of the *contrary* and particularly since fractality implies the encompassment of the *similar*,[26] I believe that the Indian caste system, as Dumont (1980) himself describes it and, for example, as saliently emerges from Parry's (1979) or Moffatt's (1975) excellent studies, represents a classic example of a social fractal. This is evinced by the caste system's ability to replicate itself in its various sections, levels of purity, or magnitudes of population. This also applies, though to a lesser extent, to the rather "broken" Nepalese "caste system" as implied, for example, in Macdonald (1975) and Höfer (1979) and (as I shall discuss later) is also evident in Thamghar.

Hindu mythology provides many examples of how Hindus describe the world and imagine it using the fractal concept. One clear example of fractalic self-similarity is evinced by the myth of Kṛṣṇa's mother, who breastfeeds him and later sees the entire world he has swallowed, including her milk and herself, within his mouth (O'Flaherty 1980a: 268).[27] Indeed, the fundamental Hindu perception of divinity involves fractalic imagery; the notion that divinity pervades everything, that it is in essence self-similar, that *īśwar* (god) is omnipresent and that the entire world is to be found within the great gods, certain ritual paraphernalia, and so on. Similarly, the idea that the gods and goddesses are or may be united in the form of a single deity and that Śiva is not merely represented by, but also resides or is embodied in his *liṅgam*, to mention a few frequently referred-to examples from Thamghar, are all based on an implicit notion of fractal divinity.

Hindu architecture and predominantly the structure of ancient Indian cities[28] and temples[29] provide perhaps the most striking visual examples of three-dimensional fractals that could well replace Mandelbrot's computer-generated geometrical contrivances. This salient aspect of Hindu architecture, which does not appear to have been sufficiently heeded in the past,[30] is slowly gaining recognition among contemporary scholars.[31] One notable Nepalese example is the Buddhist shrine of Mahaboudha of Patan, in the Katmandu valley, completed in 1600 A.D. This temple is a replica of the Mahabodhi temple of Bodhgaya which dates to the fifth to sixth century A.D. It is built in the Hindu Śikara style and features various architectonic elements and numerous Buddha clay figurines that exhibit self-similarity across their scales.

More abstractly, Marriott's ethno-sociological approach,[32] a highly significant and influential interpretive perspective in the anthropology of Hinduism, shares the present work's emphasis upon iterative, self-similar relationships on various scales. Viewing Hindus and Hinduism via six analytical components of popular Indian biology and physics, it thus implies an inherent fractality dominating the Hindu world.[33]

In an attempt to capture Hindu perceptions of reality, where no clear-cut boundaries exist between cosmological levels, a whole and its parts evince

*Mahaboudha temple in Patan, the Katmandu valley.*

spatial continuity and shared qualities, and the former is imagined to be found within each of the latter in the form of a miniature condensed image, a number of scholars such as Trawick-Egnor (1978: passim and particularly 179–81),[34] Ramanujan (1989a, 1989b)[35] and Handelman and Shulman (1997: 193–98),[36] employ the metaphor of a hologram.

Admittedly, little appears to differentiate between the notions of holography and fractality, indeed, the terms have been employed almost interchangeably (e.g., Wagner 1991). Nonetheless, in my view, the fractal concept seems more appropriate for a description and discussion of Hindu perception and thought, primarily since the equation of part and whole in a hologram appears only *after* the latter is cut into fragments, while the notion of fractality conveys self-similarity within an *undivided entirety*. In addition, unlike holography, a fractal often implies a state of recursive encompassment and this is, to my mind, highly pertinent both to general Hindu intuitions and understandings of reality and particularly to the local notions prevalent in the Brāhmaṇ village, which form the ethnographic basis for the present book.

The above brief discussion, although far from exhaustive in this regard, makes it apparent how the fractal concept is ingrained within Hindu thought and practice, infusing almost all spheres of life and imagery. As already mentioned, this book makes the case for an additional *fractal dimension* as implied by the way Thamgharian Brāhmaṇs perceive human marriage, ritual life, cosmology, the self, and the optimal modes of action and creation in the world. This fractal, unlike other fractalic aspects of the Hindu world already mentioned above is unique in being saliently *gendered*.

The significance and implications of the image of a male encompassing a female via marriage in Thamghar touch upon a number of fundamental aspects that are related to the manner in which people constitute themselves and construe their society and the world around them. In effect, it embodies the template for the ideal modality of existence in the Thamgharian cosmos, providing a highly powerful mental meta-frame that serves as a vehicle for reflection within which much experience, action, and interaction gain their form and meaning and are translated into cognition and emotion. One of the crucial arguments of the present work is that this gendered fractalic image is not merely a unique local notion representing one of the principle building blocks of local perceptions of self, the other, and the cosmos, as may ostensibly be the case, but that it embodies a previously neglected Hindu fractal dimension that is also implicit and crucial for our understanding of Hindu perceptions, life, and society elsewhere. The remainder of the present chapter will present the structure of the book, followed by a short introduction to Thamghar and its people, and will conclude with a brief note about the fieldwork conducted.

## GENERAL OUTLINE OF THE BOOK

The book is divided into three parts according to its three cardinal areas of inquiry and levels of analysis which follow the main realms of life where fractalic thinking and imagery appear to play a decisive role in the life of Thamgharian Brāhmaṇs, namely Man, Cosmos, and Rice.

Part I—Man, commences with chapter 2, which provides an evocative description of a rural Brāhmaṇ world consisting of a closed social system, located in the midst of an inhospitable universe, embattled from both within and without. Analyzing the dynamics of social relationships, sorcery, and mistrust, this chapter attempts to understand the local self-perpetuating vicious cycle of mistrust and hostility, thus adding color and depth to the picture portrayed in the brief ethnographic note provided at the end of the present chapter.

Interpretations of high-caste Nepalese kinship usually argue that while the sister is perceived to be sacred, the wife is dangerous and ambivalent. Chapter 3 argues that this structuralist-oriented view is too simplistic, demonstrating how Brāhmaṇ women are fundamentally ambivalent in their various kinship roles *throughout* their lives. This chapter provides an innovative reinterpretation of the major life stages of Hindu women, the ambivalent "female core" of high-caste Nepalese kinship. It further sets the scene for introducing Thamgharian marriage imagery, which is governed by the image of the wife being encompassed within her husband's body (matrimonial encompassment). Why do women need to be encompassed? What is the significance and implications of this imagery? An incisive analysis of fundamental local notions and prominent mental images, which seem to appear vividly in Thamgharians' minds and dominate their perceptions of personhood, social relationships, body, house, marriage, kinship, sorcery, and cosmology, casts new light on Brāhmaṇ social and kinship milieus. Furthermore, it is suggested that it is the dialectics of these notions and images that, in part at least, lie behind the ongoing intensification of the village's vicious cycle of mistrust and hostility. The chapter ends with a comparative analysis of the background, workings, and implications of various other ways in which Hindus conceptualize the conjugal bond elsewhere. In particular, the juxtaposition with southern India will demonstrate how a similar mental image may be imbued with a very distinct significance when placed within a somewhat different, albeit Hindu, cultural context.

Chapter 4, which opens Part II—Cosmos, attempts to locate the Thamgharian image of matrimonial encompassment within a broader metaphysical context. This is achieved through an analytical review of the prominent landmarks in the development of sexual imagery and the major fluctuations experienced by the notion of divine marriage in Hindu thought and practice. These are traced from Vedic cosmogonies, the fire and other ancient sacrifices, through Puranic mythology, ending with contemporary Hindu myth and ritual.

Against this theoretical backdrop, chapter 5 reflects upon the way in which Thamgharians present the world they inhabit to themselves. This presentation is dominated by the notion of a gendered fractal of a male encompassing a female, as expressed by the complex way in which they establish and perform a village *jagya* (an elaborated ritual enclosure and the complex ritual performed therein) including the performance of *hom*—the climactic fire ceremony. This is the most dramatic and picturesque of all local rituals, paradigmatic of Brāhmaṇ elaborated *pūjās* and *saṁskāras*. As a prerequisite for the discussion and analysis of the *jagya*, this chapter commences with a general introduction to the concept of *jagya* followed by an analysis of the nature of Hindu "symbols" and highlights the shortcomings of previous attempts to conceptualize these. Thus, it introduces the notion of "embodied representation," which not only faithfully conveys Thamgharian local notions but proves applicable for other Hindu contexts and promises to be highly pertinent for the analysis of ritual beyond the borders of south Asia.

Chapter 6 ends Part II with an in-depth study of the *jagya* and *hom*, providing an insight into the complex way in which, through the manipulation of an intricate web of identifications and analogies between micro- and macrocosms at work, the superimposition of the kinship notion of matrimonial encompassment on the cosmos within the ritual realm sets into motion and effects the regeneration of its participants through gestation and rebirth. Most anthropological and Indological theoretical engagements with the fractal concept[37] employ it as a useful metaphor for primarily descriptive purposes, as part of the discussion of complex indigenous perceptions, material culture, and architecture, and explain its appearance mainly in terms of a rather arbitrary aesthetic or symbolic preference. The crucial questions *why* many people in traditional societies, and Hindus in particular, find it appealing to think with, create, and build fractals, and *what* are the deep significance, role, and imports of this concept, remain largely unanswered. In a similar vein Hindu rituals have long been an anthropological staple, yet little attention has so far been paid to the nature and principles behind their efficacy. Many scholars mention the mutual influence between similar micro- and macrocosms, and the identification and, at times, merger of the *kartā* (the main actor and subject of a ritual operation) with both the microcosmic enclosure he fashions and with one or several of the major Hindu gods. What all previous scholarly discussions fail to explain is *why* this should form a conduit for mutual influence and effect. Drawing upon Gell's (1998) interpretation of image sorcery, chapter 6 analyzes the Thamgharian *jagya* and *hom* thus offering an innovative and convincing approach for understanding the above queries. In addition, it also engages in a comprehensive comparison with a range of similar Vedic, medieval, and contemporary Hindu rituals, all of which are characterized by the mutual manipulation of micro- and macrocosms, thus providing an insight into the complex trajectories of agency and efficacy that dominate them, shedding new light on the striving for personal and general alignment with the cosmos in Hindu life. In this way, this chapter demonstrates how cosmic fractality is the backbone of the Hindu world's causality, efficacy, and interconnectivity and considerably enhances our understanding of general ritual efficacy elsewhere. In effect this chapter exposes why, from a Hindu point of view, the cosmos *must* be organized as a fractal. By and large, studies of Hindu culture and society pay little attention to the significance of gender within the organic Hindu universe and overlook the manner in which nonhuman entities (objects) exercise (mediate) social agency, particularly within the ritual context. This chapter provides new insights by focusing on local gender sensitivities and the import and implication of gender attribution to nonhuman entities. It does so by directly tackling the intersubjective relationships and the sexual imports of action and interaction between and with them.

Anthropologists have long emphasized the crucial role of rice, the Nepalese staple, in the social and ritual life of high-castes, and its symbolic meaning has been the subject of fierce debate. However, the complex, prolonged cultivation process, the significance of the grain's gender, and the manner in which it mediates and exerts agency throughout its life trajectory have been ignored. Chapter 7 which opens Part III—Rice, analyzes rice cultivation as an agricultural *jagya*, designed to harness the divine cosmic forces for the mass production of virginal rice, providing a key example of the employment of matrimonial encompassment as a template for creation and production. A detailed analysis of the notion of the gendered rice grain (comprising a male husk and female grain), throughout its life trajectory from its monsoon sowing to the cooking pot or use as an auspicious forehead mark (*ṭikā*) at the end of a *pūjā*, casts new light on some of the prominent roles played by rice, not only in Thamghar but possibly also elsewhere in Nepal and south Asia. The new ethnography presented in this chapter and the gender-sensitive analysis of both the village *jagya* and the *kunyā*, its three-dimensional agricultural projection (taken as the earth goddess, Bhume's, ripened temple of rice) established during the harvest, which represents the climax of Thamgharian ritual technology, greatly illuminate the architectural development of the north Indian Śikara Hindu temple, providing an additional, gendered perspective for its comprehension, which has a crucial implication for our understanding of Hindu architecture more generally.

Chapter 8, which concludes Part III, brings us back to our point of departure: the Thamgharian social universe. How are the mutual animosity and mistrust kept in check without spiraling toward complete collapse and fragmentation of the Thamgharian social world? The answer appears to come from a most unexpected direction. Agriculture is usually treated as a separate domain of human activity, a major source of subsistence, and a fundamental economic pursuit. This view is too narrow and fails to appreciate the holistic nature and inherent interconnectedness of the different realms of the Hindu fractalic world, life, and experience. Chapter 7 demonstrates the way in which rice agriculture is intimately interwoven with cosmology and ritual and how it cannot be disentangled from them for analytical purposes. Chapter 8 demonstrates how the final stage of rice cultivation, when rice is threshed and brought home, is an inversion event that serves as a complex cultural arena in which imaginings of self, the other, and collectivity are operated upon, manipulated, transformed, and reestablished. In effect, it is argued that the rice fields embody an ideal social landscape, demonstrating the crucial implications that rice agriculture, being the ultimate lever of sociability and trust, has for the Thamgharian social universe, while offering a radical reinterpretation of inversion in social life.

The final chapter of this work, chapter 9, reflects on the themes discussed throughout, albeit in a more abstract and theoretical manner. It ties together the various threads laid out thus far into a whole and proposes potential avenues for future research.

## A HIGH-CASTE VILLAGE IN CENTRAL NEPAL

The main inspiration for this book, its themes and major arguments, derives from and is the consequence of a reflection upon the data collected during two years of field research in Nepal (1995–1997), most of which were spent in the remote village of Thamghar, supplemented by a number of subsequent short fieldtrips and brief investigations, which concluded in early 2005. The village of Thamghar lies on the slopes of a mountain range in central Nepal, amidst natural and planted forest patches and endless rows of rice terraces situated on the more moderate escarpments reclining toward the river bends some 1,500 feet below, with the Himalayan snow-capped peaks towering above. The village, accommodating 221 families, is dispersed over a large area, taking up to forty minutes to cross by foot.

The high-caste Brāhmaṇs I studied belong to the more general stock of Indo-Aryan[38] people who, despite their rather obscure historical background, view themselves and are generally believed to have immigrated to Nepal from India over the course of the last ten centuries.[39] Of the latter's relatively simple caste system,[40] compared with that of Hindu Newārs[41] or the "classical" Indian caste system,[42] it is mainly the extremes that are represented in Thamghar; most local inhabitants are high-caste[43] orthodox Hindus, the vast

*Thamghar—general view of the village's upper part.*

majority of them Brāhmaṇs (Bāhuns in Nepali),[44] with a minority of "Untouchable" artisan castes (the tailor-musicians Damāi, Sārki leatherworkers, and Kāmi blacksmiths),[45] and a handful of families of other ethnic origins (Baram, Gharti). The various differences in conduct and lifestyle between the majority of Upādhyāya and minority of Jaisi Brāhmaṇs living in the village are mainly related to purity; Upādhyāya Brāhmaṇs will not eat rice cooked by a Jaisi, and only the former, being at the apex of ritual purity, may serve as priests (*purohit* or *puret*). Otherwise, the disparity between them is negligible, hardly significant in everyday life, and has very little if any bearing on the subjects discussed in this work.

To designate their habitat, local people mainly mention the name of their clan (*kul*) village (*gāũ*). A clan village is a cluster of usually no more than ten houses, which ideally at least, belong to the same clan, and is normally the result of land partition between the descendants of one patriarch. More often than not, clan villages are situated far away from each other and even within them houses are dispersed as far as possible. This is locally glossed as a measure for avoiding the incessant mutual sorcery accusations and frequent conflicts that proximity breeds. The younger generations living in Thamghar may often include the name of their ward as part of their address. A ward is a larger political unit that contains a number of clan villages or parts of them. Twelve such wards make up the local VDC (village development committee), which includes a number of additional, adjacent settlements. It thus becomes apparent that the title "the village of Thamghar" employed here reflects neither the local usage of "village" nor the government's (rather arbitrary and recent) official political division. Notwithstanding, "Thamghar" includes two named geographical areas that are locally viewed as inseparable and are clearly detached from other hamlets in the region. Moreover, this geographical unit encapsulates the most significant indigenous social networks and reflects the limits of most people's mutual personal familiarity. Yet it should be noted that as may be obvious, local family and clan (*kul*) ties often diffuse into adjacent territories and Thamgharian priests serve clients (*jajmāns*) from neighboring hamlets and vice versa.

A well-known and frequently recounted story in Thamghar narrates the adventures of and praises the extraordinary personality of Damodar Raj Upadhya, a mysterious and gifted priest, royal astrologer, and *siddha* (locally glossed as "the one whose words materialize," "successful in all he does"),[46] who lived in the area about three hundred years ago. As a reward for his loyal services, miracle-making, and extraordinary assistance to King Ram Shah (one of Prithvi Narayan Shah's ancestors)[47] in his conquests of the then Magar kingdoms in the area, the king offered to give his pious astrologer whatever area the latter could envelop, using a thread, while the king smoked one hookah. Being a *siddha*, Damodar managed to draw the thread not only around the area of what is contemporary Thamghar's VDC but also around that of a neighboring one, returning even before much smoke came out of the king's hookah. As a result, he was given the entire area in a royal grant of land (*birtā*). At the summit of the hill above Thamghar, Damodar established a Kāli temple, which still serves as the regional and central locus of Kāli worship during the major autumn festival of Dasaī[48] (Dashera in India). Following his mysterious disappearance and to this day, Damodar himself is worshipped in an adjacent modest shrine, in a cave-like niche under a large stone.[49] A number of the area's main clans, the principal actors in the aforementioned Dasaī celebrations, the region's highest priest and the ten people who had held the position of royal land-revenue collectors (*thari* or *mukhiyā*) until it was abolished fifteen years ago all proudly claim Damodar as their ancestor.

While Damodar's origins are said to be Indian, most local clans cannot clearly trace the historical immigration of their forefathers to Nepal, and as the oral histories I collected testify, they arrived from other Himalayan areas, mostly from eastern Nepal.[50] The minority of village "Untouchables" are said to have come along with their patron Brāhmaṇ families and seem to have been allocated peripheral areas on which to build their houses, surrounding the dispersed core of Brāhmaṇ clan villages. Shortage of land, the prevalent inheritance customs where the patriarch's land is equally divided among his sons and the anxiety provoked by living in close proximity, coerced Brāhmaṇs to settle in new forest areas; thus, while quite a few Brāhmaṇ houses are situated on the outskirts of the inhabited area, one "Untouchable" clan village can be found closer to the "center." Every Brāhmaṇ household still maintains the hereditary, age-old mutual service/patronage relationship with a number of "Untouchable" families, as found elsewhere in Nepal,[51] yet the intensity of these is on the decline along with its diminishing economic justification and beneficence.[52] The rules of "Untouchability" are strictly adhered to within the village,[53] and the two revolts by the local low-castes that occurred during my fieldwork were only concerned with the level of their annual wages.

Most Thamgharian Brāhmaṇs[54] do not possess a steady income, with the exception of the teachers in the local schools, the village priests, and the handful of local small business owners. An indispensable need for cash drives a majority of men, at least at one stage of their lives, to seek work outside the village, mainly in India or elsewhere in Nepal, for periods that range from a few months up to thirty years. Whatever their source of income, all Thamgharian Brāhmaṇs are also farmers and the majority of them rely solely on their rice harvest for subsistence. A number of village families would sell their surplus grain for cash if they had the opportunity to do so, but even barter is very rarely viable. Primarily for the purpose of rice cultivation, most families own two bulls and a plough. At least one cow is essential for the daily supply of fresh dung that is used, mixed with mud, for purifying the kitchen and dining areas following every rice meal, while the urine and milk are required for ritual usage. For its supply of fresh drinking milk, every household relies on its female buffalo's (*bhaĩsi*) well-being, and this is reflected in the animal's high price (12,000 NRs[55] in 1997). A number of goats are needed for the sacrifice of Dasaī, the feast of marriage, or, rarely, for ordinary consumption and are usually kept (at night) in a protected shed situated alongside the ground floor's veranda or within the house in order to prevent them from falling pray to a passing tiger. Vegetables and flowers for worship are provided by a modest kitchen garden during the rainy season.

Its remote geographical location makes living conditions in Thamghar relatively harsh; as in other hill areas of Nepal, the village has neither running water nor electricity, passage to the outside world is by footpath only,[56] and access to medical help is greatly restricted. Newspapers are nonexistent and mail, if it arrives at all, is delivered by hand no more than once a week. Almost every village house is equipped with a radio receiver, but the price of batteries is often prohibitive for many households and the reception is usually poor. The village is off Nepal's main trekking routes and during my fieldwork no other non-Nepali person visited the village. The majority of village women and a number of men have never visited the capital Katmandu. Moreover, throughout the area there seems to be a clear vacuum of political and state authority; the central government is too weak and remote and the police force is regarded neither as an agent of law and order nor as a source of assistance or refuge. Until thirty years ago, most householders who were sufficiently well-off would send their sons to complete their Sanskrit studies in Banaras or Katmandu, hence I found no shortage of practicing priests with a formal Sanskrit education. More recently, if they could afford to do so, householders began sending their sons to study in regional schools, which emphasize the study of mathematics and English. Few, if any, present-day students whose father serves as a priest are willing or indeed expected to return to Thamghar to take over this role. Education is very high on the villagers' agenda; almost all boys and the vast majority of girls (except for the poorest) are enrolled in one of the three local schools, and immense pressure is placed on achieving academic excellence. When away from home, particularly while living in Katmandu, most young village men openly oppose the caste system; they often fail to adhere to many of the strict purity rules vigilantly observed by their parents and some even neglect wearing the sacred thread (*janaï*) received during their initiation (*bartoon*). Once in the village, however, they behave according to the local norms and sons of priests may occasionally officiate the performance of worship (*pūjās*) alongside their fathers.

Finally, it has become customary for anthropological essays, whose main purpose is often to put solitary occurrences into a wider context, to further allude to and often make the claim for wide yet often rather unsound generalizations on the basis of what is, always, rather *limited* data, not only in geographical terms but also due to the limited span of research, the community involved, or the sample of people studied. The present work makes no such claim. On the contrary, as may be expected, while a few ethnographic elements mentioned herein may be found among some other high-caste Nepalese or in India, it must be stressed that many cannot. Hence, while the data drawn out of a few Thamgharian ethnographic moments is employed throughout as a basis upon which major debates and various themes in Indic studies could be reconsidered, discussed, and theorized upon, it would be a mistake and a distortion to exploit it for the sake of any sweeping generalization regarding Nepalese and Indian Brāhmaṇs.

## UNPREDICTABILITY AND FIELDWORK IN THAMGHAR

The long span of field research that forms the basis of this book exemplifies the unpredictability of anthropological work as notably highlighted by Strathern (1999: 1–11). To begin with, though my original research proposal was entitled "Education and socialization processes among Brāhmaṇs in the Katmandu valley," I ended up conducting an expansive study of the intricacies of social relationships, kinship, ritual, cosmology, and rice agriculture in a rural, remote, high-caste community. Furthermore, retrospectively, while "in the field," there seemed to be little to suggest that this fundamental anthropological practice of sharing the worlds of the people I lived with or as it is more formally called "participant observation" will result in a book about a particular Hindu fractal, predominantly since at the time fractality was still mainly recognized by scientists and was yet to become strongly associated with Hinduism.

In addition, as may be expected, the high-caste Brāhmaṇs in the Himalayan settlement of Thamghar are not aware of fractal theory and Hindu fractals, nor do they ordinarily articulate their perceptions of self and the other, the nature of personhood, the gender of objects, or the agency of the various animated entities in their vicinity. In a similar vein, they do not wonder about the efficacy of the numerous acts of worship (*pūjās*) they regularly perform and philosophical speculations about the nature of existence and the cosmos are not normally at the forefront of people's concerns. The ongoing flow of life and the world's inherent duality, namely the existence of an apparent, manifested reality enmeshed with or concealing an invisible, indirectly perceived one, are all taken for granted, merit little special attention, and thus largely "go without saying" to borrow from Bloch (1992). These themes, which are far from self-evident, are nonetheless anchored, rather implicitly, in people's everyday discourse, experience, behavior, and social intercourse. Thus, in many ways, fieldwork for me was like participating in a mystery play whose true plot revealed itself only at a later stage of the drama.

In other words, the fractality of the Thamgharian Hindu universe, with its primary emphasis on a fundamentally gendered fractal, gradually revealed itself during fieldwork as well as the vast library work, the hours spent in front of the computer screen, and while debating the issues discussed herein with various scholars and colleagues in the years that followed. Thus in a way, the manner in which this book is written is like replaying my individual intellectual journey in reverse.

In the first three months of research I lived in the household of Sharmananda Subedi (71),[57] an experienced and highly knowledgeable priest, together with his wife, daughter-in-law (*buhāri*), and two young grandchildren. Sharmananda is probably the most popular village *jānne mānche* (literally, "the one who knows;" that is, a healer) in Thamghar, and not just because his medicinal services are provided free of charge. As he once told me, he became a healer at the age of twenty-five, following an encounter with a mysterious

*Sharmananda Subedi's house.*

wandering and silent *yogī* (ascetic), who stayed in his house for several days without uttering a word. Before departing, never to return, he gave Sharmananda a long scroll of sacred mantras, which became the basis of a remarkable career. Although outwardly he is a strict patriarch, harsh toward his wife (twenty years his junior) and quite suspicious of his daughter-in-law, who is not permitted to cook in his house, he nevertheless acts as the confidant of many village women who secretly come to deposit their money with him. Sharmananda's wife, Radika (52), testified that she never suffered at the hands of her parents-in-law, as Sharmananda took the place of his deceased father at a young age, and she treats Renuka, her sixteen-year-old daughter-in-law, with much love and tenderness. Taking very little care of herself, Radika works from morning to evening and would not miss what she views as the highly meritorious act of the twice-daily drinking a few drops of *khuṭṭa pāni* (the water she uses to wash her husband's feet before every rice meal), although Sharmananda himself cares little about this matter. Notwithstanding, she is quite a strong-minded woman who does not refrain from expressing her opinion, often putting her husband in his place, and she is endowed with a fantastic, rare sense of humor and a good-natured rolling laugh.

The early mornings before breakfast (5:00–9:00 a.m.) in Sharmananda's house, were often spent attending the endless visits by village men, women, and children seeking therapeutic advice, solace, diagnosis, and relief from their various medical conditions. Usually, the latter were found to be the result of an evil, malignant attack by a ghost or demon, elicited by envious neighbors, competitors, or foes. Later in the day I often accompanied Sharmananda on his house calls, which were frequently performed for healing the family's female buffalo (*bhaĩsi*), whose milk had dried up due to an evil spell or vicious stare. Although probably the most popular local healer, Sharmananda claimed no exclusivity; the supply of first-aid accessories and medicines I brought with me proved indispensable and soon the aforementioned morning public clinic turned into a holistic institution. Thus villagers were first treated by Sharmananda, who sacrificed pigeons, blessed rice grains, or blew air over their foreheads (*phuknu*), in order to exorcise the evil spirits that had entered their bodies. Following this, they turned to the alternative treatment I offered in the form of paracetamol, ear drops, antibiotic creams, salts for their diarrhea-affected children, bandages, and so on. Often Sharmananda would add a number of blows or recite a powerful sacred mantra over a medicine before I dispensed it, to magnify its healing potential. At the time, the nearest health-post was about two hours walk away from the village, and my ability to alleviate some of the people's daily suffering opened many hearts and made my presence in Thamghar more comprehensible, not only from a local point of view but also from mine.

In the fourth month of my stay in the village I met Eknath Lamichane (24, then an MBA student in Katmandu and the eldest of seven children), who later became my research assistant. I began spending time in his home and finally divided my time in the village between the two families. Muktinath, Eknath's father, a respectable landowner and a veteran teacher working in the local high school, became a part-time practicing priest during my fieldwork. When he realized that he could not answer some of my questions, he didn't hesitate to admit so and was very helpful in initiating and facilitating two extended learning sessions held with Mr. Karananda Raj Upadhya (86), the most respected priest in the region. Mr. Upadhya kindly agreed to come to Thamghar to stay in Muktinath's house for a few days each time for this purpose. Muktinath's wife, Sita (42), a most diligent and gentle woman, took care of me as if I were one of her own seven surviving children (two others had died in infancy).

The two households provided me with a habitat full of *māyā* (love) and warmth. Belonging to two of the largest local clans (*kuls*), situated about thirty-five to forty minutes walk apart, they constituted a flexible and convenient base for research, allowing me to maintain a dual vantage point for the examination of Thamgharian life. Being associated with two central local families assisted me in expanding my intimate circle of acquaintances to include many households belonging to their huge clans.

Much of my time was spent in standard "participant observatory" activities such as chatting, following members of my families in their daily activities, informal interviews with individuals, attending the daily morning and evening male gatherings at the three village tea shops, participating in ad-hoc village councils, accompanying village men to a regional festival (*jātrā*), or joining a family visit to Chitwan,[58] as well as recording genealogies and personal histories, and collecting children's songs and drawings. In addition, I also performed a comprehensive survey of all village households, closely inspected the construction process of new houses, and made a number of detailed studies of the domestic use of space. My data regarding illness and sorcery was complemented by a series of sessions I held with students of the four highest classes in the local high school. In each session, students were asked to write, anonymously, about illness, incidents of sorcery, and attacks of evil spirits against members of their family, and include details such as the symptoms, diagnosis, treatment, number of occurrences, and the like. My knowledge and understanding of local rituals draws upon the ample opportunities I had to attend religious ceremonies. Apart from those that were readily observable in the priestly households I lived with, I accompanied my two householders, as well as other priests, on their various functions among their client families (*jajmāns*). When I began my fieldwork no one owned a camera in Thamghar, thus I soon became known as the American photographer (*amrikane khicne mānche*) and in this capacity I was invited to attend numerous family and other major rites. Recording and filming became almost an integral part of village rituals (*pūjās*), and I was given all necessary assistance in documenting these events. The two priests in my host families proved to be an invaluable mine of knowledge regarding Hindu thought, ritual, and myth, and never tired of answering my questions.

Following my initial five months in Thamghar, invitations to eat and sleep in various households became very frequent (and, with various fluctuations, remained this way throughout my fieldwork), and steering between people's expectations without disappointing or offending a number of them was far from easy. These prolonged visits provided me with an excellent opportunity to become closely acquainted with many village families, who otherwise would have remained relative strangers, and considerably expanded the scope of my research. Once the relationship with the women in both my host households became closer, the latter became an excellent mine of information, knowledge, and wisdom. Many questions that could not be asked directly in other circumstances were answered quite freely during two women's gatherings I organized, away from village men. Although I often managed to converse with many other village women, it is as well to bear in mind that the obvious difficulties a male researcher faces in gaining access to women in an orthodox, patriarchal Hindu society (such as that of Thamghar) mean that throughout my fieldwork I was mainly exposed to the male point of view. Apart from intimate moments in the company of my two host families, it was among groups of workers in the midst of the rice fields, where I spent many days throughout the rice cultivation season, and particularly on a number of nights (during threshing as described in chapter 8), that I came close to transcending the boundary between "participation" and "observation."

Finally, the ongoing "People's War," the extremely violent confrontation between the radical communist CPN (Community Party of Nepal) or Maobadis, as they are locally called, and the government-royal security forces began during the course of my fieldwork. One of the first incidents of the "People's War"[59]—an attack on the vice-chairman of the local VDC (a prominent local Congress politician) just outside Thamghar—took place on January 1, 1996, yet the conflict had no significant impact on local affairs throughout the principal research period and no bearing on the subjects discussed below.

However, although Thamghar has not been the scene of a bloody battlefield and was largely spared from much of the brutality of war experienced by many other rural villages throughout Nepal, it could not escape from the general intensification of the "People's War," particularly during the last five years. Hence, just like most of the kingdom's hilly and rural areas, ways of life have changed considerably since my time in Thamghar, and many of the practices and customs mentioned in this book are nowadays performed differently or not at all. Furthermore, sadly, in recent years Thamgharians have found themselves in the impossible situation in which most other rural Nepalese are situated: being unwillingly caught between the wrath of both the royal armed forces and the Maoists. This situation drove almost all young Thamgharians to migrate from the village, mainly to the capital Katmandu and to India, leaving their old parents and wives behind and turning the village into a semi-phantom settlement mirroring the current state of much of rural Nepal. Hence the "ethnographic present" employed throughout should not be taken as anything other than what it is, namely an anthropological literary convention. Further details about my integration into the village and personal experiences there are provided in appendix A.

## NOTES

1. See Mandelbrot, B. B. 1982. *The Fractal Geometry of Nature*.
2. Mandelbrot (1982, 2005); Mandelbrot and Hudson (2004).
3. For a number of notable exceptions see Wagner (1991), Gell (1998, 1999), Eglash (1999), Gaddis (2002), and Jackson (2004).

4. As both noun and adjective.

5. Mandelbrot (1982: 33). Gleick is particularly illuminating on this point (1987: 94–98).

6. Also reproduced, for example, in Gleick's *Chaos*, 1987.

7. The term fractal is an appropriation of the Latin adjective *fractus* that derives from the verb *frangere*—to break (Gleick 1987: 98).

8. Mandelbrot (1982: 17–18).

9. Mandelbrot (1982: 17–18).

10. See, among others, Strathern (1988), Wagner (1991), Gell (1998, 1999), and Eglash (1999).

11. See, for example, Turner (1964: 16), Hocart (1970: 60–71, 250–61), Wayman (1982), Gaborieau (1992), Hugh-Jones (1995: 226–52), and Coomaraswamy (1995).

12. One of the contentions of this book, to be elaborated upon later, is that fractality is the basis for ritual efficacy in the Hindu world and may hold the key for understanding ritual efficacy elsewhere.

13. This obviously applies, in varying degrees, to a range of Hindu avatārs; about Buddhism and Taoism see, for example, Jackson (2004), about Jainism see Trivedi (1989) and White (2000).

14. The quest for *mokṣa* or *nirvāṇa* can be seen as the aspiration of the part to realize its unity with the whole in a fracatlic cosmos.

15. On this much-discussed notion see also Shulman (1980: 41) and Gupta (1988: 34).

16. *Ṛg Veda* 1.164.20.

17. Das (1985: 189); Fuller (1988: 50, 57).

18. Kramrisch (1991).

19. Gonda (1957: 24).

20. Beck (1976: 237–40); Coomaraswamy (1995).

21. Good (1987: 40).

22. Volwahsen (1969: 43–88); Kramrisch (1976: 132, 154); Snodgrass (1994: 126–56); Hardy (1998: 107).

23. Beck (1976: 240–1); Kramrisch (1991: 102).

24. Trawick-Egnor (1978: 122).

25. Pintchman (1994: 120–21); Handelman and Shulman (1997: 56–57).

26. At this stage suffice it to note that "encompassment" is employed herein in its ordinary sense of to include or to surround (*The Collins English Dictionary*, 1987) and that the term matrimonial encompassment bears only superficial similarity with Dumont's (1980) notion of the "encompassment of the contrary" in the context of the Indian caste system. The disparity between the present and the Dumontian notion of "encompassment" will be discussed in detail in chapter 3.

27. *Bhāgavata Purāṇa* 10.8.21–45. Jackson (2004: 29–31) finds some of the earliest fractal visions in Hinduism within the *Athharva Veda* and *Bhagavad Gītā*.

28. Namely, the south Indian temple city of Shrīrangam (Volwahsen 1969: 56).

29. This is well exemplified in plates I, III, and IV in Kramrisch (1976). See also the photos in Volwahsen (1969: 75, 83, 121, 129).

30. Volwahsen (1969); Brown (1971); Kramrisch (1976); Michell (1977); Malville (1991); Meister (1991).

31. Trivedi (1989); Hardy (1995, 1998)—although using an alternative terminology drawn from architectural theory; Ben Dov (2002); Jackson (2004).

32. Marriott and Inden (1977); Marriott (1989, 1998).

33. Nevertheless, as Thamgharians were totally unaware of the cardinal analytical categories and conceptual components of this approach, its employment would make little sense and it is not helpful for discussing the Nepalese material presented herein.

34. The Tamil universe, Trawick-Egnor contends, is taken to be infinitely divisible and, for example, the eye is both like and at the same time also a part of a person.

35. Ramanujan mainly discusses myths where the microcosm is both within and like the macrocosm, exemplifying what in the present terms is taken to be a literary fractal. Thus, he maintains that within the great Hindu epics, every story is enveloped within a meta-story so that one tale is the context for another within it, while the latter acts as a microcosmic replica of the former.

36. Handelman and Shulman discuss the nature of Śaiva cosmology arguing that it evinces self-similarity, the presence of the whole in each of its parts and interconnectedness. As an example (which is not, however, visually self-similar) they bring the *liṅga*, which is the sign for, a part of, but also contains the whole god (Śiva).

37. For example, Trivedi (1989), Wagner (1991), Gell (1998, 1999), Eglash (1999), and Jackson (2004).

38. Or more generally, the Indo-Nepalese people (including Muslims) who share their Indian origin. Following notably Doherty (1974: 40, n1) and Bennett (1983: 10, 15), I use the terms Brāhmaṇ-Chetris or high-castes to refer to the upper strata of Nepali-speaking Hindu people of Nepal, which constitute a separate, coherent cultural unit. I shall avoid the title of Parbtiyā or its derivatives (Parbat or Parbate) because in Thamghar, at least, these terms have negative connotations.

39. Bista (1972: 5, 12–15); Bennett (1983: 9); Ramirez (2000a: 12). According to Fürer-Haimendorf (1960: 17–18), the Nepalese Brāhmaṇs entered Nepal without their menial "Untouchables" (perhaps quite like their Kashmiri counterparts [Madan 1995: 110]), which is the reason why they became land cultivators (as Thamgharians are), in sharp contrast with Indian Brāhmaṇs.

40. See Höfer (1979), Bennett (1983: 8–16), and Ramirez (2000a: 30–35).

41. For an excellent discussion of the Newār caste system see Gellner and Quigley's (1995) edited volume.

42. Dumont (1980).

43. A "caste" is locally referred to as *jāt*. As is the case elsewhere in India or Nepal (see the recent eloquent discussion in Cameron (1998: 9–14)), *jāt* or *jāti* refers to classification of anything in the world, namely, women are taken as female, women (*āimāi*) *jāt*. For the sake of the Western reader, I shall mainly employ the term "caste" to denote human *jāts*.

44. Of which one hundred households are Upādhyāya Brāhmaṇs, sixty-seven households are Jaisi Brāhmaṇs (the offspring of union between Upādhyāya men and an Upādhyāya widow or divorcee), and only six households are Chetris (Kṣatriya).

45. A total of forty-three households.

46. *Siddhas* were "perfect *yogīs* (*jogis*)" (literally, practitioners of yoga, extreme Tantrics, ascetics) belonging to a late medieval movement that formed a complex amalgamation of various religious ideas, alchemical traditions, and magic, and was predominantly influenced by Śaivaistic and Tantric ideas. Its founders, Gorakhnāth and Matsyendranāth, were deified in Nepal and to this day their legends are still well known in the northern parts of the Indian subcontinent (Eliade 1973: 301–18).

47. Prithvi Narayan Shah is considered to be the founder of modern Nepal.

48. Also called Durgā *pūjā*, this is the climax of high-castes' annual festivals, held in honor of the goddess Durgā (Kāli) during the light fortnight of Asoj (September–October).

49. This is one of the stones carried by Damodar for building the Kāli temple nearby, reminiscent of the worship of Goraknāth in a cave shrine in Gorkha (Tingey 1990: 182).

50. Those Thamgharian Brāhmaṇs who are familiar with the division between eastern (Purbiyā) and western (Kumaī) Brāhmaṇs associate themselves with the former.

51. Blustain (1977); Fortier (1995); Cameron (1998).

52. In the Katmandu valley, according to Bista (1972: 26), these relationships have totally dissolved.

53. Entering an "Untouchable" house and vice versa, taking food and water or having intercourse with an "Untouchable" renders the high-caste person involved or the entire clan (*kul*) in the latter instance, "Untouchables." Concomitantly, a Brāhmaṇ touched by an "Untouchable" must be purified via *chiṭo hālne*, where a few drops of pure water are sprinkled on him/her by another high-caste person.

54. Almost needless to mention, each person in Thamghar is unique and people's perspectives are not monolithic; young differ from old, women often (but not always) have different views from men, and so on. In short, endless variations in view and action may be found regarding most issues, and almost any "rule" has its exceptions, yet common tendencies and general moods can be traced. Unless otherwise stated, I refer to the latter by using the apparently all-inclusive terms "villagers" or "Thamgharians." I attempt to present the main lines of diversity and the range of the different opinions and perspectives I was exposed to, when these are of particular relevance and significance.

55. Nepalese Rupees. For comparison, in 1997, the monthly salary of a senior local school teacher was 2600 NRs.

56. The nearest paved road, leading to Katmandu, is about twenty-six kilometers of mountainous walk away.

57. All figures in parentheses refer to a person's age.

58. One of the main settlements in the Terai.

59. February 13, 1996, is generally accepted as the starting date of the "People's War."

# I

## MAN: UNDERSTANDING MISTRUST

# 2

## Himalayan Uncertainties

### INTRODUCTION

One may be inclined to believe that those who share and endure objectively harsh living conditions, particularly in the absence of a true centralized authority,[1] would cling to each other in order to overcome obvious hardships together. This, however, is not the case in the Brāhmaṇ village of Thamghar, where the atmosphere is characterized by mutual animosity, suspicion, fundamental mistrust, and anxiety. The latter seem to percolate through almost all facets of life, infusing every level of society and interaction, abating any motivation to join forces and paralyzing cooperation, which is thus limited to the inevitable.

At this initial stage it is crucial to recall Strathern's (1999: 18–19) caution regarding the usage of terminology and the evaluative and other unwanted overtones and implications that are part and parcel of "the Euro-American language on which anthropologists draw." In the main, Strathern points out that, quite unjustly, sociality is frequently understood as implying sociability, reciprocity is thought to entail altruism, and relationship is seen as solidarity.[2] This issue proves highly germane to the Thamgharian social milieu.

Indeed, in Thamghar sociality means that in order to deal with major personal conflicts or public disputes, an ad hoc council whose members are never officially appointed, voted for, or known in advance (each is a local *ṭhulo mānche*, literally "big man," a senior respectable and influential householder), is summoned. Sociality may also mean the existence of perfunctory, mutual social obligations, a strictly regulated system of work exchange and domestic reciprocity, as well as structured ways for distributing the meager allocation of government funds[3] and for dealing with issues such as death and local and national politics. Sociality also involves providing help to an old widow whose roof has caved in or to a householder whose haystack has been burned to ashes. Yet it should be borne in mind that these expressions of sociality do not imply much in the way of sociability, altruism, solidarity, interpersonal harmony, rapport, or trust.

Rather than focusing on the aforementioned expressions or "by-products" of sociality, the present analysis, which can be thought of as an extended introduction to the Thamgharian social world, will be concerned with what is undoubtedly the dominating and, from a local point of view, the most significant and disturbing feature of the Thamgharian social universe, namely its embroilment within a self-perpetuating vicious circle of hostility and mistrust. Following a brief note on Hindu "evil," a subject highly relevant to the subsequent discussion of the dynamics of the Thamgharian social world, I will examine the pattern of the relationships maintained with the various cosmic forces and nonhuman entities, dominating what appears to be a hungry and emotive cosmos.

Life in Thamghar seems fraught with uncertainty and appears to be a rather precarious affair. Although ostensibly this may be predominantly due to multiple external dangers, local concern lies particularly with the inherent evil that is believed to lurk within fellow villagers' hearts, behind their mask of amiability. In the main, I shall argue that the inability to contain the Thamgharian social vicious circle stems primarily from the inevitable misappropriation of the rather mechanistic pattern of interaction with the nonhuman entities of the cosmos into the human social realm and the absence of the restraining mechanism of commensality in the latter.

Next, I will turn to examine the various expressions of the Thamgharian tendency to construct a false surface of uniformity enshrouding an intriguing diversity. In particular, this is highlighted by the effort to conceal the measurement or scale of householders' wealth and prestige, namely the family's rice supply. Concomitantly, it will be demonstrated how, rather than serving as a dwelling place and a means for objectifying one's prestige, the Thamgharian house primarily embodies an embattled shrine of rice, which is yet unable to protect the latter from the animated evil intentions of fellow villagers.

Finally, I will suggest that the dynamics of social relationship discussed hitherto cannot fully account for the intensity of doubt and mistrust evinced in Thamghar. This conclusion leads to chapter 3, which examines how elements belonging to the realm of kinship, particularly the dialectics of cardinal local perspectives and mental images elicited by the notion of a male encompassing a female in marriage, may yield a better understanding of the predicament of the Thamgharian social world.

Much of the ensuing discussion is closely linked to what in the West may be termed "evil," yet local Thamgharian and general Hindu ideas of "evil" differ considerably from Western notions, hence a short elaboration is in order.

## A Note on Hindu "Evil"

"Evil" forms an integral part of the Hindu empirical world where *saṁsāra*, the cosmic cycle of death and rebirth in itself is regarded as evil. Evil is part of god, is inevitable, often necessary and may be beneficial (O'Flaherty 1980b; Shulman 1980: 347, 1993; Inden 1985a: 148). Sharmananda Subedi (71) expressed this notion lucidly when he said "*Īśwar*' (god) is everywhere, in every tree, bird, stone, house, man, *bhūt* (an evil spirit, ghost) and *boksi* (a sorcerer, witch). Everything on earth should *samjhinu paryo*—be remembered or taken as a *mandir* (temple), since *deutā* (*deotā*) *bascha*—god—"sits" and dwells inside. Our words are *īśwar*'s words, and when we lie, it is due to the present Kāli *yug* (an era of darkness that is destined to end in total destruction)."

This statement overrides the Western dichotomy between "good" and "evil," accepting evil as an integral part of the fractalic world of temples filled with god's presence and regarding lying as an external, situational evil rather than a moral issue or a character flaw. This, of course, is not to say that Hindus in general and Thamgharians in particular do not distinguish between good and bad, but mainly that for them the "problem of evil" is not a moral but a major *practical* one (O'Flaherty 1980b; Inden 1985a: 143, note 3). Indeed, Thamgharian Brāhmaṇs do not engage in philosophical speculations regarding the origins of or justification for evil but are primarily preoccupied with warding it off and avoiding it.

This is considerably complicated by the presence of two related elements. First is the fact that good and evil, truth and lie, surface and core, "moral evil" and "natural evil" are not easily distinguishable. This is not, as Macfarlane argues regarding capitalist society, a situation where the opposition of good and evil has been eliminated and where good and evil have become indistinguishable (Macfarlane 1985), but is more an issue of actuality and potentiality.

Stated differently, as Inden (1985a: 146) suggests, good and evil are but a mask of the problem of power (*śakti*), which in Hindu perception is fundamentally ambivalent and has the potential for both benefit and harm. Secondly, any manifestation of power involves fundamental uncertainty, particularly when it is beyond one's control, and (usually) presents its true nature (harmful or benevolent, reassuring or ominous) *a posteriori*—after it has made its mark.

Much remains to be said regarding the dialectics and management of power as it is perceived in Thamghar and I shall revert to these issues later. First I shall consider villagers' relationships with some of the nonhuman forces inhabiting their world. These relationships are pertinent primarily since they seem to provide villagers with a number of fundamental notions or a prototypical pattern for implementation in the sphere of human interaction in social life. Their logic derives from a particular ontology and epistemology entrenched in one of Hinduism's ancient "gastro-cosmological" visions, as discussed below.

### Metaphysics of Hunger

As unfolded in the *Upaniṣad*, "this world was fashioned by the hunger of the immensity," "from food all things are born . . . all are eaten and eat, hence everything is food."[4] "All existence could be brought back to the fundamental dualism of two factors: food (*anna*) and the devourer (*annāda*)." The all-pervading divinity is both the food and the eater;[5] fire is life and it is through fire that man can partake in the cosmic life (Daniélou 1964: 63–66). According to Shulman (1993: 38–39), "Hindu notions of divinity frequently include the aspect of the all-devouring Absolute," and in the *Bhagavad Gītā*, Arjuna is faced by the ultimate truth of existence where the "whole ultimately consumes the beings it has created."

Alternatively, in ancient Hindu texts the world is seen as the food to be cooked or recooked by the Brāhmaṇ priest through ritual means (Malamoud 1996: 23–53). Malamoud interprets the *Brāhmaṇas* as portraying a world full of holes on each and every scale of being, be it the gods, semidivine entities, man, and so on. Since these voids are "nothing less than death itself" (ibid. 53) they must be constantly refilled with food, otherwise they may turn to devour humanity itself. Empty cavities and chasms are an invitation for disorder, chaos, and evil to creep inside and dwell therein, hence the continuous, harmonious progression of the cosmos through time is based on an eternal filling and feeding in a ritual process (Malamoud 1996: 54–58). Thus, this ancient vision of a hungry cosmic fractal explicates what lies at the root of the constant need for worship.

## THE CONSEQUENCE OF MISAPPROPRIATION

### Food for Emotion

As will be demonstrated, the above Hindu "gastro-cosmology" seems to inform the Thamgharian notions of the cosmos and the sort of relationships that must be maintained with its nonhuman forces. In addition, for Thamgharian Brāhmaṇs the cosmos is not merely hungry but is also fundamentally *emotive*. As will be explicated below, villagers believe that to a large extent, man's cardinal fortunes depend on the satiety of the animated powers found in the universe, be they ghosts and spirits, *nāgs* and *nāginis* (the semidivine snake deity and his consort), gods and goddesses, planets, or other celestial bodies, which are fundamentally hungry and emotional.

These entities are believed to inhabit worlds that parallel or intersect our own or, in the case of the major Hindu divinities as well as the stars and planets, even reflect and encompass our world altogether, forming part of what in the present terms is seen as an existential fractal. Man's relationships with these entities, predicated upon the exchange of food, take the form of what Babb (1991: 69–70 and passim) calls "fluid and humble intimacy" and are characterized by their underlying hierarchical implications, impersonal and instrumental nature, the irrelevance of morality, and the mechanics of causality.

Hence, for example, misfortune almost always strikes without warning and is believed to be due to uncontrolled anger evoked by an unintended offence committed toward a deity, or simply results from its insatiated hunger. Gods are said to "rise" (*uṭhcha*, *ris uṭhāunu*) in anger while stars are believed to "break" (*bigrincha*) in rage, and both are reputed to use every means at their disposal, often in a rather capricious and merciless manner and with no regard for human loyalty or past experience, in order to receive prompt attention from their ardent devotees. A swift pacification or filling of the cosmic void, fashioned through a lavish *pūjā*,[6] overturns this harmful chain of events, returning man, deity, or planet back into their harmonious orbit.

The clear knowledge gap between these higher entities, often equipped with encompassing consciousness and infinite erudition (which puts them in a permanently advantageous position), and humans instills great uncertainty into their relationships. This can be minimized through the periodic performance of "prophylactic" *pūjās*, but can never be fully obviated. Nevertheless, once it is performed, a *pūjā* seems to draw these cosmic forces into a contractual interaction, in which man exercises ultimate control over the situation. If it is performed appropriately the nonhuman entities cannot help but accept the offerings and fulfill the intention of the rite, in the spirit of the Vedic sacrifice with its automatic determinant effects (Malamoud 1996: 50). Hence, with the exception of one Vaiṣṇavite family, Thamgharian worship is far removed from the devotion of *bhakti*, where a deeply emotional and truly intimate bond is formed between man and his divinities.[7]

Although the wrath of the major gods and planets may turn a man's life into total calamity, Thamgharian Brāhmaṇs do not view these in awe since they are relatively easy to negotiate and interact with and do not form part of the horizons of villagers' mundane reality. It is mainly the "lower" semidivine entities, in the form of demons, ghosts, and other evil creatures, which are believed to lurk in almost every corner in and out of the village, that elicit consternation and make the Thamgharian landscape appear, to borrow from Heesterman (1992), rather "inauspicious." Villagers speak of an overwhelming list of anonymous entities and whimsical malevolent agencies,[8] all of whom are highly sensitive, easily offended and manipulated, suffer from insatiated hunger, and often mistake humans for food. These malicious beings are frequently warded off or tempted to leave by offering food, and only the village *boksis* are said to maintain enduring feeding relationships with them.

Danger appears to be almost ubiquitous but its intensity varies according to time and place. During the day, the village with its clusters of *kul* (clan) "villages" is considered relatively safe, except for a number of specific places where devouring *bhūts* (evil spirits) dwell, such as large trees or *dobāṭo* (path intersections)—their favorite haunts. Danger is particularly intense in the forest just beyond the loosely demarcated village boundaries,[9] yet peril tends to diminish as one proceeds away from the village and completely dissolves beyond the auspicious rivulets and one large river that almost surround the ridge of Thamghar.

Within this "no man's land" outside the village, evil agencies reportedly dwell in deep cracks in the land or cavities under large rocks, and almost any unusual feature in the landscape is associated with menace; even simply sitting on the bare ground may entail danger. There are, however, a number of notable exceptions within this region, namely the rice fields with their adjacent Bhūme (Bhūmī)[10] temples, which together with the forest form a patchwork quilt surrounding the inhabited area, and the regional Kāli temple on top of the hill overlooking the village. Similarly, the local footpaths with their *cautāros* (*cautāra*)[11] are also viewed as loci of relative safety. The village's *masān ghāṭ* (cremation ground) situated by the largest river in the region, is a highly inauspicious place but poses little hazard during the day.

Twilight time, or *kāl samaya* (*samae*, literally a time of death), is considered a particularly dangerous time, when no work, travel, or even reading may be accomplished safely. The perils of the night that follows are not as paralyzing, yet the night is believed to be ruled by evil; all places considered relatively safe during the day are haunted by nocturnal malevolent forces after dark. People tend to stay confined in their houses until daybreak, and if night travel is unavoidable, one must be accompanied and equipped with a lit torch or an oil lamp—reputable deterrents against evil spirits.

Many people who for various reasons had to walk alone or sleep outside at night are said to have suffered evil attacks, which resulted in their immediate or delayed but unnatural death, insanity, or other personal calamity. Only the house is regarded as a relative locus of refuge from the array of evil agents roaming the village under the shroud of darkness. I noticed that many village children are too scared to go out of their house, even to the yard in front of the main door, during the night. Sleeping alone is considered dangerous and frightening, thus children sleep near their parents, grandparents near their grandchildren. Fear is not perceived to be shameful and is perhaps the sole exception among all other emotions, which are normally strictly concealed.

The pervasiveness of evil forces dictates vigilance and careful planning of travel and almost all activities, and often requires astrological consultation. Nonetheless, the certainty that, when encountered, it will be possible to enfeeble the malevolent effects of these impersonal and unpredictable forces of the cosmos, and the belief that via relatively simple, almost mechanistic procedures one may assume control over them or hold them at bay, instills confidence and allows Thamgharian Brāhmaṇs to lead their lives in relative peace of mind.

The image portrayed thus far may convey the impression that Thamghar is a community embattled by forces from *without*, yet it is particularly the play of dark uncontrolled forces beneath the benign surface appearances *within* the fragile capsule of interpersonal relationships that is permeated by intense, ominous uncertainty and gives rise to much unremitting anxiety, making social navigation a complex endeavor indeed.

## Himalayan Moods in Turmoil

Commenting on the successful appropriation of the manipulative Indian strategy based on panegyric, flattery, and gratitude within the ritual realm onto everyday public behavior ("coercive subordination"), Appadurai notes that the projection of one mode of action or a certain behavioral paradigm from one realm of life onto another is a prevalent cultural strategy (Appadurai 1985, 1990). Thamgharian Brāhmaṇs' attempts to relocate the elements governing their interactive framework with the various nonhuman emotional forces of the cosmos and implement them upon the reality of a vibrant human social world proves more complicated.

This is since, in contrast with hungry spirits and deities, human hearts and bodies seethe with diversified emotions and feelings, generating complex motives and real intentions in a nonmechanistic manner. Furthermore, human sentiments tend to be endemic, passionate, and self-reinforcing, and are often strongly affected by past experience and subjective dispositions.

Himalayan quotidian reality in the remote village of Thamghar is harsh and existence often seems a rather precarious affair, brimming with misery, hardship, and misfortune, with only short glimpses of brightness. While villagers tend to take personal credit for the latter, seeing it as the positive fruition of their own actions, they deflect blame for the former onto external situational circumstances, primarily within the social realm. By doing so, they adopt what social psychologists call "the observer's point of view" and are inclined to attribute fellow social actors' behavior to their *internal* personal dispositions.[12]

Indeed, in line with Stone (1983), the personal testimonies and case histories I collected in Thamghar as well as the anonymous written reports of the local high school students regarding illness, evil spirits, and sorcery, confirm the belief that the majority of personal, financial, and physical adversities one experiences through life are realizations of fellow villagers' evil intentions, embodying their unleashed dark emotions of jealousy, hatred, and greed. Thus it becomes clear that Thamgharians employ the template governing their relationship with the animated cosmic forces as a guiding framework for the *interpretation* of social life.

However, while the considerable unpredictability that permeates the former is regularly restrained and is almost completely obviated through the efficient mechanism of the *pūjā*, villagers are unable to implement this culinary manipulation of emotions within social life due to the low level of solidarity and ever-present mutual suspicion reverberating through the village social milieu. This may gain clarity following a brief elaboration about food and commensality in Thamghar.

As may already be apparent from the Thamgharian "gastro-cosmology," the prominent place of food in Brāhmaṇ life and the centrality of its prototype, rice (*khānā*, literally meaning food, is a synonym for *bhāt*, cooked rice), cannot be overemphasized.[13] Limitations of space preclude covering every facet of the complex meaning and significance of food in Thamghar and its function as the prime tool for defining the household unit, or discussing its rich imagery and prolific idiomatic usage in everyday discourse. Hence my present commentary will be limited to its critical role as a primary medium of *trust* (*biśwās*, *patyār*, to believe/trust—*patyāunu*).

Hindu food is a moral construct, synonymous with life itself (Khare 1976a: 2–3, 1992: 1) and is a highly condensed "social fact" used as an instrument for encoding myriad social and other messages (Appadurai 1981: 484). Many of these are concerned with the (mainly ritual and purity) status of the persons involved in or excluded from food transactions or commensality; (mainly cooked) food is an instrument for marking hierarchies and making them explicit, creating and sharpening social divisions.[14]

However, these notions have only marginal social significance in the rather closed high-caste Brāhmaṇ community in Thamghar, and are of little relevance to the village's culinary manipulative strategies and "gastro-politics," to borrow from Appadurai (1981). Whenever the need arises, for example on the relatively rare occasions of village *bhāters* (public feasts), the higher ritual status Upādhyāya Brāhmaṇs assume what Malamoud sees as Brāhmaṇs' fundamental role of cooks (Malamoud 1996: 27), a noble task that underscores their status, thus rendering the issue of ritual purity in the main village public transactions of food almost irrelevant. In Thamghar, the chief social significance of sharing food lies in its inherent assumption and implication of trust, as well as its ability to generate and maintain social relations.

As often noted regarding grains in India (Marriott 1968: 143–45; Raheja 1988; Malamoud 1996: 52–53), *cāmal*, uncooked husked rice is regarded by villagers as a "superior medium," a highly efficacious conductor or vehicle for the transference of various qualities and moral substances. These capacities are believed to augment considerably through cooking, but while general cooked foodstuffs or rice cooked/fried in pure *ghyu* (clarified butter) are considered *cokho* (pure), *bhāt*—rice cooked in water, is *biṭulo* (mildly impure). As such, *bhāt* is extremely vulnerable to pollution and evil manipulation (sorcery) and cannot be offered in a *pūjā*.[15] Hence it is no wonder that the process of cooking and consuming Thamghar's main staple, essential for sustaining life "is not a leisurely, convivial event" to borrow from Bennett (1983: 41), and is deemed highly perilous. This is the reason why it must be confined to the dim and hidden interior of the ground floor of the house, away from the potentially malignant gaze of fellow villagers, where it is sheltered from ocular aggression, as Babb (1991: 215) terms it.

These notions rest upon the idea that any type of food, once internalized, has the capacity to merge with, transform, and operate upon the consumer and affect him/her positively (confer blessing, auspiciousness, or purity) or negatively (pollute, inflict inauspiciousness and illness, poison, or even kill) according to its particular moral and qualitative load. As Parry writes:

> In Hindu culture, a man *is* what he eats.[16] Not only is his bodily substance created out of food, but so is his moral disposition. . . . Just as butter is formed by churning milk, so the mind is formed by churning food, [thus] eating [. . .] is a dangerous pastime. It lays one open to malign influences and the risk of infection. (1985: 613–14)

Therefore, unaware of the teachings of the *Bhagavad Gītā*, which warn that "he who takes food alone eats nothing but sin" (Gonda 1965: 216), many village householders prefer to eat in near solitude with only their wife (whose qualities permeate the food she cooks and serves) for company. This enables them to avoid even the minuscule danger of *choka lāgne*, the unintentional evil eye that may be mutually cast by family members through staring plainly at each other's plates during a meal. For this reason, *bhāters*, where the food is prepared and consumed in the open, are viewed as highly hazardous events.

Eating together is a trust-building action and a prologue for social intimacy. This is well exemplified by the fact that on the rare occasions when relatives or respected guests from outside the village are invited to sleep in village homes, it is imperative that they first eat *khānā* before retiring to sleep. While commensality is emblematic of trust but does not necessarily imply much solidarity, withdrawal from it is an overt sign of mistrust, is deemed an explicit act of hostility and aggression, and implies the accusation of sorcery.

Ramakrishna (26) expressed a general village view when he told me "if people had not to eat together, they would probably cease to speak to each other," yet even when verbal communication is completely absent, commensality may still exist, representing the last thread of sociality; I know of two brothers who, though they have avoided verbal communication for the last twenty-seven years, still strictly participate (in person or through a family representative) in mutual *bhāters*, lest they become true enemies. Likewise, the first symbolic expression of a joint family's fragmentation is the "division of the *culo*" (*chulo*, hearth), that is, the denouncement of commensality; frequently, the *buhāri*, the stranger incoming daughter-in-law, is barred from cooking for her suspicious in-laws.

Lack of trust forestalls commensality and generates further mistrust, resulting in a situation where social visiting and sharing food are virtually nonexistent in Thamghar; overly anxious about the risk of being bewitched or poisoned, villagers strictly avoid eating in each others' houses and unexpected visitors are viewed with much suspicion. Therefore, wherever possible, people take circuitous paths around others' houses in order to avoid the habitual meal invitations, which although expected in accordance with village etiquette (according to which failure to make such invitations entices sorcery), are *always* refused. Even priests, who are obliged to partake in the lavish meal prepared for them at their *jajmān*'s house following the performance of a *pūjā*, try to avoid it by sending their *kanyā* (virgin) daughter whose divine status and powers grant her immunity from evil[17] as a representative. In a similar vein, participation in *bhāters*, a social imperative in certain events during the year, is always done reluctantly; people eat hastily and complain about the menace and trepidation involved.

Even the supposedly solidarity-producing *kul deutā pūjās* (the joint festive worship of the clan's tutelary deity) are unable to generate much cohesion and, as Fürer-Haimendorf (1966a: 40–41) and Ramirez (2000b: 173) note regarding other high-caste Nepalese, a Thamgharian *kul* is not akin to a cohesive "corporate kin group" but is fraught with rivalry, competition, and antagonism.[18] Conflicts between *kuls* seem to be extremely rare and in fact, as in Sicily where "your enemy is the man in your own trade" (Franchetti 1974 [1876]: 71),[19] Thamgharians expect to find their worst rivals within their own *kul*. Therefore, it is little wonder that *kul deutā pūjās* are permeated with much tension centered around villagers' main concern—preventing a rival faction from stealing one of the precious *murtis* (stone manifestations of the gods) or other sacred paraphernalia, which would enable them to celebrate the next *pūjā* separately, as is indeed the case among a number of the village *kuls*.

Thus it may be concluded that while sharing food may yield trust, the latter is also a necessary prerequisite for the former, hence Thamgharian Brāhmaṇs appear to be embroiled in a self-perpetuating vicious cycle of mistrust. As a result, and as will be demonstrated below, evil seems to creep into the "yawning gaps of hunger" opened within the Thamgharian social cosmos. The absence of the significant restrictive mechanism of food transactions and commensality allows subversive uncertainty to spiral high, shattering and throwing into question any kernel of sociability and harmony in the village.

## All of Them are *Titro*

In order to forestall fellow villagers from objectifying their negative sentiments, yet being unable to negotiate, mitigate, or manipulate them otherwise, Thamgharians opt for construing an overall false appearance of evenness and uniformity in almost all realms of life. A critical token for this strategy is the tendency to strictly conceal all feelings and emotions, particularly negative ones, and to eschew direct verbal confrontation and physical violence. Even within the family nucleus interpersonal relations are, as Fürer-Haimendorf (1966b: 172) finds among Nepalese Chetris, rather formal and "there is not much room for spontaneity and the expression of warm emotions," with the sole exception of small children who are often hugged and pampered.

Concomitantly, Thamgharians maintain that the desired social strategy is one of *kāne-khusi garnu* or *khusi banāune*, literally, "making the ears/another person happy," that is, one should keep in mind another's expectations and preferences and calculate his speech accordingly since *tito satya bhandā miṭho asatya rāmro huncha*—a sweet lie is better than a bitter truth.

Numerous testimonies and rich local exegesis evince a certain daunting "geology" of self and personhood in Thamghar; villagers believe that the affable tongue and the calm and friendly appearance of their neighbors is merely a mask under which lurks a hidden, fearsome inner-self (heart).

A common saying in the village is that *mukhmā Ram Ram, bagalimā churā*—people [smile and] enunciate god's name (Ram, an avatār of Viṣṇu) while hiding a sharpened knife or a dagger in their pocket. Villagers who are *sojo mānche* (literally, straight, direct), that is, naive and genuine, are ridiculed and considered stupid. Instead, one should navigate one's life in a *bāṅgo-ṭiṅgo* manner (literally, convoluted, curved), using indirect and roundabout strategies. The notion of *cālāk* (astute, clever, prudent, artful), despite its otherwise considerably questionable reputation in general Nepalese public discourse, is one of the village's culturally intimate *ideals*.[20]

As found by Sharma (2001: 121) in a north-Indian Himalayan village so also in Thamghar (as will be further demonstrated later) "the politics of reputation become all-important"; honor, which is a major component of one's prestige, is the most precious possession a villager may have. This was tragically evident twice during the course of my fieldwork when those who had lost these decided to put an end to their lives. Thus, one of the most common local indirect methods of harming others is through violation of their privacy, by dispersion of true or false information that will subject them to loss of face and destroy their prestige and good name. This is often termed *mukh phornu*, meaning exposing the secrets and smashing one's "face."

Thamgharians' continuous attempts to spy, reveal, and expose information about others in order to harm them is also evident in the rich local vocabulary describing the different methods for obtaining and employing such information and those whose expertise it is, along with the acknowledgment of the different status accorded to information, depending on the kind of evidence provided for it and its authenticity.[21] Men and women alike are engaged in constant management of information, prevarication and dissemination of misinformation about themselves and others, and, as far as possible, concealment of personal details. The combination of the maintenance of silence, surveillance, and the manipulation of information, as well as the capacity for emotion and impression management as described by Gambetta (1993: 34–43) regarding Sicily's mistrust-saturated social milieu, seem to be essential social skills in Thamghar.

In such a state of affairs, it is no wonder that villagers often say they can believe *no one* but themselves, but this should not be taken too literally as will be further discussed in the following chapter regarding a husband and wife's relationship, for example. Thamgharians depict themselves in the words *hāami śaṅkālu mānche* [*haru*]—we are wary people; *sadhaī manmā ciso pascha*—negative feelings and suspicion, imagined as chill and moisture, constantly infiltrate the (otherwise solid, dry, and static) heart. Accordingly, the majority of villagers I addressed on this issue explained that they do not have any real friends in Thamghar. *Mit* (ritual friendship) is usually formed with people from other villages. Ram Saran Adhikari (16) told me he has many friends at school. Answering my question whether to some he may "open his heart" and divulge personal details and thoughts (*manmā kurā*), he told me: "How can I do that? They might use what I have told them against me." From a young age children are taught not to trust others and are instructed to avoid playing in courtyards of houses belonging to other *kuls*. Failure to adhere to this results in strong rebuke.

The many case histories and direct evidence I collected[22] lend support to the following grievance, often expressed by Thamgharian householders. A widespread complaint in the village is that "all people here are *titros*." A *titro (titrā)* is a common partridge of the surrounding jungle, known for its excellent ability to camouflage itself, and for its fantastic speed. It rarely enters the village, but when the corn in the terraces between the houses is high enough, its shrill screams are clearly heard from within the long leaves, yet it is very rarely, if ever, seen.

People say that such a bird, befriended by a hunter, would help him pursue fellow *titros*. To do this the bird lies on the ground in the middle of the forest, crying as if seriously wounded. Other birds that come to assist the wounded one are easy prey for the hunter hiding nearby. With this allegory in mind, it should come as no surprise that people usually limit their social contacts and cooperation to the inevitable, or that the nature of relationships is mostly instrumental and contractual.

The latter is well exemplified by the nature of the village's *parma*, which is a system of calculated mutual aid, and *paīco* (*paincho*), a well regulated manner of reciprocity, which every village household is "enmeshed" in, particularly since rice agriculture is too large a task for a single household to handle by itself.[23] In these highly symmetrical systems, work and goods are meticulously exchanged in a strict manner and are returned, as soon as possible, in identical terms of quantity, time, and workforce. Alternatively, an exact equivalence may simply be paid in cash or kind (mainly rice grains).

Hence, inherent within the fragile Thamgharian fabric of social life is an element that dissolves and paralyses it from within. This becomes more apparent if we consider the major social axiom recapitulated by Gell as follows:

> Social relationships, to endure over time, have to be founded on "unfinished business." The essence of exchange as a binding social force, is the delay, or lag, between transactions which, if the exchange relation is to endure, should never result in perfect reciprocation, but always in some renewed, residual, imbalance." (1998: 80–81)

Thamghar's rigid, symmetrical, contractual pattern of cooperation leaves no room for imbalance or laxity that may foster the development of social relationships or "entrap" one into friendship. Irregularity, if introduced at all, is difficult to contain and tends to spiral out of control; failure to reciprocate is regarded as *pāp* (sin) and is expected to elicit much anger, triggering the orchestration of an evil performance.

Furthermore, as Hart (1988: 189) observes, friends are free and friendship depends on the maintenance of such freedom, thus there is little wonder that in a social milieu dominated by coercion, anxiety, and threat, amity is only very rarely able to flourish.

## HOUSE, RICE, AND SELF

### Veiled Prestige

The Thamgharian tendency to contrive a uniform façade, enshrouding an intriguing diversity, is best illustrated by the emphasis placed on the concealment of wealth and scale of prestige. This practice appears at odds with the basic Euro-American notion of prestige as a social attribute, status, or reputation, based upon, negotiated, and constantly contested through public recognition. Moreover, in sharp contrast with north Indian praxis where a dowry is still a prominent means for assessing and representing social status, honor, and prestige (Roulet 1996), a Thamgharian bride's *dāijo* (dowry) is concealed within a trunk and its monetary component is given to the groom in a sealed envelope. What is the meaning of prestige and what are the reasons for and social implications of keeping its scale hidden? These issues, central to the dynamics of the Thamgharian social ethos of mistrust and hostility, will now be considered.

To a large extent, Ortner's assertion (1981: 360) that prestige criteria "provide the largest framework and the ultimate reference point for the organization of almost every aspect of social life," holds true in Thamghar, where *ijjat* (prestige and honor) seems to be of utmost importance, playing a central role in people's individual and collective perceptions in social relations. Yet, unlike what may be true of other Hindu communities (Ortner, ibid.), the "key" to high social status and prestige within the rather homogenous high-caste community in Thamghar is not caste, since the small minority of "Untouchables" does not constitute a reference group for Brāhmaṇs.

According to the Laws of Manu: "The seniority [prestige] of priests comes from knowledge; of rulers, from manly power; of commoners, from wealth in grain; and of servants alone, from birth" (Doniger and Smith 1991: 33)

Consistent with Manu, Thamgharian Brāhmaṇs, many of whom combine priesthood with the earthly vocation of farming, view rice grains as wealth and as the main scale of prestige,[24] while the rather obscure notion of "priestly knowledge" affords individual priests a further increment in social status. A man's prestige is also augmented by begetting male offspring, by his personal moral conduct, and, although not directly measured in these terms, his reputation and prestige also depend on the fidelity, chastity, and morality of his wife and daughters, who constitute what Das (1976b: 15) calls "a repository of [the] family['s] honor," as well as his conformity to various religious and social rules.

The importance of rice is tied to its status as the staple diet on which every family is completely dependent and this is probably the reason why, as in many other rice growing societies (Ohnuki-Tierney 1993: 63–80), it has become the measure or scale of wealth and prestige. To this day rice is still used as currency to pay for work and certain commodities. In Thamghar the scale of wealth and prestige (rice) is kept well hidden from the mutual inspection of villagers for 364 days of the year.[25] This is easily achieved by preventing all except family members from entering the *talo* (or *tala*, first floor) of the house, where the rice is stored.

Apart from embodying the scale of wealth and prestige, the amount of rice a householder is able to produce is also a major constituent of his perception of self and forms the basis for his self-esteem and confidence. This is particularly evident in villagers' reflections regarding the possibility that their rice may not meet their family's needs, despite the fact that there are usually sufficient surpluses. These notions are also evinced by the way Thamgharians imagine their house; stated differently, the architecture of their personhood is embedded in the manner in which Brāhmaṇs build and operate within their house, which is, first and foremost, a sanctuary and a shrine of rice. A short elaboration on the nature of the Thamgharian house will render the significance of rice in the life of a Thamgharian householder clearer.

## Man and His House: Man as His House

As will be demonstrated below, for the Brāhmaṇs of Thamghar the house is a focus of pivotal and complex processes, meanings, and perceptions of self, society, and the world in which they live. The house is a silent emblem symbolizing the continuous, precarious process of human struggle for survival in a hostile, "inauspicious" world, the inherent existential dualism and the inevitable immanent danger in this world. It is believed to be a relatively safe and secure place for living, for storing the family food supplies, for sleeping and having sex, as well as for cooking, eating, and feeding the gods.

The house is simultaneously a living entity and a *mandir* (temple), the dwelling of various gods, mainly the *pañchāyan deutā* (usually: Viṣṇu, Śiva, Gaṇeś, Sūrya, and Devī) and Bāstu Puruṣa, the virile god of the house, represented by an upright tree trunk thrust in the floor and attached to one of the main beams. The Thamgharian house represents a microcosm of which humans are believed to be part, a prominent segment in a series of nested microcosms (as is the case in many other places in India[26] and the world over),[27] which together appear to embody an existential fractal. Hence, as Karananda

Raj Upadhya (86), the most distinguished priest in the area[28] explains, all entities in the world should go through (the Hindu) *saṁskāras* (*sanskār*, life cycle rituals) but apart from humans, these are strictly performed only for a limited number of "living enshrined entities," most notably for the house.

Thamgharian Brāhmaṇs celebrate the conception or birth of their house, when the (male) foundation stone is placed inside Bhūme (the earth goddess, land),[29] and since the family residing in a house is regarded as its *ātmā* (*ātman*) or soul (self, mind, core of being, and God's *aṅsa*),[30] the house is said to be reborn (the second time, similarly to Brāhmaṇs who are "twice born"),[31] when they take up residence. The exact moment when new life is infused into the house is conceptualized as the moment when the fire in the *culo* (hearth) is ignited and the householder's wife cooks *khānā* for the first time.

The *culo*'s fire parallels the body's internal fire, which recooks and digests the ingested food, burning for as long as the person is alive. When this internal fire is weak and cannot digest the food properly, Pramod Bhatta (21) explains, "*hāmro peṭ lagne*" (we experience stomach-ache or diarrhea). Both fires must never be extinguished.[32]

The *saṁskāras* a house enjoys during its life span culminate in its marriage, which renders it *purā*, that is, "complete" or "whole" in a state of ultimate auspiciousness, purity, and security.[33] Finally, the house is believed to die and accordingly also become polluted like a human corpse when its *ātmā*, the family who dwells in it, leaves and moves elsewhere. Likewise, the human body is viewed as a house, a "living temple" and a microcosm in itself, the habitat of a divine couple—a male *mālik* (ruler) and a subordinate female *nokar* (servant). Both are nourished or "worshipped" twice daily in the rice meals, often referred to as *peṭ pūjā* (abdomen, internal *pūjā*).

Knowledgeable villagers explain how the entire universe "can be found" within (is analogous to) the human body, and how a house is in fact a person; the roof being the head, with walls for clothes and so on. The parallelism between the micro- and macrocosms of the house, the human body, and the universe is also manifested by the view that all are made of similar substances and share the same basic elements, be it *ḍuṅga-māṭo* (stone and soil, mud), which are forms of Pṛthivī (the earth goddess), or the *pañca tattwa* (five elements).[34] The relationship between man and his house is seen as one of complementary parts, albeit of different orders of being, which are analogous to each other.

Therefore, in Thamgharian cognition we find three fractal dimensions or images that seem to overlap and fuse together; each living entity, be it the cosmos, house, man, or any other, possesses an energetic, radiating burning core of life and is both a temple inhabited by god(s) and a body encompassing an *ātmā*, the rather elusive fraction of the distributed cosmic Brahman. Moreover, *ātmā* (mind) and body, spirit, and matter are not perceived as absolute categories but are taken in relational and relative terms. The body of one is the *ātmā* of another, thus each gains its ultimate significance from its particular scale of existence. In this way, Thamgharian Brāhmaṇs are able to steer clear of the mind-body dilemma that has been haunting Western philosophy for generations. Indeed, the question arises only if we assume that mind and body, spirit and matter are contradictory or at least fundamentally distinct, but this perception is usually overridden by Hindu definitions of reality (Marriott 1989). The above Thamgharian perception also seems able to lay to rest what Western scholars often depict as the perennial Hindu "paradox," presented by the *Upaniṣhadic* equivalence of Brahman-*ātman* and their simultaneous transcendence and immanence in the world.[35]

Language highlights the notional analogy between man and his house; the same term, *ghar*, is used metonymically to denote both the house and the family dwelling inside. Furthermore, the house is not merely the body and skin of its human "*ātmā*," or the embodiment of the agency of the householder responsible for its construction; Thamgharian householders actually identify with their house and, asked why they would not let fellow villagers into the *talo* (first floor), they answer: "We don't know what is in other people's hearts so how can we let them into our heart?" Thus a householder dwells within the universe he himself embodies, in line with the transcendence and immanence of the universal Brahman and the major Hindu gods.

Thamgharians' tendency to lead their lives according to similar standards of living and their attempts to engineer a façade of uniformity are expressed in terms of clothing, education, or the use of consumer products, as well as in the avoidance of any material ostentation.[36] It is also evident from the simple and rather standard structure of their houses, which do not serve as vehicles for the objectification of rank, wealth, power, or prestige but as an instrument for the latter's concealment.[37]

The layout of a Thamgharian house is relatively simple; two to three stories high with one large room on each floor,[38] it is organized according to and embodies the same cardinal biaxial concerns that govern the lives of its inhabitants: purity-sacredness (which also implies auspiciousness) and trust. In theory, every villager (apart from the village "Untouchables") may enter the house and, once inside, is subject to the numerous constraints marked by the intricate inner division of the ground floor space along the axis of purity.

At one extreme, at the peak of purity and sacredness stands the *culo* (with the family's modest shrine close at hand) situated at the apex of a series of leveled mud platforms dividing the floor of the *bhānsā* (the kitchen, cooking and eating area—the universe of women). Each level demarcates the designated area one is allowed to occupy during a meal, according to one's level of ritual purity. The *culo* is often situated at the farthest corner away from the only door and is hidden behind a wall, thwarting any potentially evil gaze that may sabotage the cooking and eating processes.[39] The only window in the ground floor, made of a dense wooden lattice, is small and a relatively low one (sometimes located no more than half a meter above floor level), which doesn't disturb the almost complete darkness of the *bhānsā*. Preventing any view of the interior, it enables the householder's wife, busy inside, to listen to outside discussions.

While the solid mud and stone structure of the house, together with its thatched (or stone-tiled) roof, protects its inner domains and inhabitants from the elements, wild animals (mainly tigers), and from fellow villagers' inquisitive and, at times, malevolent gaze, it is only the more fluid, virtually invisible *balini* (*baleṇḍāri*)—the sphere fashioned on the ground by the raindrops falling off the roof—that embodies its potent "shield" against most evil spirits.[40] Despite its ephemeral and unidimensional appearance, the *balini* is conceptualized as enveloping and protecting the house from all sides.

The Thamgharian house, embattled architecturally and ritually, undergoes continuous ritual fortification. Moreover, inside, along its solid structure and particularly above its main orifice (the door) dwell powerful guardian gods and deities assisted by their *senā* (army) of protecting spirits. Nonetheless, this seems to fall short of instilling the desired feelings of safety and confidence in villagers' hearts, as attested by the general propensity to sleep outside the walls, on the lower or upper verandas surrounding the house on three sides, in order to guard it.[41] Villagers are not inclined to break this habit despite their great fear of the wild tigers that often roam the village after dark in search of prey.[42]

The second prominent axis dividing the Thamgharian house—the axis of trust, dividing the entire structure vertically—will now be considered. This axis draws upon the imagery of the house as the embodiment of the householder himself, where the *talo* (the first floor), analogous to his inner-self or heart, is the locus of intention and emotion. It is there that the prominent scale of prestige and wealth (rice) is held in *bhakaris* (woven bamboo containers, large enough for a young boy to stand or sit inside), together with other forms of wealth such as cash, jewelry, and other special family possessions.

As implied above, a strict taboo precludes anyone but family members who share the *culo* from wandering into the *talo*, which is accessible only via the ground floor.[43] Even *kul* members are not normally invited to cross the threshold into the house without particular reason and, along with other fellow villagers, are entertained on the outside veranda, sitting on *gundris* (rice straw mats woven by the householder's wife).

Why, in light of their highly competitive spirit and preoccupation with making mutual comparisons and judgments, should people feel the need to conceal the scale of wealth and prestige and avoid its inspection throughout the year? No doubt, this tendency creates further tension and anxiety in the community since assessment is hard to make while the scale is veiled behind walls. At least part of the answer may already be apparent.

One reason for keeping both the scale and one's position on it hidden is that this prevents the possibility of public recognition of the derogatory state of poverty, which entails disgrace and loss of face, prestige, and almost all social status. This happens when no more rice is left on the first floor and the householder must resort to buying rice in order to feed his family. Villagers will do everything in their power to avoid this, even at the cost of their children going hungry. Another reason for hiding the rice is the general belief that jealousy and mutual competition would incite villagers to use spells, the evil eye, and other forms of sorcery to harm their neighbors. Rice, one of the most important possessions of a family, should therefore be hidden away. The extreme mutual suspicion and anxiety in Thamghar seem to be epitomized and objectified by (the communal belief in) *birs*—terrible, destructive demons, more malicious than any other evil spirits or tools at the hands of the village *boksis*.[44]

People are said to harbor a *bir* in order to destroy fellow villagers and their houses by sending it to invade and demolish, or to literally devour them from within. Thus a Brāhmaṇ householder together with his family may find themselves "sacrificed" by others and consumed in their own shrine. While such a *bir* is active there is nothing to protect the house or its inhabitants from the terrible invasion and the series of complicated *pūjās* that are performed in order to exorcise *birs* are only rarely effective.

Villagers explain that *birs* are kept on the *talo* in a special box (*madus*) or inside one of the empty *bhakaris* and demand careful, daily care and occasional blood sacrifices. Unless kept content, an angry *bir* might harm and even kill one of the members of the household where he resides. One of the common and milder uses of a *bir* is to send it to other people's houses to take away the *saha* and steal the rice hidden on the first floor, grain by grain, until no rice is left for the family to consume. In Thamghar, as in other high-caste Nepalese communities (e.g., Rutter 1993: 122, 212), *saha*[45] is referred to as the mysterious ability of rice (but sometimes also of other foodstuffs and money) to multiply infinitely and the potency of everlasting, unlimited sufficiency. Villagers are convinced that without *saha*, the rice supply would not suffice for a family's subsistence. Hence, concealed from malevolent eyes and out of direct reach, the scale of wealth and prestige is thought to be subject to evil manipulation and destruction. Consequently, while some (financially) modest families are surprisingly able to hold lavish, festive meals during religious ceremonies, others of similar status must clandestinely buy food for their families. This is indeed a dire scenario.

On other occasions, *birs* are said to be unleashed in order to invade the bodies of (mainly) young *buhāris* and possess them. As a result they may become completely barren or merely "fail" to give birth to a male offspring, and might finally die in pain when the *bir*, devouring their bodies from within, reaches the heart.

People also note that others do not allow them into the *talo* since they are afraid to reveal the presence of a *bir* therein. Accusations of keeping *birs* (sorcery) are among the gravest that can be made in Thamghar, hence all available means are employed to avoid false accusations. This explains why householders keep their "heart" and its contents hidden and do not allow it to become exposed to common knowledge and malign use.

In short, a *bir* embodies villagers' dark emotions and represents direct and indirect evil attack on others' prestige, primarily by stealing their rice and its *saha*, but also through invasion and possession of women's bodies, with the obvious sexual overtones and potential lethal consequences this may entail.[46] Thamgharians say that when a *bir* attacks a household, the main beam of the house (to which the manifestation of Bāstu is attached) and its entire structure starts to *kāmnu* (tremble violently) and indeed, when the house becomes possessed, the householder himself is fundamentally shaken and undermined.

Unlike the hierarchies of ritual purity, caste, age, and *kul* seniority, one's position on the hidden scale of wealth is not stable but is subject to constant change. Primarily this change is perceived to come about through the evil manipulation of *birs*, but it may also be generated by the acquisition of wealth through ordinary means such as employment elsewhere in Nepal or in India. Any such change necessarily affects all other positions on the scale, potentially undermining one's own, implying danger and giving rise to feelings of uncertainty and grave anxiety. Unlike Webber's Calvinist Protestants, for Thamgharian Brāhmaṇs there is nothing divine about the acquisition of wealth—it is the principle of hierarchy that is deemed sacred.

Since it is hidden from the public eye, scale triggers the imagination and releases doubt on the one hand, while providing an explanatory framework and justification for personal misery as well as others' success on the other hand. The outcome is a state where hierarchy is obscured and believed to be subject to demonic, uncontrolled alteration and manipulation.

## A NUMBER OF CONCLUDING REMARKS

All facets of personal and social life in Thamghar are pervaded by acute uncertainty, endemic anxiety, and mistrust that seem to build upon the perception of a fundamental duality in the world; the notion that, for example, there are two distinct aspects to both houses and humans: the benign façade and the potentially "evil" interior. Put differently, Thamgharian Brāhmaṇs believe that they all wear masks in a macabre social play on Goffman's back and front stages (Goffman 1959), which draws upon the equivalence of the exterior social persona = moral and amiable but inauthentic, and inner-self = authentic but negative and treacherous. The strategies employed to overcome the paucity of trust in social life inevitably fail, further aggravating the situation and proving the truism that lack of information destroys trust, as noted by the eighteenth-century Italian thinkers Doria and Genovesi (Pagden 1988).

Unlike the relationships Brāhmaṇs maintain with the nonhuman forces of the cosmos, where the intrinsic unpredictability is kept in check through instrumental "gastro-mechanical" means, within the human social world, uncertainty spirals sky-high in a rather unrestrained manner. As Gambetta (1988b: 234) writes, distrust "has the capacity to be *self-fulfilling*, to generate a reality consistent with itself."[47] Indeed, as we have seen, mistrust is associated with complex dynamic processes involving mutually affecting elements, which are intimately embroiled in people's emotional apparatus and may generate particular moods. These can affect villagers' judgment of the world and other people in it and have clear endemic, self-perpetuating properties.

Thus mistrust may gain momentum, creating a vicious cycle of deep-seated suspicion, hostility, and an ethos of doubt. Even in the absence of other forms of misfortune or incidents indicative of covert masked hostility, the inevitable, gradual annual decline in the quantity of rice seems to provide sufficient grounds for generating "cognitive and emotional inertia" (following Good 1988: 41–43), and thus fostering mutual resentment and suspicion. Put differently, the lacunae of hunger fashioned gradually in householders' "hearts" (within the *bhakaris* of rice) beckon *bir*s to creep in and dwell inside, eventually leading to the downfall of sociability in an entropic process of dissolution and paralysis.

In general, it seems questionable whether a vicious cycle of mistrust and hostility, even when bound to what appears to be an infinite escalation (either in the presence or the absence of apparent counterbalancing social elements) in a certain community, would necessarily draw the latter into irredeemable fragmentation. This is since just like expressions of sociality do not necessarily imply much in the way of sociability, altruism, solidarity, or trust, endemic mistrust and covert hostility do not necessarily exclude all cooperation or the possibility of strong interdependency.

Likewise, as will be demonstrated in Part III (which analyzes rice agriculture) and particularly in chapter 8 (focusing on the final stages of the cultivation process), an increase in cooperation and expression of sociability does not necessarily imply a deep-seated interpersonal attachment and long-term trust. Nonetheless, the analysis of the complex rice agricultural process in Part III refines the picture of the social universe portrayed hitherto, suggesting that rather than being bound to an irreversible upward spiral, levels of mistrust and hostility in Thamghar are subject to a fluctuating annual process.

Although Thamgharian Brāhmaṇs draw upon a similar pool of cultural ideas and approaches to misfortune shared by various other high-castes,[48] various Nepalese ethnic groups,[49] and many people in India,[50] and though the discourse of evil spirits, sorcery, and witchcraft undoubtedly involves a certain level of interpersonal mistrust, it does not necessarily follow that a community where such discourse exists will have a social milieu akin to that of Thamghar.

Clearly, as Strathern (1980) argues regarding the ostensibly universal nature-culture dichotomy, I would suggest that evil spirits, sorcery, and witchcraft beliefs, as well as endemic mistrust, although they may seem universal or pan-Hindu, can only gain their true meaning and significance within their specific context. Moreover, the analytical expositions of the socio-historical dynamics in a number of relevant social circumstances, for example, Doria and Genovesi,[51] Gambetta (1993),[52] and Macfarlane (1985),[53] demonstrate that these involve extremely complex processes and a combination of elements whose significance derives from their specific social and cultural contexts.

This further highlights the limited applicability of general explanatory frameworks, such as the allegedly universal struggle for "the limited good" in "peasant society" (Foster 1965), for explaining specific social circumstances.[54] Therefore, Stone's (1988: 76–78) suggestion (following Foster, ibid.) that it is the persistent fear of shortages and the "ideology of scarcity" that place pressure and lead to ill, instrumental interpersonal relationships in the particular high-caste central Nepalese community she studies, is to my mind, unsatisfactory and cannot account for the situation in Thamghar.

I hope that the discussion hitherto has clarified that even if Foster's definition of a peasant community was applicable to Thamghar, viewing the village's social milieu, which exhibits some of the interpersonal relationships Foster (and Stone) aims to account for, solely or mainly in terms of "the image of the limited good," is indeed limiting and misleading. Furthermore, most Thamgharian households do not suffer from an annual shortage of foodgrains, including rice.

As far as historical explanations for the emergence of the particular social circumstances in Thamghar are concerned, the lack of relevant historical documents and the limited vantage point provided by a relatively short-term anthropological fieldwork, preclude me from going much beyond Thamgharian villagers' own justification for the *dukkha* associated with living in a deeply fragmented and atomized social world. Drawing upon the general template of Hindu historical determinism, villagers often maintain that they are experiencing the impending closure (in catastrophic destruction) of the present Kāli *yug*—the era of darkness and amorality.[55]

I have thus confined my discussion to the exposure, extrapolation, and consideration of some of the salient, mutually reinforcing and perpetuating elements that seem to foster villagers' constant rearrangement of their social realities in a manner that reproduces their mutual mistrust and resentment.

I now wish to take a further, complementary step to complete the ethnographic picture of the Thamgharian social milieu portrayed thus far and discuss the fluctuating life stages and various kinship roles Thamgharian women—the ambivalent "female core" of the family—assume throughout their lives. This will set the scene for introducing the particular way villagers imagine the unity fashioned between husband and wife during the marriage ceremony, namely the (fractalic) image of "matrimonial encompassment," which forms the focal point of the entire book.

## NOTES

1. Reflecting on cohesion and trust in traditional Islamic society and in the Soviet Union, Gellner (1988) advances the idea that anarchy and a lack of centralized authority may produce trust and cohesion.

2. See also Ramirez (2000a: 252).

3. At the time, the newly elected UML (Communist) government introduced a 500,000 NR budget for every VDC (Village Development Committee), each comprising more than ten villages.

4. *Taittiriya Upaniṣad* 2.2.1.

5. *Taittiriya Upaniṣad* 3.10.6.

6. This often follows the prescription of a *jānne mānche*, usually a priest, astrologer, and healer. Diagnosis is based on a series of signs, cosmic indexes, and physical examination of the victim.

7. Fuller (1992: 155–81).

8. Among them are *bhūt*, *pret*, *masān*, *picās* (*pichas*), *tarhsāune*, *rāches, batās*, *lagan*, *hidowā*, *pañchābahia*, which all come under the general category of *ched*.

9. These are marked by the tree-temples of *ban* (forest) Devīs (fierce territorial goddesses). Each *ban* Devī is worshiped annually by the inhabitants of the village area she is believed to protect.

10. The earth goddess, also called *Annapurna*—full of grain.

11. This famous feature of the Nepali landscape is found along mountainous footpaths throughout the hilly areas of the country. A *cautāro* consists of a rectangular stone-coated ramp about 1.5 meters high, with steps surrounding it at a level convenient for porters to support their load and for people to sit. At its center, two sacred

trees are planted and married ceremoniously, making it an auspicious, safe resting place for travelers.

12. Michener, DeLamater, and Schwartz (1990: 114–45). See also Appadurai (1985: 245) regarding the Western tendency to conceptualize "gratitude" as referring to some inner disposition of the actor.

13. The meanings and significance of rice, which stretch far beyond its obvious role as a staple food, and its unique place in the life of Thamgharian Brāhmaṇs, will be further discussed below and in subsequent chapters.

14. Regarding India see Marriott (1968, 1976), Khare (1976a), Dumont (1980), and Selwyn (1980). In Nepal this mainly applies to cooked rice, as discussed by Stone (1978, 1983), Höfer (1979), Löwdin (1986), and Ramirez (2000a: 31).

15. Highly reminiscent of Vaiṣṇav Assamese (Cantlie 1981: 46), the widely discussed (i.e., Khare 1976a) north Indian distinction between *kaccā* (inferior food boiled in water) and *pakkā* (superior food fried in ghee), is unknown in Thamghar, where the former simply denotes raw, uncooked food and the latter refers to any type of cooked food, in keeping with the literal meaning of these terms noted by Dumont (1980: 384, n64b). In its wider and mainly idiomatic usage, *pakkā* also means being ripe and mature, with strong sexual overtones.

16. Original emphasis.

17. This will be elaborated upon in chapter 3.

18. This appears to be the case among many high-castes in India (Mandelbaum 1968), but compare Bennett (1983: 135–36) and Lecomte-Tilouine (1993a) who find the *kul* and its communal *pūjā* to be the epitome of agnatic solidarity.

19. Cited in Gambetta (1988a: 163).

20. Following Herzfeld (1997: 3).

21. For example, *kuro* (*kura*) *lāunu*: to reveal the secrets of another, *gaigwi garnu*: to spread a false and offensive rumor about someone who is not present, *ciyo garnu*: to invest much effort for obtaining secret information, often via a discreet channel.

22. I refrain from mentioning specific cases in this regard as the people involved may be too easily recognized.

23. Only a number of Brāhmaṇ householders can afford to hire workers on a daily salary (*nivek*) basis.

24. This follows Strathern's (2000: 51–53) discussion of scale, its meanings and implications, observing that the notion of culture as the workings of human activity and their varied effects and consequences, implies a measure or scale. In Thamghar, these consequences are scaled by the amount of rice a household produces.

25. The sole annual occasion when it is briefly displayed forms the focus of chapter 8.

26. For example, Moore (1983: 220, 1989), Daniel (1984: 105–62).

27. For example, Bourdieu (1977: 89–91, 1992, particularly 277), Howe (1983), Carsten and Hugh-Jones (1995), Hugh-Jones (1995), Rivière (1995), and Waterson (1997).

28. Who proudly displays his genealogy, showing that he is a descendant of the *siddha* Damodar's firstborn son.

29. From this moment on and throughout the entire construction process, the householder remains at the site day and night, guarding it to prevent others from inflicting evil on his house during this critical stage. Later, ideally at least, the house is never left unattended.

30. *Aṅgsa*—one of God's many equal, shared parts or limbs.

31. *Bartoon*, the second major "rite of passage" for a Brāhmaṇ male (the first being *nuwāran*—name-giving ceremony), is a male initiation rite usually undertaken between the ages of eight to fifteen (or later, though it must be performed before marriage), where the boy is given his *gāyatrī* (*gāyantrī*) mantra (sacred Vedic verse to be recited every morning) and his *janaï* (sacred thread)—the hallmark of high-caste status, preparing him (in theory at least) for adult life and marriage. It is also called *bratabandha* or *bartaman* (Bennett 1983: 59–70). A woman is said to be reborn for the second time in her marriage.

32. Although this is the most fundamental and commonly presented local perception of the body, a few knowledgeable persons are familiar with a more complex perception where in addition to the *ātmā* the human body also contains the *jīv*, which is the essence of life or personal life-impulse. The *jīv* and *ātmā* are commonly held to be a couple (yet with no specific gender attribution), and quite similarly to a husband and wife (as elaborated in the following chapter) are in a clear hierarchical relationship; one is a *nokar* (slave) while the other is the *mālik* (master). Both reside within the body (abdomen, thorax) rather than the head. Additionally, the abdomen also contains the *aat*, a special location where all bodily energy is concentrated and stems from.

33. The marriage of the house in Bāstu *pūjā* is further discussed in chapter 6.

34. These are *Pṛthivī* (land, earth), *akās* (space, sky), *jal* (water), *tej* (energy, power, ray), and *bāyu* (wind).

35. On this theme see, for example, Renou (1953: 69) and Hiriyanna (1993: 48–72).

36. In sharp contrast with many other Nepalese women who often wear much gold in the form of numerous earrings and necklaces, most Thamgharian women wear the bare minimum to mark their married status, except when they embark on a rare trip out of the village.

37. There are of course minor differences, but most of them are not evident from the outside.

38. On the Brāhmaṇ-Chetri house see also the meticulous accounts of Gaborieau (1991) and Bouillier (1991).

39. The level of a household's orthodoxy is often expressed in this intricate division; a simple inner division of space is a sign of low concern about the issue of purity and pollution and vice versa.

40. During particular rituals the *balini* is reinforced by circles of water and sometimes also sacrificial blood.

41. In some households women may sleep inside.

42. During my fieldwork, tigers tragically killed two girls sleeping outside in a neighboring village.

43. Elsewhere in central Nepal, the ill-trusted daughter-in-law is barred such access (Gaborieau 1991: 49), yet in Thamghar, although the *buhāri* is often not trusted, I never heard of such practice.

44. About *birs* in other high-caste communities see Höfer and Shrestha (1973) and Stone (1977: 68–71, 1983: 974–75).

45. *Saha* derives from the ancient Vedic concept of *sāhas* (often associated with *Agni*, the god of fire), which denotes potency and energy, the combination of power-substance with an autonomous independent character (or agency) in the possession of gods and other powerful beings, objects and substances, enabling them to conquer and rendering them effective and influential (Gonda 1957: 1–25). Turner (1996: 593) adds that *saha* also means help, plenty, the sexual desire of female animals or sexual heat, and is used to denote animal copulation. The *Tharus* in the Dang valley see *saha* as "the beautiful season of abundance—the time when the rice is ripe and ready to be harvested" and also associate the word with desire and love (Krauskopff 2003: 78, 82).

46. In south India possession also means demonic sexual abuse (Nabokov 1997). Thamgharian householders' approach to the possession of their wives by *birs* implies this, yet they avoid explicit reference to it.

47. Original emphasis.

48. Höfer and Shrestha (1973), Blustain (1977), Stone (1977), Bennett (1983), Rutter (1993: 214–16), and Gray (1995).

49. Gellner and Shrestha (1993); Reinhard (1994: 263–92); Macdonald (1994: 376–81); Gellner (1994); and the comparative discussion in Höfer and Shrestha (1973: 62–67).

50. Epstein (1959); Pocock (1973: 25–40); Babb (1975: 200–8); Standing (1981); Caplan (1985); Fuller (1992: 236–40); and Kapferer (1997).

51. Regarding the deep-seated mistrust in eighteenth-century Naples, discussed by Pagden (1988).

52. About the rise of the Sicilian mafia, which encourages and makes profit from mistrust.

53. Regarding the break-up of witchcraft and the disappearance of evil from English social discourse between the sixteenth and eighteenth centuries.

54. The limitations of Foster's thesis have been discussed by many; see, for example, Piker (1966) and Kaplan and Saler (1966).

55. See Good (2000: 281) for an eloquent portrayal of the Hindu perception of the cyclic progression of time.

# 3

## Encompassing the Ambivalent Female Core

### INTRODUCTION

Focusing on various aspects of Hindu kinship and gender, recent accounts by Trawick (1990), Raheja and Gold (1994), Lamb (2000), and Osella and Osella (2000),[1] provide prominent examples of the complexity and multivocality of the Hindu modes of thinking. The latter, as argued by many, are characterized by the absence of absolute scales, the fluidity of categories, and the capacity to maintain a relativistic approach, allowing contradictory points of view to coexist. These notions seem far removed from Western systems of reasoning, which incessantly attempt to unearth coherence and hidden logic, and to divide and compartmentalize reality.[2]

Arguably one of the finest works compiled on high-caste Brāhmaṇs-Chetris and their womenfolk, Bennett's *Dangerous Wives and Sacred Sisters* (1983)—a significant landmark in the anthropology of Nepal in its time—still seems to inform, and even to a large extent dominate, the scholarly discourse on high-caste Hindu Nepalese women. Nonetheless, its main structurally oriented thesis, portraying the way in which, from a Hindu male standpoint, women are both socially and symbolically split between the roles of wife and sister in the respective "partrifocal" and "filiafocal" contexts, and the implied dichotomy between *danger* and *sacredness* are, to my mind and as I shall attempt to demonstrate below, somewhat misleading. I therefore wish to distance myself from this approach on a number of major issues.

First, if as Bennett argues the goddess (Devī) with her dual gentle and fierce aspects, often identified with the two logically opposed personae and iconographies of the dark, malevolent Kāli and the docile, benevolent Gaurī (or Pārvati),[3] is indeed a core symbol that informs the social roles of actual women, sacredness and danger cannot be viewed as separate throughout women's various life stages and their different roles as daughters, sisters, wives, and mothers. This is since, as Bennett herself skillfully elaborates (and as is also true in Thamghar), the various aspects of the goddess are fundamentally intertwined and embedded *together* in her image as it is fashioned in villagers' minds (Bennett 1983: 261–308). Viewed another way, the range of Devī's personalities and forms—complex, varied, fluid, and transmutable as they are, which reproduces the basic female paradigm as Hindus see it—reflects the mixture of these notions found in actual women (Harlan and Courtright 1995: 8–15).

As I shall argue later, not only do women's roles and positions change throughout their lives and in accordance with specific contexts, but they are viewed from multiple and shifting perspectives *simultaneously*. Consequently, the "male perspective" of their womenfolk (or the "patrifocal ideology" to use Bennett's terminology) and the realities of women's lives cannot simply be depicted along the lines of their existence as "pure *or* polluting, sacred *or* dangerous" (Bennett 1983: 316, [my emphasis]), but are more complex. Furthermore, to my mind, Bennett's claim that the dual aspects of Devī are in agreement with the different roles women embody, equating the virgin, unmarried consanguinal woman with sacredness and the affinal woman with danger, entails an intrinsic contradiction, since it is the *single* virgin Kāli who is regarded as dangerous and often "out of control," while the *married* Pārvati is the gentle and benevolent one.

Marriott's (1998: passim, particularly 288–84) critique of previous scholarly discussions of the Hindu woman,[4] whom he aptly terms "the female family core," is highly pertinent here. Marriott argues that no single value, perspective, or line of analysis, no matter how accurately depicted, can be fully adequate; instead, multiple contextualized views and larger syntheses are needed. He further maintains that "not all [female] roles need be conceived as contradictory, for they are intercommunicating and share components that are easily joined" (ibid.).

Nevertheless, Bennett's analytical account remains one of the richest to date, an invaluable source of reference on high-caste women and Brāhmaṇ-Chetri culture which, given its unquestionable scholarship and insight, will no doubt remain a prominent anchor for further debate.

The approach adopted herein is governed by the notion that Thamgharian kinship ideas, and in particular the perception of women, should be imagined as a conglomerate of multiple perspectives and (frequently) contradictory dispositions that embody a mental hybrid vivid in villagers' minds. Presentation of a comprehensive picture of this hybridity is beyond the scope of the present chapter, thus I shall attempt to expose only those key elements governing the local perceptions of kinship and gender, which are crucial for the comprehension of the workings and implications of the fractalic gendered image governing Thamgharian life or serve to illuminate the daunting Thamgharian social ethos and emotional milieu described in the previous chapter.

This chapter opens with an examination of women's main life-stages from birth to marriage in order to provide the necessary basis for discussing Thamgharian marriage imagery and explicating its significance, which is of fundamental importance to the remainder of this work. Among various central kinship notions, the logic of the transference of *dān*, the ominous "Indian gift" imbued with the giver's negative "moral substances" such as sins and various sinister influences and qualities, will be discussed. This is highly pertinent to the manner in which the bride, who is regarded as the ultimate (virgin) "gift" or *kanyādān*, is conceptualized in Thamghar. Concomitantly, the cardinal multiple perspectives on the in-married wife will be discussed and elaborated upon. The Thamgharian marriage imagery and the key notion of "matrimonial encompassment" will be presented toward the end of the chapter and will be followed by an analysis of its significance.

The concluding discussion will suggest that it is particularly the dialectics of various cardinal concepts, kinship notions, and prominent mental images elicited by the image of matrimonial encompassment that may provide the grounds for an ongoing intensification of the vicious cycle of mistrust and hostility characterizing the village social milieu. Finally, a short comparative note will examine other important manners in which Hindus tend to imagine matrimony. The juxtaposition of the Thamgharian matrimonial encompassment with an almost identical marriage imagery set within the very different southern Indian Tamil kinship milieu will, I hope, prove particularly illuminating. In both (Hindu) contexts the mental images of the conjugal bond are highly significant and pertinent to the perceptions of self and relationship, but these perceptions are, however, very different in nature.

The general Thamgharian kinship notions discussed herein are obviously not unique, sharing many cardinal elements with other high-caste Hindu communities in Nepal[5] and northern India.[6] Likewise, to a large extent, Thamgharian ideas of kinship and cardinal perceptions of women[7] vis-à-vis men,[8] particularly those that place differences in action, capacity, energy (*śakti*), and causation at the core of the unalterable disparity between the sexes and provide the primary reason for their need to be merged in marriage, echo earlier notions that may be found throughout Hindu tradition as is well inscribed in ancient Hindu scriptures starting from the Vedas.[9] Concomitantly, the final comparison with northern and southern India will demonstrate how the concept of matrimonial encompassment is far more common throughout the Hindu world than may first appear to be the case.

## WOMEN IN FLUX

For both clarity and brevity, the ensuing discussion will follow the main stages in Thamgharian women's lives—the primary "role-types" they embody throughout their fluctuating life-trajectory.[10]

### *Kanyā*—the Virgin Goddess

As will be demonstrated below, Thamgharian attitudes toward children in general and young girls in particular are fraught with contradictory notions and practices, as well as numerous irregularities, reflecting the fundamental ambivalence associated with *śakti*.[11] Particularly salient are the inconsistencies associated with ritual vis-à-vis purity status and age seniority. For boys, "ambivalence" dissolves early on in childhood together with their aura of divinity, while girls remain ambivalent to varying degrees throughout their lives.

All babies are likened to *deutās* (*deotās*, divinities), thus they are resistant to birth and any other forms of pollution. Young children are free to wander into the purest corners of the house; like local royalty or divinities, they are the first to receive their meals and all their whims are satisfied. Their immunity to pollution allows them to associate freely with menstruating women and the village "Untouchables," and they are sometimes even encouraged to eat from the hands of the latter as a ploy against the evil eye.

However, children are also said to be born Śūdras (a general title for "Untouchables"), thus their food leftovers are not considered auspicious *prasād*,[12] but are regarded as highly polluting. In the same vein, notwithstanding their divinity, children's lower purity status precludes pre-*bartoon* boys or (unmarried) *kanyās* from cooking in their natal home. This is due to the Thamgharian view that, as also noted by Pandey (1993: 1–35), the Hindu *saṁskāras* are mainly purificatory rites, facilitating one's gradual ascent up the scale of ritual purity. While boys attain their utmost ritual purity through *bartoon*, girls, whose purity status in adulthood remains inferior, become ultimately pure only when united with a man in marriage.

The title *kanyā* (or *konyā*), attributed to all girls from birth until they reach the age of ten, embraces the idea that the divine auspicious status of a young virgin depends on her intense *śakti* being controlled and bounded, that is, well *contained* and sealed within her body; a *kanyā* in her father's house parallels the south Indian goddess sealed within her shrine or a box (Shulman 1980: 192–211, 349–50).

Once her *śakti* is "unleashed" and inauspiciously "displayed" when the girl reaches menarche or is no longer a *virgo intacta*, not only is it depleted, but it also becomes extremely polluting and harmful to men, particularly her own kinsmen, brothers, and father. Given that most village girls do not experience puberty before the age of fifteen, the large five-year safety margin adopted by villagers reflects their deep-seated anxiety regarding uncontrolled female *śakti*.

A *kanyā*'s *śakti*, most intense in early childhood, ranks the younger *kanyā* above her older sisters in terms of ritual status. It also renders her immune to evil spirits or sorcery yet magnifies her potential danger, if born inauspiciously (i.e., under *mul nakṣatra* [*nachetra*]—inauspicious star),[13] compared with the relatively minor danger posed by a boy born in similar circumstances. I know about one village girl who was born inauspiciously under a particularly ominous celestial situation. In order to avoid death as a result of his daughter's gaze, the father took the drastic measure of sending the baby together with her mother to the latter's *māiti* (natal home), where they stayed until the child was six years old. In the meantime, the father's second wife had given birth to two sons.

Despite her divine status and enormous *śakti* during early childhood, a sister's notable ability to bless and ensure the life of her brothers in Bhai (younger brother's)[14] Ṭikā,[15] is referred to with a pinch of salt by her older brothers. It is only following her marriage, long after her *kanyā* status has expired, that village men regard their sister's ability to keep death at bay with the utmost seriousness. Similarly, the patriarch (a girl's father or grandfather) often jokingly performs *ḍhognu*[16] to her.

Like the other female members of the household (including her mother), a *kanyā* is strictly excluded from various major ritual contexts, notably the *kul deutā pūjā* (sacrifice to the lineage's deity), but is nevertheless worshipped as a goddess in others, to which I now turn.

A young girl's divine status is epitomized in her ability to both personify the goddess in various family and village rituals (prominently in the central Dasaī *pūjā* held at the regional Kāli temple), and embody the family's "field of *dān*"[17]—a role she retains throughout her life. This terminology merges together the general Hindu view of a woman as a "field,"[18] with that of Thai Buddhist monks as embodying their laymen's "field of merit" (Tambiah 1984: 16, 54). A girl's prominent role as a "field of *dān*" is underscored in her marriage, conceptualized as *kanyādān*, literally "the 'gift' of a virgin." Emblematic of high-caste Nepalese and northern Indian marriages,[19] and despite its various local adaptations, this concept seems to underlie *all* Hindu Indian marriages as Trautmann (1981: 277–315) cogently argues.

Raheja (1988) may have perhaps overstretched her point, arguing that the giving and receiving of *dān* should be at the *center* of all interpretations of Hindu life,[20] yet the significance of the concept of *dān* in the Hindu universe of thought and practice is beyond doubt.[21] The following elaboration on the Thamgharian notion of *dān* is designed to provide the necessary background information for the ensuing analysis of the Thamgharian marriage.[22]

### *Dān* and *Kanyādān*

Previous research on the high-caste Nepalese system of prestations and salutations,[23] which was particularly concerned with the hierarchical principles governing the exchange of *ṭikā*,[24] the giving of *dakṣiṇā* (*dacchinā*, see below) and the performance of *ḍhognu*, did not focus on *dān* and its menacing aspects. In Thamghar, *dān* transference is a multifarious, somewhat fluid issue, with its own ambiguities, inconsistencies, specific and elaborate rules of performance, and plethora of exceptions.

Thamgharians distinguish between two different types of *dān*. The first is the "ideal gift" (Burghart 1987: 244, n4) given to the poor, free from thoughts of personal gain or immediate return; the donor expects to find the object donated (be it, for example, a bed or a Bhūme-*dān*—a piece of land or property) in *swarga* (the abode of the gods, heaven) as an "insurance," to borrow from Mauss (1967: 55), after his death. This type of *dān* is of little relevance here and very few villagers possess the necessary financial resources required for giving such a rare and prestigious "pure" gift, uncontaminated by evil.

The present discussion will focus on the second type of *dān*, which rests on the idea that by giving, one gives oneself (Mauss 1967: 57–59). More specifically, as Parry writes, *dān* is the dubious Indian "gift" containing part of the spiritual essence of the donor, given with the underlying goal of disposal of his or her "ethical burdens by ritualistic means" (Parry 1986: 459–63, 1989: 66–77, 1994: 132).[25] Indeed, in Thamghar, giving *dān* is a primary mechanism for moral cleansing through the transference of "moral substances," namely inauspiciousness, *dos* (blame, fault), *pāp* (sin), and other sinister influences and harmful qualities (but not impurity), which are gradually accumulated in everyday life. Giving *dān* is an indispensable habitual practice not only due to its highly meritorious nature, enhancing the donor's well-being in *this* life, but also since it saves one from enduring *narka* (hell) after death, since as Thamgharians emphasize, each *pāp* and *puṇya* (merit) one accumulates are translated into future (postmortem) sojourns in *narka* or *swarga* respectively.

The Thamgharian *dakṣiṇā*, as is also the case in India (Heesterman 1959; Parry 1994: 130), is a highly ambiguous concept that denotes the nonnegotiable monetary priestly fee given at the conclusion of a *pūjā*. It may be accompanied by or in itself embody *dān*, thus it is often referred to in the village as *dān-dakṣiṇā*. This ambiguity allows villagers to employ the term *dakṣiṇā* to refer to the *dān* they habitually give to the *kanyā* in their family, thus shrouding the *dān*'s sinister aspects. In this way they are able to mitigate or altogether avoid the apparent paradox implied by the logic of *dān*, discussed below, entailed by giving *dān* to one's *own* family member.

The logic of the *dān* transference in Thamghar can be summarized as follows:

(a) *Dān* should only be transferred to or by a person or an entity, such as a *kanyā*, priest, "wife-taking" affines, or a deity, capable of *pāp kaṭāunu* (i.e., "eliminating" and cleansing one of his or her menacing qualities), who is in a position to bless and confer auspiciousness to the donor by virtue of his or her superior ritual status. This hierarchical aspect lies at the root of the recipient's immunity to the potential ill effects encapsulated within the *dān*, and must be mutually acknowledged by the parties involved through the superior giving an auspicious *ṭikā* to the inferior, and the latter performing *ḍhognu*.

However, immunity is not always taken for granted, and in order to prevent the evil-laden *dān* from clinging to their bodies, priests use Sanskrit mantras and avoid taking *dān* at particularly inauspicious times, such as during an eclipse. Overall, Thamgharian *purohits* (priests) maintain that they are able to deliver all the *dān* collected onward to their own priest, a deity, or the family's *kanyā*, without internalizing or being affected by it; they enjoy consuming *dān*'s empty vehicle—the pure rice grains—in their homes, without suffering any moral peril. The funeral *pandas* of Banaras, rotting in a state of "perpetual moral crisis" due to the *dān* they collect (Parry 1994: 123), would probably deem this an impossible ideal.

(b) In order for *dān* to be transferred effectively, not only must its recipient be of high ritual and purity status, but the vehicle used in the transference should also be the worthiest, and preferably *śaktiṣali* (full of energy). Impurity renders both vehicle and cargo highly vulnerable to the latter's sinister effects and makes the entire transference highly inauspicious.

In Thamghar, the mediums deemed by villagers to be the most efficacious for the conveyance of *dān* are *godān* (the holy cow), *kanyādān* (a maiden), *annadān* (mainly *cāmal*—uncooked husked rice gains),[26] money (mainly copper coins),[27] and the water of a flowing river.[28] Hence, in Thamghar at least, it seems that preferably, not only are *dān* "gifts" *female* but they are also *divine*. This combination of divine and female *śakti* seems to endow the above entities (as was the case in ancient India [Mauss 1967: 55] where they were taken as personified beings) with a remarkable capacity to transfer moral substances and I shall revert to this issue later.

(c) *Dān* must *never* be reciprocated and must be transferred outside the *kul*, lest it endanger the well-being of its members.[29] Yet, this does not imply that in Thamghar *dān* is "a free gift which makes no friends" (Laidlaw 2000). On the contrary, as Mauss (1967: 58) finds regarding *dān* and Shulman (1985: 142–43) notes regarding *dakṣiṇā*, Thamgharian *dān* provides a hold over its recipient and establishes a resilient fabric of relationship between the parties involved, which is inherited and maintained across generations.

Furthermore, as was implied above, in addition to acting as the vehicle, *dān-dakṣiṇā* also embodies a form of wealth, possessing an intrinsic worth, which suggests that it is also a sort of payment, given to whoever is willing to relieve one of his or her moral burdens. Village priests often remark that it is the value of both the rice grains and monetary *dakṣiṇā* that are their main incentives for maintaining their profession. Therefore, I believe that in Thamghar, *dān* and its transference may best be conceptualized not as a "gift" but as a paid *service*—indeed there are no "free (rice) meals" in the village.

(d) Any transference of *dān*, however minimal, always takes place as part of, or in itself forms, a ritual (*pūjā*), endowing it with an auspicious, divine seal.

Two principal *dān* "collectors" habitually assist in alleviating the daunting moral load of a Thamgharian Brāhmaṇ household—the family's *purohit* (priest) and its young *kanyā*. At the conclusion of every *pūjā* where he officiates, the priest collects all the grains (mainly rice grains) offered to the gods (imbued with the family's *dān*) and takes them home, where they form a welcome contribution to his household's wealth. Village householders zealously ensure that not even a single grain, saturated with evil qualities, is left behind. In addition, a priest visits all his *jajmān*'s households monthly, on *saṅkrānti* (*sokrati*, the first day of a lunar month) to collect the *dān* accumulated during the previous month, locally termed *siddhadān*.[30]

The priestly function of *dān* collection should not be underestimated, yet it is the family's own *kanyā* who seems to serve as its principal "field of *dān*." Accepting *dān* on all occasions when it is also given to the family priest, she is the sole recipient on many other occasions throughout the year, such as *pūrṇimā* (full moon) and *aūsi* (*aunsi*, no moon), the winter and summer solstices, and the equinox.

Marriage epitomizes a *kanyā*'s role as a "field of *dān*" when, as *kanyādān*, she personifies her entire natal family's *dān*, and is transformed from a mere recipient of *dān* to the embodiment of its primary vehicle.[31] The fundamental significance of *kanyādān* is captured well by the Sanskrit mantra uttered by the priest in the name of the bride's father, while the latter performs *pradakṣiṇā* (*pardacchinā*, circumambulation) around the groom: *yāni kāni ca pāpāni Brāhmaṇ hatya śatāni ca tāni-tāni pra·asyantu pradakṣiṇam-padepade*; whatever *pāp* I [and my family] may have committed, such as the killing of *hundreds* of Brāhmaṇs, will be eliminated in each and every step of the circumambulation.[32]

Next, in the climactic stage of the ceremony (called *kanyādān*), the father "transfers" his daughter to the groom. Thus, giving his daughter away in marriage emerges as the ultimate act of *dān* transference a householder can make during his lifetime, and for this reason alone, having at least one daughter seems highly desired in every family.

The exposition of the governing principles behind the transference of *dān* affords a number of insights. First, it provides an explanation for the past emphasis placed on marrying girls off before their puberty, prior to the abolishment of child marriages in the 1960s. Second, not only does the logic of *dān* correspond to the superior status a *kanyā* has in her *māiti* (natal home), but it also implies the higher status of wife-takers vis-à-vis wife-givers, which was often termed, perhaps somewhat misleadingly, as "hypergamy." This is misleading since, as Bennett (1977: 262–63) notes, what is involved here is in fact an "*ad hoc* hypergamy"; a result of the arrangement of marriage itself, which has very few ramifications outside the context of affinal relationships.[33]

Furthermore, the logic of *dān* clearly demonstrates why Thamgharian and north Indian weddings always take place at the *bride*'s natal house, since only the bride's (newly) inferior family can act as a "giver" of anything that may contain *dān*, including food. Finally, the divine *kanyā* appears to be viewed as belonging to a different *kul* even prior to her marriage, while still living in her *māiti*, and this seems to facilitate her ability to accept *dān* from her own *kul* members (natal family). The logic of *dān* transference further permeates Thamgharian notions of kinship and affects local frameworks of self and subjectivity, as will be discussed later.

As children grow up and their adult teeth emerge (around the age of seven to nine), marking the end of their immunity to pollution, the positions of boys and girls undergo a gradual but dramatic transformation. Both are expected to comply with the strictly adult gender roles, culinary etiquette, and social norms. A girl approaching the age of ten loses her particular *kanyā* status and is transformed from ritual asset to burden, a potential locus of pollution and danger, and a source of constant anxiety. Concomitantly, she must become submissive and modest and is "harnessed" (together with her mother and older sisters) for sharing the responsibility for the household chores.

In contrast, as early as eight years of age, her brother may go through his *bartoon*, permitting him to actively participate in *pūjās* and (theoretically at least) to get married. The main expectation from boys during this phase of childhood is to excel in their studies and be obedient to elders. A twelve-year-old, post *bartoon* boy enjoys a higher ritual and purity status than his sister, who at the same age appears to enter what may perhaps be the most inauspicious stage of her life as will be described below.

## "Betwixt and Between": From the Age of Ten until Marriage

Ordinary Thamgharian aversion and fear of menstrual blood is immeasurably exacerbated at the onset of a girl's menstruation, which epitomizes male anxiety regarding female *śakti* running loose and thus becoming inauspicious and virulent. Menarche seems to evoke connotations of a vicious internal defloration, as reflected in the belief that at her menarche, a girl's mere gaze upon one of her agnates, particularly her father, may be lethal, in keeping with the mythical power of the virgin goddess who can cause blindness, castration, or death by her mere vision (Shulman 1980: 211).

To avert this, the girl is marginalized, secluded, *gupta* (hidden), and confined to a dark room[34] in a distant house belonging to another *kul*, for a period that, as Thamgharians recall, used until recently to last for up to eighteen days, but is currently about seven days. As may be expected, in her situation a girl is never a welcome guest and the entire experience as I heard from the girls in my host family, is a most unpleasant one.

As Arjun Neopane, a middle-aged man, adamantly states, menstruation is believed to make *kanyādān biṭulo*, that is, the girl and the *dān* she embodies become polluted, considerably reducing the potential merit gained through *kanyādān* and inflicting peril on all involved in her marriage. Hence, a primary grievance held by householders concerns the prohibition of child marriages, which could prevent the above situation as will be further discussed below.

In ethno-sociological terms, Marriott's (1998) statement regarding Hindu women in general, holds true for the mature Thamgharian girl and woman:

> Females . . . as a class are more participant in mixing flows than males, and also more subject to fluid markings and unmatching flux. [They have] "greater liquidity" . . . [and] greater inner-bodily space."

Thus mature women are porous, more permeable to pollution and evil than men who are supposedly equipped with harder, solid, evil-resistant skin, and who, conceptually at least (though not ordinarily in everyday village practice), cannot be defiled. Furthermore, a sexually mature, unmarried daughter posits an acute danger to the honor and prestige of the family, her modesty and chastity generating deep concern, particularly in her agnates' minds. Thus, as related by northern Indian Gujars (Raheja 1995: 29), a menstruating girl in her father's house is perceived to be "the greatest danger."

It is therefore little wonder that the indulgence and the pampering of young age is now turned upside down; not only must a mature girl attend to others' whims, but channeling her labor to the benefit of the household now ranks much higher on the family's priority list than her education.[35] Her chastity is under constant suspicion and she is carefully monitored. A girl's lower status vis-à-vis her brothers, is now underlined by her position as the penultimate recipient of food in the family.

A number of scholarly accounts portray what it is like to "grow up female" in a Nepalese high-caste household in even more dramatic terms than I have used (Skinner 1990: 71–75; Kondos 1991; Skinner and Holland 1998). All seem to agree that the experiences of the adolescent girl are far from pleasant, in clear contrast with Bennett's (1978; 1983: viii–ix, 142–45) unequivocal emphasis on the sacredness, auspiciousness, purity, and high status girls enjoy in their *māiti*. In a critical note on Bennett (1983), which is highly relevant to the issue under discussion, Gold argues that "the voices of her informants, whom she often allows to speak at length, do not always convince us of the stark contrasts drawn in the analysis" (1994: 36).

Indeed, on the one hand Bennett states that "in her *maita* [*māiti*] a woman is always perceived as the pure, virgin *kanyā*" (Bennett 1978: 38), while on the other she writes that "all the women I interviewed seemed to have perceived the superior position of boys from an early age" (1983: 166–67).[36]

The basic issue at stake, namely, the position of an unmarried menstruating girl in her *māiti*, seems to be best illustrated by the debate over the interpretation of a girl's seclusion at her menarche. Bennett argues that a girl's seclusion is, first and foremost, in order to protect *her* categorical virginity, sacredness, and purity from all men (Bennett 1978, 1983: 239, 242). However, in Thamghar, as in a number of other high-caste communities,[37] it is the girl who is regarded as the source of destructive potency and her seclusion is, in the main, for the protection of her agnates and not vice versa.[38] Furthermore, her "*kanyā*hood" is clearly blemished by *her* menstrual blood, but not by the potential gaze of her agnates. Hence, in Thamghar at least, the view of a premarried girl as simply "sacred" appears too limited and does not hold true.

I shall now turn to a discussion of the somewhat forgotten, nowadays obsolete, but previously ideal state in a high-caste woman's life trajectory, namely, the married prepubescent *kanyā*, which seems to correspond well with Bennett's notion of the "all pure and sacred virgin" in her *māiti*.

## A "Married *Kanyā*"—from Marriage in Infancy to Menarche

The descriptions given by elderly lay householders and priests alike, exude the impression that a "married *kanyā*, was indeed the epitome of purity and earthly divinity, combining the ultimate control of female *śakti*, through both the natural (divine) somatic boundary fashioned by virtue of her juvenile age and marriage. As Shulman (1980: 211, 349) notes, virginity is "an archetypal image of power sealed within limits" and rather like the Tamil virgin goddess, marriage does not seem to destroy a *kanyā*'s intense *śakti*, guaranteed by her virginity, but makes it auspicious, accessible, pure, and thus all benevolent.

A census I conducted in Thamghar revealed that the vast majority of women and men over the age of forty were usually married at four and seven years of age respectively, or at the very latest by the age of eleven. This clearly shows that child marriages were the norm in Thamghar, as in many other places in high-caste Nepal and India.[39] In those days, villagers recalled, following her marriage the young *kanyā* immediately returned to her *māiti* and remained there until puberty.

Thamgharians explain that once married, a girl's menarche does not pose much danger since her *śakti* is imagined as controlled and channeled toward her husband. Hence, at the onset of menarche, a married *kanyā* was immediately sent to her husband's house, where she was not marginalized or secluded any more rigorously than other menstruating married woman.

The notion of the married *kanyā* seems to be an attempt to solve the perennial Hindu dilemma of harnessing the female *śakti* while avoiding its ominous aspects, by providing an additional measure for controlling the impending outburst of dangerous female sexuality (menarche). It also provides a bridge mediating between the two salient aspects of the goddess as reflected upon by villagers: the extreme *śakti* of the *virgin* Kāli and the gentleness, benevolence, and purity of the *married*, faithful Pārvati. This is reminiscent of the Punjabi mythical construct of the virgin-mother goddess *Mātā* described by Hershman (1977), whose impure and dangerous sexuality seems to be obviated by viewing the goddess as already converted into the auspicious state of motherhood.

## Life as a *Buhāri*—a Daughter-in-Law and Wife

North Indian kinship marriage prescriptions are strictly adhered to in Thamghar; villagers recite the following basic requirements for an appropriate match (set up with the assistance of a *lami* (*lomi*)—a professional mediator): the bride and groom must belong to separate *gotras*[40] and *thars*,[41] their *kuls* must not be related by marriage in the previous fourteen generations,[42] and they should come from *equal* background in terms of *jāt*, financial and social status. As a rule, village exogamy should be carefully maintained, lest a girl's *māiti* be too close to her *ghar* (conjugal place), in order to avoid the involvement of her relatives, who are expected to side with her at a time of crisis. Hence, prior to the arrangement of marriage, equal status must be assumed between the parties involved, who more often than not are total strangers.

What for a girl is often a traumatic transition from *māiti* to *ghar* is also a perplexing experience for her affines, since from their point of view she is a highly ambivalent, daunting, female stranger,[43] penetrating the core of their otherwise well-guarded and fortified microcosmic sanctuary; a potential threat to the unity of the household, she is also the main hope for its perpetuation.

The general lot of contemporary Thamgharian *buhāris* may be much better than in previous generations, since nowadays their in-laws are well aware that if mistreated, a *buhāri* is likely to make a swift return to her *māiti* and may soon compel her husband to separate from the joint family. Yet her education, greater independence, and perhaps rebelliousness clearly do not promote her in-laws' trust and she is regarded as a potential immanent threat to the well-being of the household and its members. Furthermore, as a rule, the girl is married without her consent, far earlier than she would have liked, and thus she may be depressed, moody, or simply overstressed and anxious about her new life and husband, whom she did not choose. Consequently, she may well lack the energy and enthusiasm to work from morning till night in her *ghar*, thus realizing her in-laws' fears. This intimidating image of the in-married young wife seems to last until she gives birth, ideally to a boy, thus proving her loyalty to her *ghar*.[44]

Besides being viewed as a "dangerous stranger" by her affines, a *buhāri* is mainly identified with the concept of *kanyādān*, yet nowadays, a bride is inevitably a *biṭulo kanyādān* (i.e., a menstruating one). This identification has a number of unfavorable implications, adding to her already ambivalent image. First, as *kanyādān* she is imagined to personify her *māiti*'s ripened, ultimate *dān*—an enormous, latently perilous freight of inauspiciousness and *pāp*. Therefore her affines, conscious of their ad hoc, *temporal* superiority, may seriously doubt their immunity against the menacing aspects of the (*kanyā*) *dān* they have received. The perils inherent in the *dān* and its transference are exacerbated by its pollution since the supposedly pure *kanyā* is, in fact, a mature menstruating woman. Hence, rather than merely serving as its unadulterated vessel or vehicle, the *buhāri* may be intrinsically affected and "saturated" with her *māiti*'s *dān*, which hinders its onward transference. In addition, the doubt cast over the bride's honor and virginity by postpubescent marriage is amplified by the rare love affairs involving unmarried village girls. Furthermore, marriage inevitably draws two equal-status families *up* and *down* the otherwise stable, divine hierarchy of ritual status that lies at the crucial nexus of Brāhmaṇs' personal and social life. Thus, a major daunting aspect associated with *kanyādān* marriages concerns the ill feelings the bride's family is believed to harbor against their ad hoc superiors and the measures they are expected to take in order to avoid or mitigate their inferior position.

One alternative open for wife-givers is to give their daughter to the son of their family priest (or another member of his *kul*), an ideal recipient of *dān* with whom hierarchical relationships and regular transference of *dān*, conceptually at least, already exist, and whose superiority is beyond question.[45] Yet this negates the emphasis on village exogamy and is rarely chosen. Another manner through which wife-givers attempt to take the edge off their lowly public status, is through the use of jocular humiliation and ridiculing of the groom and his party, as seems to be common among northern Indians (Kolenda 1984, 1990).

The above intertwined, emotionally and cognitively disturbing aspects of present-day *kanyādān* marriages, seem to transform the stranger *buhāri* into an emblem of danger and uncertainty—a potential locus of destruction at the core of the household she marries into. This is epitomized by the following prevalent belief, consistent with the local notions of causality in the "emotional universe," which seem to come to the fore particularly when the aforementioned public farce does not take place. I often heard that among the other "gifts" and various items inside the bride's *dāijo* trunk, her potentially humiliated and angered parents may place, without her knowledge, a pitcher containing a capricious predatory *bir*.[46] This, I was told, is done in order to *dukkha dinu* (*dine*), to assign sorrow and ill fortune onto the groom's family. Once again we see that as with the scale of wealth and prestige, the demoniac realm is imagined to be strongly associated either with the alteration of social hierarchy or with its aftermath. Taken together, these elements illustrate how the Thamgharian *buhāri* is clearly not viewed as "an auspicious bringer of fortune to her husband's family" (Raheja 1995: 29), nor is she the pure *kanyā* that parents-in-law could have hoped to receive in previous generations.

In the Hindu world "actors and their actions are held to be aspects of each other" (Marriott 1976: 123; Daniel 1984: 211–13), affliction and its relief, the disease and its cure are often identical (Bayly 1989: 133) and, similarly, clothes retain the qualities of the people who fashioned and exchanged them (Bayly 1986). This is well exemplified in Banaras, where the Māhābrāhmaṇs (funeral priests) embody and are poisoned by the *dān* they collect, and rather than merely impersonating the deceased with whom they associate, they are made cosubstantial with it and become the *pret* (ghost) or *pitṛ* (ancestor spirit) themselves (Parry 1980: 93). Similarly, although I never heard this explicitly in Thamghar, I believe that it may be legitimate to surmise that in the minds of many villagers, a *buhāri* is conceptually identified with the archetype of danger she innocently carries—a ferocious *bir*.

In Thamghar's public male discourse, the in-coming wife is presented as the *nokar* (servant) of her husband or *mālik* (master, ruler, or god).[47] Indeed, the wife is supposedly blessed by sipping her husband's *khuṭṭa pāni* (the water she uses to wash his legs prior to every rice meal) and by eating from his plate, where both the water and leftovers are likened to auspicious *prasād*. While many young village women refuse to perform the former, most still eat what otherwise would be the extremely polluted (*juṭho*) leftovers of their man. A general village saying, which captures well the notion that women have "slippery loose ends" that must be firmly controlled, states that women can't be trusted since they lack Rudra *ghaṇṭi* (an Adam's apple)—seen as a masculine device for keeping secrets locked in the heart.

The daunting image of the *buhāri* is clearly reflected in the local male ethos, according to which women are blamed for almost everything: for lack of male offspring or children in general, the tensions rife within the joint family and its inevitable early fragmentation, the death of a son or husband (this is taken as a valid sign that the poor woman is a *boksi* [sorcerer] in training), and at times, she may even be blamed, posthumously, for bringing about her own death.[48] Likewise, many village widows are suspected of being *boksis*.[49] Due to a woman's greater vulnerability to impurity, a promiscuous wife who sleeps with an "Untouchable" may set off what Fürer-Haimendorf (1966b: 154) calls a chain-effect, an "avalanche" or swift degradation of her *kul*'s purity to the lowest possible level. A Brāhmaṇ male who behaves similarly, would only cause his immediate family to become "Untouchable." Furthermore, in a local *baṁśāvali* (genealogy) diagram I examined, prepared by a young village student, I could not find even a single woman's name. Thus, from the "patriline point of view" to paraphrase Bennett (1983), women appear as "guests" in their own houses (Madan 1965: 77), a "non-endurable" element (Kondos 1991: 118), an ephemeral and transient phenomenon in a male-dominated world.

## MULTIPLE PERSPECTIVES AND PERCEPTUAL RESOLUTION

The aforementioned discussion may give the impression that a young *buhāri* is imagined as utterly dangerous and inauspicious in her new *ghar* and village. Yet I would like to suggest that this view, which is one of the cornerstones of Bennett's work (1983) is somewhat limited, and at least regarding Thamghar would even be misleading. "The dangerous stranger wife" is perhaps an image contrived in the minds of many parents-in-law and reflected in high-caste male discourse, where it becomes an "ideal," but as Hart argues, real people and empirical life rarely conform to any such ideal (Hart 1988: 191).

Moreover, I would like to argue that Thamgharian in-laws, for whom the *buhāri* may indeed seem dangerous and untrustworthy, would not identify their chosen potential contributor to the perpetuation of their household with a predatory *bir*. This would mean pointing a blaming finger inwardly, an option rarely taken in Thamghar, creating a patent cognitive dissonance. Furthermore, the points of view expressed by the parents-in-law and the "public male discourse" regarding the incoming *buhāri* are not the only existing perspectives, but are part of a wider array of views. Of these, the most prominent and highly relevant for the present discussion is the husband's standpoint, to which I now turn.

By adhering to a number of simple, habitual rules of conduct, men easily avoid the dangers inherent within the ambivalent aspects of women's *śakti*, evinced in their periodical involvement in extremely polluting organic processes such as menstruation and birth. In addition, as will be demonstrated below, the particular manner in which marriage is conceptualized in Thamghar allows village men to feel secure vis-à-vis their wives. Indeed, every Thamgharian man with whom I privately discussed this was adamant that his wife was *not* dangerous and regarded such a possibility with disdain, no matter how lowly he may have viewed her or how good or problematic their relationship was.

Seen through the eyes of her husband, a Thamgharian wife possesses a number of gentle and benevolent virtues in addition to a woman's obvious positive aspects associated with motherhood. First and foremost, Harlan and Courtright's (1995: 12) remark regarding Hindu women in general also holds true for Thamgharian wives, whose *śakti* is "understood as encompassing the ability to preserve the lives of [their] husbands."

This notion is predicated upon the perceived discrepancy in the *śakti* possessed by the two sexes; in contrast with a woman's unlimited *śakti*, evinced through its fluctuating overflows (menstruation) and transformative processes of internal growth (gestation), a mature man is the focus of finite *śakti* condensed in his bones, which is believed to be depleted through use in the form of semen (*birya*, *śukra, bij*, *biu*).[50] Copious ejaculation, householders warn, leads to somatic collapse and even death, but the wife is normally believed to be able to replenish her husband's *śakti*, and thus it is lonely men (widowers and bachelors) who are expected to die before their time. Therefore, blaming a widow for her husband's death is not only part of the rhetoric of village male ethos, but derives from the recognition of a wife's crucial responsibility for the longevity of her husband.

Bennett (1983: 246–54) rightly underscores the sister's role in ensuring her *brother*'s life during Tihār's Bhai Ṭikā, yet she (1976, particularly 1983: 218–34, 251) underestimates the equally significant effect of the wife on her *husband*'s life, as epitomized in the festival of Tij, the exclusive festival of married women.[51] Moreover, emphasizing the clear aspects of overt eroticism associated with Tij, she does not heed all of its dimensions of inversion and remarkable discourse of resistance.[52] Thamgharian men cherish the *ṭikā* given to them by their *married* sisters in Tihār, yet also believe that through the rigorous fast and other austerities their *wife* performs in the true spirit of *pativratādharma* (Courtright 1995: 187) as part of Tij,[53] she is able to preserve and enhance their well-being and *āyu* (long life), as well as bestow blessing on to them.

In my view, the complex inversion of a woman's behavior, position, and relationship vis-à-vis her husband is central to Tij's logic of design and is not merely a "steam valve" as Gluckman (1954) and Turner (1969: 172–203) would argue. Instead, inversion is crucial for the efficacy of the entire performance so that the wife can assume a higher position than her husband, since blessing and protection can only be conferred from above. I have noticed that a number of Thamgharian husbands seem reluctant to cook the pre- and post-fast meals for their wives, thus attempting to tone down this inversion. In addition, the female *śakti* is only auspicious and benevolent if bounded and controlled, but this inevitably reduces its power. Thus by allowing women to unleash their *śakti* "negatively" (eroticism and resistance), away from the village and for a limited duration, Tij provides a "safe" cultural arena in which the wife can exercise her utmost power for the benefit of her husband. It is thus clear that a married woman's benevolent and destructive aspects are inseparably intertwined. Furthermore, Thamgharian ideas follow Hindu religious imagination in viewing a wife's fidelity and chastity as forming an indispensable moral shield, protecting her husband. This is the main theme in the famous myth told in the *Swasthāni*,[54] where Brinda, the demon Jalandhar's faithful wife, is seduced by Viṣṇu disguised as Jalandhar, thus enabling Śiva to kill the demon.[55] The enormous power of a woman's fidelity is shown here not only by Brinda's ability to protect her husband's life (as long as she remained chaste), but also by her terrible curse, turning Viṣṇu into a plant, tree, stone, and weed, which are indeed the main forms in which he is worshipped by Hindus to this day, as a *tulasi* (basil),[56] *pipal* (tree),[57] *śāligrām* (*śālingrām*, a fossil, ammonite) and *kuś* (grass).[58] Finally, if a daughter is the family's "field of *dān*," a dumping ground for its sins and inauspiciousness, the wife acts as the husband's "field of *puṇya*," where his *merit* is augmented and cultivated; every *puṇya* she is able to generate and accumulate through her rigorous *dharma* (religious work, devotion, fasts, and the performance of *pūjās*) is shared with her husband. Thus, young village men often exempt themselves from everyday *dharma*, maintaining that it is mainly a woman's task and that the actions of their devoted wives suffice for them both.

It is now time to tie together the various ethnographic threads mentioned thus far. My basic argument is that "public male discourse" or "the ideology of the patriline" should not be taken at face value. Therefore, holding the *buhāri* responsible for the agonizing disreputable fragmentation of the household via manipulation and alienation of her husband (which may often be the case), also implies that the husband's outlook and relationship with his wife may fundamentally differ from what is projected in public discourse. In fact, this common accusation often enshrouds the bitter internal tensions within the household and provides a convenient way for a son to present his (long overdue) intentions to separate from his parents, or is part of the latter's strategy to save face. Likewise, a patriarch's deep mistrust of his *buhāri* is often expressed through his refusal to permit her to cook for him, demanding that his elderly wife carry on with this heavy burden until his last day. However, this also implies his *trust* toward his *own* wife in this cardinal matter of cooking rice. Furthermore, as will be recalled, almost every Thamgharian householder spends extensive periods of time working in other parts of the country or in India, during which he entrusts all household responsibilities to his wife, never to his father, brothers, or a neighbor.[59] Certainly, a man may adopt his parents' viewpoint or the harsh conduct called for in public male ethos, but this is only rarely the case. Particularly after separation from the extended family, village men seem to treat their wives much more positively than most of them would ever publicly express in the village teashops or confess in a private conversation with an anthropologist.[60]

The above discussion seems to draw attention to one fundamental aspect that may "go without saying" in much everyday experience and scholarly writing, namely, that people's manifold perspectives on an object, a person, or a situation often involve an additional element related to their distinct social distance from the latter. In Thamghar, three main perspectives of the *buhāri* were analyzed: that of her husband, his parents, and fellow villagers. I would like to suggest that the differences between them are not merely because of *the way in which they look* (their distinct perspectives), or the role they assume when articulating their views, but also due to their discrepant *perceptual resolution*—what their respective social distance from the *buhāri* allows them to *see*.

What is at stake is not whether a married woman's image is split or how it oscillates between the contexts of *māiti* and *ghar*. Rather, it is the various perspectives through which she is simultaneously viewed by different people or the way she is perceived by the same people in various circumstances within a given social context; the existence and significance of the gap between expression and conduct, face value, and essence. Thus, it may be concluded that a Thamgharian wife is not simply dangerous, just as a *kanyā* is not strictly "sacred" or, more aptly, divine. Similarly, Thamgharian "ideology of the patriline" does not exist as a unified notion, although this may be the impression often gained from villagers' oral testimonies. This supports Raheja and Gold who argue that their research of kinship and women in north India "provides little evidence for unitary male or female perspectives on kinship relations" (Raheja 1994a: 18).[61]

Thamgharian men's acknowledgment of their wives' virtues does not mean that their relationships are not based on fundamental inequality and domination. On the contrary, women's positive aspects may *only* be viewed as such against the backdrop of strict control and extreme subordination. A Thamgharian husband may feel secure vis-à-vis his wife not due to her lack of menacing aspects, but as a result of his ultimate control over her. I would further like to suggest that the notion of one's wife being "dangerous" is rendered almost inconceivable by the particular manner in which male and female bodies are imagined to become intertwined in the marriage ceremony, as will be illustrated below.

## MATRIMONIAL ENCOMPASSMENT

Marriage in the Hindu world, whether among humans or divinities, in earthly life or mythology, is generally regarded as the preferred, ideal, and inescapable manner of coming to terms with, taming, channeling, and taking advantage of the female's ambivalent forces or *śakti*, which thus become auspicious and prosperous.[62] Thamgharian marriage imagery, which appears to epitomize these notions, will now be discussed.

Not surprisingly, it is particularly the village men who state that marriage (*bihā*, *bibāha*) marks the bride's *nayā jīvanko suruwāt*—the beginning of her new life; the complex marriage rituals, labeled as the final *saṁskāra* (life-cycle ritual)[63] and centered upon the fire ceremony of *hom* (performed in the *jagya*)[64] are believed to effect the auspicious second birth of the bride, as a purer person, within her husband's body. As I often heard, particularly from Thamgharian men, *dui* (*dwi*) *sarir bhae pani yeoṭey ho*—although apparently having two separate bodies, a husband and wife become one, while the wife is made *puruṣko bāya aṅga*—her husband's "left organ."

Therefore, the wife's entire body is imagined as being *encompassed*, occupying some space within the left side of her husband's body. This seems to correspond well with one of the readings of *mithuna*, in which "Prakṛti as Supreme Śakti is within God, hidden in His own qualities" (Kramrisch 1976: 346–47).[65] Stated differently, Thamgharians perceive the insoluble bond between husband and wife as *matrimonial encompassment*.

While men talk about the wife becoming an integral part of their body in marriage as if it was a natural fact of life and beyond any question, women (with the exception of a number of young in-married *buhāris* who had little awareness of it), seem to accept this notion, but the extent to which they actually endorse it is obviously open to debate. The women with whom I discussed it tended to respond along the lines of Mina Aryal (50) who, answering my question whether the wife becomes part of her husband in marriage, said, "Yes indeed, this is what is said [to be the case], but what do/how would we [I/women] know?"

Prior to discussing the various implications of matrimonial encompassment, I wish to illustrate how this imagery is expressed in the embodied symbolism of a Thamgharian wedding's "performative grammar." The climax of the marriage ceremony is the crucial auspicious stage of *kanyādān* when, following a series of elaborate gestures and various ritual elements, the virgin girl is "given" or "transferred" (in)to her husband('s body). In the *Swasthāni Vrata Kathā*, Śiva's marriage to Sati is established by the irreversible act of Daksa (Sati's father) giving his daughter's hand to Śiva, while Birani (her mother) pours water from a golden vase onto their hands (Buddhi 1994: 87).[66] As I witnessed a number of times and in accordance with the mythic marriage, in the critical *kanyādān* stage of contemporary village marriages, the bride's father gives her right thumb to the groom, who grasps it in his right hand while her mother pours pure water through a *saṅkha* (a conch). The latter is one of the manifestations of the Lord Viṣṇu, whose divine presence is glossed by his role as a witness while the water is said to ensure the auspiciousness of the union thus fashioned.

A comparison with similar symbolic gestures performed during *dān* transference will render the significance of the groom's grasping of the bride's thumb more salient. In Thamghar, *dān* is given and accepted in a variety of manners depending on the type of entity that transfers it. For example, when *dān* is given to a priest in a *pūjā*, normally he need only touch the rice filled leaf plate to acknowledge acceptance. When the *dān* is embodied within a living cow (*godān*) it is pulled away by the recipient, and similarly, when a leaf plate containing a coin in lieu of a cow is given as *godān*, the recipient should pull the plate away on the ground, as if it were a real cow. By the same token, through grasping the bride's thumb, the groom symbolically encompasses her within his body. From here on, forming an integral part of her husband's body, the wife who until now has worn her *rakṣā-bandhan* (a protective thread imbued with intense *śakti*) like all boys, unmarried girls, and men, on her auspicious right wrist, must now wear it on her left one.[67]

## Imagining Domination

Below I shall attempt to clarify the meaning and significance of matrimonial encompassment, as well as the actual and conceptual implications this notion entails. First, especially due to the rather specific, confusing, and, to my mind, also misleading manner in which Dumont uses the term "encompassment," as well as the remarkable career this term has had in south Asian studies following the publication of *Homo Hierarchicus* in 1966, there is a need to clarify the disparity and affinity between the present and the Dumontian usage of the word.

Following his Ph.D. student Raymond Apthorpe's work on African society and his particular notion of hierarchy, Dumont appropriated the term "encompassment," assigning it a key place in his overall theory of the Indian caste system. At the heart of the latter is his definition of hierarchy as the encompassing of the contrary:

> I believe that hierarchy is not, essentially, a chain of superimposed commands, nor even a chain of beings of decreasing dignity, nor yet a taxonomic tree, but a relation that can succinctly be called the encompassing of the contrary.
>
> In the hierarchical case, according to Apthorpe, one category (the superior) includes the other (the inferior), which in turn excludes the first. (Dumont 1980: 239, 241)

Dumont's aforementioned concept of hierarchy, together with his notion of the subjection of power to authority, are generally deemed largely inadequate for describing and comprehending the Indian caste system,[68] nor are they in agreement with the way "hierarchy" is generally understood.[69] Furthermore, not only is Apthorpe's inclusion/exclusion notion never fully and consistently elucidated by Dumont, but it also seems quite remote from Hindu ways of thinking and is far too rigid for dealing with Indian categories, which are rather fluid and amorphous.

To sustain his (rather specific and narrow) definition of hierarchy Dumont draws upon the famous Genesis story of Eve's creation out of Adam's rib, which Dumont considers to be the best example available (Dumont 1980: 239–40). His use of the Genesis story may at best be regarded as the employment of a good example for the wrong case. Adam and Eve's relationship hardly exemplifies "the encompassing of the contrary,"[70] and neither provides significant insight into the *universal* meaning of "encompassment" or "hierarchy" (if indeed such a meaning exists at all) nor sheds any light on the nature of the Hindu caste system. The partial similarity between the gender relationship in Genesis and the Thamgharian Hindu image of matrimonial encompassment[71] (along with its other variations found elsewhere in Nepal and India as will be mentioned below) may only suggest the existence of an affinity in the manner in which two quite distinct ancient cultures prefer to imagine the hierarchy between the sexes cum domination of women. This however goes against the grain of much of Dumont's own work in which he attempts to draw a rather sharp dividing line between Western and Indian ways of thinking.

Furthermore, although when seen from a Euro-American point of view, the Thamgharian matrimonial encompassment may indeed come very close to Dumont's notion of "the encompassing of the contrary," this perception is not shared by Thamgharian Brāhmaṇs themselves. In their frequent references to the differences between the sexes, villagers never viewed the former as a binary opposition or as contradictory. In Thamghar, the term *ulṭo* (contradictory, opposite) is never used to describe the relationship between men and women, but is mainly reserved for relating to the differences between the Hindu and Muslim *dharma* (religion). Instead, as should already be apparent by now, the sexes, whose similarity and disparity are not thought to remain constant throughout an individual's life trajectory, are predominantly viewed in relational and complementary terms.[72]

Hence there is a need to clearly distance the present usage of "encompassment," which follows its ordinary import of "to include" or "surround," from Dumont's rather ambiguous employment of the term in the context of the Indian caste system. In addition, it seems that Dumont had opted for a biblical example particularly since he did not associate the notion of "the encompassing of the contrary" with any Hindu perception of *gender* or marriage, although it is mainly here that his particular notion of hierarchy probably makes the greatest sense. Thus, while Dumont's concept of hierarchy as "the encompassing of the contrary" is rather confusing, limiting, and of little relevance to the Indian caste system, his work does imply and may have been influenced by the general Hindu tendency to encompass (in the sense of include) a subversive element (an idea, practice, religion, or indeed one's wife), in order to overcome and dominate it instead of simply opposing it. In a similar vein, he was no doubt correct in pointing out that extreme hierarchical relationships in general are often given a striking expression via the notion of encompassment, a theme to which we now turn.

As Handelman (1990: 157–58) rightly observes, unlike the notion of hierarchy in its simplest form, that of a rank-ordering from high to low along a certain common set of criteria, entailing a limited control of high over low, encompassment is the ultimate form of hegemonic order where the encompassed is thought *to be made over* in terms of the higher encompassing order.

Indeed, in Thamghar, a woman's original name, given by a priest in her *nuwāran* according to the first letter of the *nakṣatra* (star) dominating the time of her birth, is replaced during her "second birth" in her wedding, when the husband gives his wife a new name according to the first letter of *his* name. Thus, from here on, he acts as her *nakṣatra* or domestic god, dominating her time and place; in effect he becomes her entire cosmos. Compatibility with the "cosmic design" (Das 1976a: 252) is deemed essential for maintaining the inherently fragile, balanced relationship between micro- and macrocosms, ensuring their harmonic progressive flow.

Concomitantly, as elsewhere in high-caste Nepal or northern India,[73] marriage also entails the transformation of the wife's *kul* and *gotra* to that of her husband, considerably limiting her ritual bonds and obligations to her *māiti*. Despite this, women attempt to underscore their ongoing relationships with their natal place and while men tend to minimize and belittle the latter as far as their wives are concerned, they nonetheless seem to zealously maintain connections with their own sisters; even elderly householders undertake the long and arduous journey to their sister's house to receive the cherished Bhai Ṭikā in Tihār, in case the sister is ill and unable to make the trip.

The extreme domination implied by encompassment, facilitates the cooptation of the ambivalent stranger *buhāri* and her transformation into a subservient *nokar*, an obedient devotee of her encompassing universe—her husband. In fact, the husband is on a par with the major Hindu gods, encompassing the cosmos and humanity,[74] which entitles him to be worshipped by his wife throughout his life. Hence, it is through a scale effect[75] that the husband gains ultimate control over his wife's potentially "evil" aspects and remains transcendent, beyond any danger.

This eclipse of the wife's aura of uncertainty appears to be accomplished in a characteristic Hindu manner—"a classical pattern in Hindu mythology in which a dangerous, destructive, but nevertheless indispensable force is channeled and *contained*"[76] (Shulman 1984: 42). In fact, this is the general Hindu approach toward virtually all forms of dissidence and heresy, which are bound to be assimilated (O'Flaherty 1971) or to quote Renou, "other religions select and eliminate; Hinduism incorporates" (Renou 1953: 53).

Indeed, matrimonial encompassment represents an ideal channeling of the wife's *śakti* for the sole benefit of her husband, through her absolute containment, recreating the somatic seal lost at puberty in an ultimate form; unlike women, men are not "porous" and their "ends" are not "loose," thus once the former are encompassed, conceptually, "overflows" or leakages become impossible. In fact, matrimonial encompassment provides an ideal solution to the perennial Hindu problem of the flow of male and female sexual substances;[77] the woman's potentially contaminating sexual fluids are well-controlled and thus auspicious, while the man's devastating gradual depletion of masculinity is averted as his semen flows "*internally*," toward his wife. Thus, "the flow"—essential, desirable, and indispensable for life and procreation—does not stop but instead is confined to a perpetual inherent circulation, rendering it pure and auspicious. Vitality is gained without the inextricably polluting and menacing aspects of sexuality, which otherwise impinge upon humanity.

In matrimonial encompassment, neither the overall masculine identity of the groom is called into question nor is the bride's feminine identity jeopardized. What is questioned and in fact appears to come to an end is the bride's notional existence as an autonomous person (to the extent that she actually enjoyed some measure of independence at her *māiti*). Thamgharian men seem to adhere to Manu's precept that a woman should never be independent (Manu 5.147–49, in Doniger and Smith 1991: 115) and thus, as glossed by Ram Babu Dhital (35), marriage is *jimmā lagāe*[78]—the "transference of custody/responsibility" from father to husband.

In addition, encompassment also implies the protection of the encompassed. Thus, conceptually (but rarely in reality as discussed above) bounded and covered by a masculine skin, the vulnerable woman becomes protected, more resistant to pollution, and less susceptible to evil. Anticipating what will be discussed in detail in part III, rice grains appear to posses a similar structure;[79] while still within their male husk, the grains of *dhān* are totally immune to pollution and evil and hardly capable of transferring *dān*, hence it is primarily *cāmal*, the vulnerable, feminine (husked) grains that are used for *dān* transference. Likewise, Brāhmaṇs gladly accept grains of unhusked rice from the hands of the village "Untouchables," yet would never accept *cāmal*, viewing it as polluted by the latter.

Strathern (1988: 242–60) discusses a rather different sort of encompassment between the sexes in Melanesia, where each sex encompasses elements of the other, but no permanent domination is implied (ibid. 334). According to her, encompassment is "the encapsulation of another's view-point, a containment of an anticipated outcome" (1988: 259). In my understanding this also implies "the encompassment of the future," to borrow from Dumont (1980: 240). This insight seems to lay to rest the unresolved status of the "sacred" *kanyā*, who embodies the goddess in a number of the village's prominent rituals, yet is excluded from others. Following Strathern, I suggest that viewing the Brāhmaṇ girl as a nonmember of her natal clan is an anticipation of her destined matrimonial encompassment; the potential future situation of the daughter is perceived as already present. Clearly, potentiality and actuality do not exclude each other in Thamgharian imagination; rather, they are viewed as one.[80] While in Melanesia encompassing is anticipation, in Thamghar anticipation is viewing the daughter as already encompassed.

Thamgharian Brāhmaṇs follow the ancient Hindu dictum that man and woman are born "incomplete" and may transcend their fragmented existence through marriage, which parallels the condition of *mithuna*. This seems to come very close to the notion of divine marriage (and thus also the human one) as implied by Kramrisch (1976: 346–47), where marriage is seen as an attempt to counteract the entropic devolution from the primordial state of undifferentiated unity (Kuiper 1975: 109) in order to rectify the unstable earthly separation of the genders. This desired ideal state is locally termed *purā*—"completeness." Being single, particularly if one is a single woman, may imply danger, inauspiciousness, and may at times be equated with evil and death, while marriage, I was often told, is emblematic of dynamic creativity and eternal progeny, a harmonious state of utmost safety, sanctity, and auspiciousness, where death is completely annihilated.

As Sharmananda Subedi (71) contends, the general view in Thamghar seems to be that a man and his wife in their house are said to turn it into a place of *sṛṣṭi (shristi)*—a divine emanation out of the procreative unity of male and female—a *living* temple. Thus, through matrimonial encompassment, Thamgharians seem to reinvent *mokṣa* (being too vague a concept and beyond most villagers' horizons of aspirations) within *saṁsāra*, a kind of salvation on earth, as close as one can get in the present life to the cosmic primordial ideal.[81] Consequently, marriage is the minimum requirement for every man and woman in the village, as well as for other divine and semidivine entities, such as gods and deities, houses, *cautāros*, *jagyas*, and various ritual paraphernalia; all must be married or simply made *purā* by joining their male and female aspects, as will be unfolded in the ensuing chapters.

We will now turn to consider one possible implication of the mental image of matrimonial encompassment within the social realm, namely, what may be the consequences of the *collective* interplay between matrimonial encompassment and other fractal images mentioned earlier, on the village's social milieu.

## ENDNOTE: FROM THE DIALECTICS OF IMAGES TO A VICIOUS CYCLE

Enshrouded under a divine guise, Thamgharian matrimonial encompassment places a man in a permanently advantageous position vis-à-vis his wife, exercising control over her through one of the most powerful ways of imagining almost complete male domination and female domestication. I would like to argue that this concept, ideal as it may seem from a male point of view, has other unintended, undesired, and far-reaching social implications, resulting from the effects of perceptual resolution and the dialectics between a number of cardinal images and perspectives.

The latter should all by now be familiar. First is the fearsome imaginative structure of the human "moral geology" or personhood, consisting of an evil inner-self encapsulated within an external, affable but inauthentic mask of the social person. The second derives from the fractal of *ātmās* (*ātamans*) and bodies, where one body is the *ātmā* of another in a Russian-doll-like manner, and the third image is that of matrimonial encompassment—of the wife occupying some space within the left side of her husband's body. An additional image, drawn from the fractal of living temples, where all entities are viewed as temples sheltering an inner divinity, is completely overlooked and does not participate in the dialectics of images discussed at present, for reasons that will be clarified later. A fifth image, associated with the fractal dimension of a living entity enveloping a burning core as its dynamic flame of life, seems to be of little relevance to the "cognitive turmoil" associated with the notion of matrimonial encompassment discussed below. Concomitantly, three main perspectives on matrimonial encompassment are at play—those of the husband, his parents, and his fellow villagers.

To this exhibition of images and perspectives we should add a number of cardinal village notions: the analogy between the house and the householder, their identification or the belief that the former embodies the latter, the notion of the *buhāri* as an ambivalent dangerous stranger who potentially carries, even embodies a *bir*, and the latent fluidity of bodily essence. The latter is mainly evinced in villagers' perception of possession, imagined as the intrusion and seizure of one's body by a divinity (god, goddess) or semidivine evil entity (*bhūt*, *bir*). While being possessed by a god (as often takes place during *kul deutā pūjās*) is highly auspicious, possession by a *bhūt* or *bir* is devastating—the victim, so villagers maintain, is devoured to death from within in a demoniac sacrifice.

The dialectics between these mental images and abstract notions, which all share a simple identical basic structure, form much of the local Thamgharian lore and perceptions of self, the other, and the world, and seem to be well inscribed in villager's minds and personal histories. Being imaginary, they are flexible, have slippery edges and are dynamic and half-transparent. As Strathern (2000: 53) argues in her comment on Melanesian imagery, people's imagining and analogy drawing have the ability to flow across contexts and different domains of life. Thus, these images have the ability to mutually mirror, overlap, influence, intensify, imply, validate, sharpen, or mitigate each other and are, in effect, analogous.

The crux of the matter may be simply stated as follows: a house is imagined as the body of its inhabitants, who are perceived as its essence of life or *ātmā* and clearly control it, building or demolishing it at will. By the same token, the image of matrimonial encompassment suggests that the husband is the house and body of his wife and thus she is likened to his *ātmā*.

As may be obvious, village men (and in-laws) would never explicitly admit or endorse this idea, as it assigns the wife supremacy over her husband

and brings *him* under *her* control. Yet this is clearly what haunts their minds; in public male discourse, the *buhāri* is blamed for manipulating and inducing her ingenuous husband to take drastic steps toward the undesirable fragmentation of the household, while he is absolved of any responsibility for his actions.

Concurrently, the notion of matrimonial encompassment further implies that marriage places the ambivalent and dangerous, post-menarche *buhāri* and perhaps also the horrific *bir* she embodies or at least carries, under a fellow villager's inauthentic mask of affability. Thus, in the minds of villagers who do not belong to the *buhāri's* new household, the image of matrimonial encompassment implies that marriage posits her within the left side of her husband's body where his *heart* lies, fashioning the vivid, daunting equivalence of the *buhāri* = *bir* = the husband's inner-self or heart. In effect, the young wife appears to "possess" her husband.

Again, one would not expect villagers to openly use the terms employed here, yet the aforementioned conclusion is implied by and correlates well with their understanding of the way the bride's parents negotiate their degraded status, by unleashing their turbulent emotions, embodied in the form of a predatory *bir* that is carried by their daughter into the core of her new *ghar*: into her husband's heart on the first floor of the house.

Along similar lines, a husband's parents may imagine a milder scenario where an ambivalent, polluted stranger usurps their son's inner-self. Hence, the *buhāri* may simply be thought of as "polluting" his heart. We may thus conclude that what from a male ego's perspective is a token of ultimate, almost absolute domination, reflected by his self-confidence and secure stance vis-à-vis his wife, may at best be perceived by others as the foray of a feminine *ātmā* or an ambivalent, polluted stranger on his inner-self or, far worse, his possession by an evil *bir*.

While the anxiety and suspicion toward an individual *buhāri* within her *ghar* gradually tones down and fades completely following the birth of her first son, the unintended daunting implications of matrimonial encompassment within the village social milieu appear to linger almost unchanged; most *buhāris* eventually turn into respectable matriarchs, yet the incessant flow of others into the core of the village has no foreseeable end.

Notwithstanding the influence of *dāī hālne*, an event that affords temporary relief from the intense interpersonal village tensions and that will form the focus of chapter 8, the image of matrimonial encompassment clearly remains "lurking" at the back of villagers' minds. Hence, the dialectics of self-fulfilling images, expectations, and dispositions, loaded with considerable emotional implications that are entangled with the notion of matrimonial encompassment and backed by objectively harsh village realities, appear to be indelibly imprinted on Thamgharians' social ethos and the nature of interpersonal relations.

The main idea I wish to advance here is that in Thamghar, abstract mental images, as epitomized by the construct of matrimonial encompassment, have the capacity to become densely imbued with meaning. Once made public, they appear to assume a life of their own and may bear consequences or have unintended and unexpected implications regarding people's perception of self and the other, as well as the way they make sense of social relationships. Unlike the static and clear two-dimensional demarcation of a geometrical fractal as seen in figure 1.1 and notwithstanding its infinite theoretical expansion, mental fractal images are dynamic, often know no boundaries, and infringe upon each other in a quite ambiguous yet meaningful manner to create what may be a rather complex and surprising outcome.

## A Short Comparative Note

The above argument may gain support through a comparison with a very similar image, albeit with quite different implications, dominating the imagination of rural Hindu Tamils within a Dravidian kinship milieu,[82] in the southern tip of the Indian subcontinent.

Writing about the perceptions of self, personhood, gender, and various other categories governing the Tamil universe, Trawick-Egnor (1978: 183–84) describes their notion of the human body as consisting of a soft, eternal, generative female interior enclosed within a hard, temporal male "crust." She further wrote about marriage: "To unite with someone is to include them within oneself, or to be included by them. (Trawick-Egnor 1978: 75). When a man and woman unite, it is as the union of body and soul, say ancient and modern Tamil love songs also. The man is the body and the woman is the soul. . . . The body is a protector of the soul (ibid. 149–50). The husband is the body of his wife. The body is the servant of the soul (ibid. 153).

In contrast with Thamghar, marriage partners in matrilineal Tamil Nadu are sought nearby, matrilateral cross-cousin marriages are preferred and people find mates who are very close, both geographically and in terms of the kin network (Trawick-Egnor 1978: 76).

In general, as exemplified in the words of Themozhiyar, a Tamil scholar and Trawick-Egnor's chief guru, Tamils stress that "bringing a woman into a house [after marriage] is like bringing a light into a place that is dark," women are the deities of the house, godlike and full of wisdom (ibid. 153). As will be recalled, this image is also implied by the Thamgharian fractal of living temples, and as may be apparent, it is denied in the context of matrimonial encompassment as this would imply that the incoming *buhāri* is *divine*.

Unlike the dangerous, ambivalent Thamgharian *buhāri*, in Tamil Nadu, a woman is regarded as the essential prominent person in the culture as a whole, and femininity is mainly equated with love and emotion (Trawick-Egnor 1978: 175–76). Therefore, it may be surmised that the Tamil perception of women and marriage may partially explain Trawick-Egnor's observation that (quite unlike Thamgharian Brāhmaṇs) when Tamils conceal their emotions, this is in order to "*encourage relationship* to become and remain in a dynamic mode, as it establishes tension that binds self to other"[83] (Trawick 1990: 256).[84] Indeed, in sharp contrast with Thamghar, what Tamils are so eager to conceal is mainly *anpu*, love (Trawick 1990: 254–58 and passim).[85]

Comparison between the marriage imagery in Thamghar and Tamil Nadu reveals that by drawing upon and making selective use of quite a similar pool of cultural images within markedly different kinship milieus, the same mental image (which I prefer to call matrimonial encompassment), lends itself to conveying almost opposing imports with rather disparate corollaries. Yet I would like to argue that in both contexts it plays a decisive role in the local metaphysics and is highly consequential and central to the framing of experience and production of meaning. Its ultimate significance clearly depends upon the context in which it is embedded. This conclusion is further confirmed by other southern communities where quite similar marriage imagery prevails, yet the general perception of women and husband-wife relationships are often more in agreement with those found in Thamghar.[86]

I would now like to shift the attention to an additional, prominent and intriguing Hindu manner for conceptualizing the conjugal bond. Particularly in northern India[87] and Nepal,[88] marriage is often understood along the lines of the image of *ardhāngini*—the wife becoming the left *half* of her husband's body via marriage.[89] In apparent contrast with the Thamgharian image of matrimonial encompassment or similar perceptions[90] that seem to justify and promote extreme subordination of the in-married wife by her husband and kin, *ardhāngini* may imply complementary interdependence and *equality* between the spouses, as found by Daniel (1980: 83–84) regarding its Tamil version.[91] However, unlike its Tamil variant, the northern *ardhāngini* does not denote the joining of two halves, one male and the other female, but instead, the wife becomes part (the "lesser half") of her husband, which underscores the hierarchical relationship between them.

Hence, as already implied by Nicholas (1995: 140–41) and suggested by Busby (1997: 276, n4; 2000: 222–29),[92] the north Indian *ardhāngini* seems, in effect, to be a form of encompassment. This is not only in agreement with the general subordination of the wife in much of Hindu northern India and Nepal, which according to Sax (1990: 501) has become "an ethnographic axiom," but is also in line with the view of high-caste Hindu widows as socially half dead (Parry 1994: 157), and the notion of *satī*, which suggests that a virtuous woman should die on her husband's funeral pyre (Leslie 1992: 175–91). Thus it may be concluded, that as found regarding the image of matrimonial encompassment, *ardhāngini* has the capacity to become densely imbued with meaning and significance, but these are ultimately dependent on their specific context.

Finally, in sharp contrast with the human *ardhāngini*, the parallel image of *ardhanārīśvara* (the lord who is half woman) in Hindu mythology, mainly associated with Śiva (or previously Rudra), is taken as an emblem of limitation where conventional sexual congress is impossible until the male and female halves split, or the goddess steps *out* of Śiva's body where she otherwise eternally dwells.[93] The possible background for this stark discrepancy between the notion of procreative human *ardhāngini* in general and the Thamgharian matrimonial encompassment in particular, and the asexual overtones of divine, mythological *ardhanārīśvara*, will gain some clarification in the ensuing chapter.

The present part explored the significance and implications of the image of matrimonial encompassment within the realm of Thamgharian kinship and social relationship. I would further like to argue that matrimonial encompassment is not merely the way in which Thamgharian Brāhmaṇs imagine marriage and the elementary kinship unit, but that for them it embodies a template for the ideal modality of existence. Moreover, I would like to suggest that matrimonial encompassment is a mental frame that serves as a vehicle of reflection, in which much experience and action, not only social interaction and kinship, gain their form and meaning and are converted into cognition and emotion. Put differently, matrimonial encompassment should be regarded as one of the prominent fractalic images dominating the life of Thamgharian Brāhmaṇs.

The full scope of the fractalic nature of this image with its sometime surprising incarnations in various realms of village life, together with its other avatārs found elsewhere in the Hindu world, will become apparent in subsequent chapters.

The following section will take a further step toward advancing these general ideas through an examination of the role of matrimonial encompassment within the mesocosmic ritual arena of the village *jagya* where, for the first time, matrimonial encompassment will emerge as a fractalic image.

## NOTES

1. See also Jamison (1996) regarding ancient India.
2. Marriott (1989); Ramanujan (1989a).

3. See for example Babb (1970), O'Flaherty (1980a: 90–91); and Fuller (1992: 40–48).

4. Bennett (1983) included.

5. Fürer-Haimendorf (1966a); Doherty (1974); Stone (1977); Gray (1982, 1995); Czarnecka (1986); Skinner (1990); Kondos (1991); and Rutter (1993).

6. Hershman (1977); Inden and Nicholas (1977); Raheja and Gold (1994); and Harlan and Courtright (1995).

7. Generally referred to as *nari*, *āimāi*, *sri* or *pothi*.

8. Called *puruṣ*, *logne-mānche*.

9. See, among many others, O'Flaherty (1980a); Larson and Bhattacharya (1981); Kinsley (1986); Pintchman (1994); Jamison (1996); Larson (1998); and Doniger (1999).

10. The last two stages of a woman's life, namely becoming a matriarch (mother-in-law within an extended family) and widowhood, have no direct relevance to the ensuing discussion and hence will not be examined herein. However, it should be noted that these stages share the inner contradiction, inconsistency, multivocality, and fluidity that characterize women's other life stages as well as the attitudes toward them as will become apparent below. For an excellent account of these last life stages among north Indian women see Lamb (2000).

11. As already mentioned in chapter 2.

12. The highly auspicious food leavings of a deity, full of blessing and grace by virtue of coming in contact with it, which is distributed at the end of every *pūjā*.

13. Thamgharian astrologers maintain that *nakṣatras* are stars. In Hindu astrology these are regarded as clusters of stars situated in different sections of the ecliptic zone (the apparent orbit of the sun), near which the moon passes during the month (Basham 1985: 490), that is, the twenty-seven or twenty-eight lunar mansions mentioned in the *Ṛg Veda* (Renou 1957: 76).

14. This title clearly demonstrates that every sister is conceptualized as *older* and thus of *higher* status than her brothers, regardless of biological age.

15. This takes place on the final, climactic day of the Tihār festival (celebrated during the dark half of the month of Kārtik, October/November). The day is named after its main focus, the attribution of a special colorful *ṭikā*, an auspicious mark on each brother's forehead. Bhai Ṭikā parallels the north Indian *rakṣā-bandhan* festival.

16. This is an expression of deference where one bows down, touching a superior's legs with one's forehead. Women should perform a slightly more modest gesture, which does not include physical contact.

17. The concept of *dān* will be explained below.

18. Inden and Nicholas (1977: 54); O'Flaherty (1980a: 29); and Manu 9.33 in Doniger and Smith (1991: 201).

19. Gough (1956: 841); Madan (1965: 102); Fürer-Haimendorf (1966a: 49); Campbell (1976: 88–89); Inden and Nicholas (1977: 43–44); Dumont (1980: 117); Krause (1980: 176); Fruzzetti (1982); Bennett (1983: 80–83, 144–45); Czarnecka (1986: 29); Parry (1986: 461–62); Raheja (1988: 121); Enslin (1990: 133); Sax (1990: 496); Carter 1995: (135–36); and Gray (1995: 80–81).

20. For a critique of Raheja (1988) see also Heesterman (1992) and Parry (1994: 135–39).

21. Heesterman (1959, 1992); Parry (1980, 1985, 1986, 1994); Shulman (1985); and Laidlaw (2000).

22. Limitation of space prevents doing the subject of *dān* full justice here, hence the following will only include those features deemed most salient and relevant to the present context.

23. Notably, the excellent discussions in Bennett (1983: mainly 148–62 and passim) and Ramirez (2000a: 184–94). See also Gray (1982).

24. The auspicious forehead mark made of husked rice mixed with red *abir* (*avir*) powder.

25. Mauss (1967: 126, n83) mentions the possibility that *dān* contains sins, but regards it as absurd.

26. Rice in general is a semidivine entity and *cāmal* is considered female. This is of much significance as will be demonstrated and further discussed in chapter 7.

27. Though money is not explicitly gendered, as a form of wealth it is viewed a manifestation of Lakṣmī, the goddess of wealth and is regarded as a substance full of energy. This is why in *pūjās* coins are often given as *godān*, in lieu of a cow.

28. The river is female, conceptualized as a confluence of the Ganges and thus a manifestation of the goddess Gaṅgā.

29. This is since *kul* members are seen as *hāṛ-nāto* (*hāṛ-nātā*), literally "bone relatives" that is, they are all thought to share substance (bone) and together form the *kul*'s "social body." Hence, they are mutually affected by each other's moral substances and major fluctuations in ritual and purity status.

30. *Siddhadān*, whose chief constituent is *cāmal*, is always accompanied by *dakṣiṇā* and includes all the ingredients required for cooking a single rice meal.

31. Subsequently, her husband, children, and their spouses become potential recipients for her consanguines' *dān*.

32. Translated by a local priest.

33. As implied by Gray (1980) and Czarnecka (1986), the complex issue of the status of the parties associated through marriage comprises a hybrid of conflicting notions and incompatible practices. Hence the transferal of *kanyādān*, important as it may be, should not be taken as the only cue for the status of affines.

34. Her inauspiciousness is so great that it is believed to affect or at least anger Sūrya, the male sun god.

35. This situation is currently undergoing a gradual yet rapid change, mainly since possessing the SLC (high-school leaving certificate) is becoming a primary requirement for a bride-to-be. Concomitantly, women's views of themselves, their position within the family, and role as *buhāris* (daughters-in-law), are no longer in line with Bennett's (1983) observations, but are more on par with Western feminist views, as was recently found by Enslin (1990, 1998) and Skinner and Holland (1998).

36. See Rutter (1993: 391) for an additional inconsistency in Bennett's (1983) work.

37. See, for example, Skinner (1990: 181–82), Kondos (1991: 113–17), and Rutter (1993: 104).

38. Clearly, being extremely polluted also makes the girl vulnerable to further pollution and evil, but this is rarely mentioned as a reason for her confinement and is certainly not its primary cause.

39. Gough (1956: 841); Bista (1972: 7); Babb (1975: 80–90); Parry (1979: 200); Dumont (1980: 56); and Skinner and Holland (1998: 91).

40. An exogamous agnatic unit, whose members may belong to different *kuls* and castes, and are believed to be descendent from one of the seven mythical *ṛṣis* (*rishis,* sages).

41. A *thar* is a group of people sharing the same family name, yet not necessarily belonging to the same *gotra*.

42. Since familial memory does not normally stretch back more than four or five generations, this ideal is rarely met.

43. These notions, prominent within north Indian kinship milieu, can already be found in the Vedas (Menski 1992: 56–62).

44. Later in life, a wife gains her utmost status when she herself becomes the matriarch of a joint household and the mother-in-law of a young *buhāri*.

45. This possibility is clearly mentioned in ancient texts, where a daughter (even wife) is given as *dakṣiṇā* (Heesterman 1959: 245 and Trautman 1981: 288–89).

46. Compare with Parry (1994: 128–29, 235) and Gray (1995: 82–83) on quite similar notions among north Indians and Nepalese Chetris respectively, where evil "exchange" between affines occurs albeit for somewhat different reasons.

47. Notwithstanding, violence against women (often found in societies where women's position is supposed to be equal to that of men) is viewed in Thamghar as one of the most despised acts, almost on a par with incest, and rarely takes place.

48. The tendency to place the blame on a married woman while exonerating the husband is also apparent in villagers remarks regarding the then Nepalese King Birendra and his wife Aishwarya, who were later killed in the royal massacre in June 2001. Many village men and women were adamant that it is the queen who was responsible for the ill state of the country and any negatively perceived steps her husband initiated during the predemocracy Panchayat era (until 1990), while the king was viewed as an honest benefactor. My impression is that this observation is by no means confined to Thamghar but is found throughout much of Nepal.

49. Yet the senior member of one of Thamghar's prestigious *kuls*, who keeps the stone embodiments of the *kul* gods in her house, is the widow of one of the former patriarchs. Nevertheless, despite her status and supposedly being part of the social body of her late husband's *kul*, like any other woman, this respected widow is excluded from the *kul*'s grand annual *pūjā*.

50. Thus in the local theory of embryology, as Inden and Nicholas (1977: 52–54) find in Bengal, the semen forms the "hard" parts of the body, while the uterine blood forms the "soft" parts of skin, flesh, and blood.

51. This important festival, celebrated throughout Hindu Nepal, falls on the third day of the bright half of Bhadau (August/September). Detailed descriptions of the festival can be found in Bista (1969), Bouillier (1982), Bennett (ibid.), and Holland and Skinner (1995).

52. These are not exclusive to Thamghar and are also exposed in Holland and Skinner (1995) and Skinner and Holland (1998). Raheja (1994b) reports of similar notions in northern India.

53. Thamgharian women distinguish between Tij, directed at ensuring their husbands' long life, and the adjacent activities and *pūjā* of Ṛṣi (Rishi)-Pañcami (Panchami), which is only observed by menstruating (married and unmarried) women, in order to cleanse themselves from the possible sin of touching a man during their periods.

54. A Nepalese compilation of episodes from the Hindu *Purāṇas*, found in each and every Thamgharian house. It is read, one chapter per night, during the month of Māgh (January/February).

55. A translation of this myth can be found in Bennett (1983: 293) and Ranjan (1999: 85–90). See also O'Flaherty (1980b: 175–76)

56. *Ocimum Basilicum.*

57. *Ficus Religiosa.*

58. On this prevalent Hindu theme see also Shulman (1980: 140).

59. Provided of course that he has already separated from the joint household.

60. Thus, although they would not normally admit it, I saw a number of men who went to fetch water and do the family's laundry instead of their wives, or alternatively, helped them by cleaning the house and attending to the kitchen garden.

61. See also Sax (1990: 495–96).

62. Babb (1970, 1975); Beck (1974); Fuller (1980, 1992); O'Flaherty (1980a); Shulman (1980); Sax (1991); and Good (2000).

63. Although obviously the final one is cremation.

64. This is the sacred enclosure established in both the bride and the groom's households and is the focal arena where most of the marriage ceremony takes place. Chapters 5 and 6 will discuss the *jagya* and *hom* in greater detail.

65. *Svetāśvatara Upaniṣad* 4.10. Elsewhere, *mithuna* is simply understood as a marital (Heesterman 1985: 136) or a sexual union (Malamoud 1996: 173).

66. For an English translation of this myth see Ranjan (1999: 30–32).

67. Similarly, the *rakṣā-bandhan* is worn on the left wrist of the married goddess in the Mīnākṣī temple in Madurai south India (Fuller 1980: 339).

68. Some of the best commentary and criticism of Dumont may be found in Marriott (1969, 1989), Parry (1980), Das (1982), Appadurai (1988), Macfarlane (1992), Quigley (1993), Gregory (1996), and Bayly (1999).

69. For example, Tambiah (1973: 191) and Handelman (1990: 157–58).

70. This idea seems to be based on Dumont's own idiosyncratic interpretation of the biblical text. In fact, the Genesis story does not describe a situation where Eve is encompassed within Adam's body but merely states that God created her out of Adam's rib (in order to establish a clear hierarchical relationship between them). Similarly, there is nothing to suggest that the Bible views the difference between the sexes in terms of contradiction. This appears to be Dumont's own (Western derived?) notion.

71. The similarity is partial since in the Bible, although Eve was created out of Adam's rib, she later gains semi-independence as unfolded in the Book of Genesis. However in Thamghar, the wife is believed to be reborn within her husband's body in order to provide a notional seal over her potential independence and subject whatever agency she may have enjoyed to her husband's control.

72. Thamgharian Brāhmaṇs seem to share these perceptions with Hindus elsewhere. See, for example, Lamb (2000) and Trawick-Egnor (1990), regarding North and South India respectively, as well as Narayanan (2003).

73. Inden and Nicholas (1977: 48); Bennett (1983: 74); Sax (1990); Fruzzetti, Östör, and Barnet (1995: 36).

74. This view, a clear expression of a fractalic perception of the Hindu universe, may already be found in the Vedas, where encompassing clearly means to restrain what is inside, the universe, and humankind (Miller 1985: 162–63). See also Renou (1961: 167–68), and Handelman and Shulman (1997: 75).

75. Scale here is used in the sense of the order of being, not a measurement or ladder.

76. My emphasis.

77. O'Flaherty (1980a: 53–61).

78. From *jimmāwāri* (*jimmewar*, responsibility) and *lagāe* (to give, to endow).

79. Yet it should be noted that rice embodies a *non*-matrimonial encompassment.

80. A similar notion is evinced by the Vedic idea of conception, where *garbha* means both a womb and an embryo (Kaelber 1976: 352–34). Likewise, Tamil notions of union and birth view them as the same event (Trawick-Egnor 1978: 67).

81. I shall expand on this point in part II.

82. One of the best sources on this is Trautmann (1981).

83. My emphasis.

84. Unfortunately, the aforementioned Tamil marriage imagery, central to Trawick-Egnor's thesis (1978), is not developed and only briefly mentioned in her later work (1990: 274–75, n16).

85. Trawick-Egnor does not dwell upon the potential social implications of the marriage imagery she analyses (1978). The suggested association of the latter with the Tamil management of emotions and the ideology of love she (1990) discusses, is my own.

86. Namely, Gough (1956).

87 Madan (1965: 125); Inden and Nicholas (1977: xiv, 45–48); Sax (1990: 497); Parry (1994: 174); Courtright (1995: 188); Östör and Fruzzetti (1995: xxvi); Raheja (1995: 33); and Lamb (2000: 209).

88. Kondos (1991: 122); Rutter (1993: 338); and Gray (1995: 173, n3).

89. Thamgharian Brāhmaṇs are, by and large, unfamiliar with this notion and I never heard any local description of marriage using these terms.

90. Namely, Daniel (1980: 64–71), Trawick (1990: 138), and Sax (1991: 81–82).

91. Locally exemplified in terms of *ardhanārīśvarar*: the androgynous form of Śiva and Śakti.

92. See also the state of a married couple in Lamb (2000: 232, fig. 7).

93. O'Flaherty (1980a: 317–20); Shulman (1980: 351); Kramrisch (1984: 201–13); Handelman and Shulman (1997: passim; esp. 76); and Doniger (1999: 246).

# II

## COSMOS: FROM IMAGE TO COSMOLOGY

Now that the picture of Thamghar's social and kinship milieus is in place, and having examined the meaning and significance of the notion of matrimonial encompassment and its import on people's perceptions of self and social relationships, it is time to further the understanding of this construct and explore its meaning and role in ritual and cosmology. This will be done via a detailed investigation of the inner working of the village *jagya* (including the performance of *hom*), paradigmatic of Thamgharian elaborated *pūjās* and *saṁskāras*.

Chapter 4 will provide a general overview of divine marriages in Hindu thought in order to locate the divine cosmic matrimony that takes place in a *jagya*, and the concept of matrimonial encompassment in general, within a broader metaphysical context. Against this theoretical backdrop, a detailed description of the Thamgharian *jagya* and the performance of *hom* will be presented in chapter 5, where prior to focusing on the ritual procedure itself, I shall briefly touch upon the dual meaning of the term *jagya*, the nature of Hindu "symbols" and the notion of "embodied representation."

Chapter 6 will offer a detailed analysis of the various stages of the *jagya* and *hom*, elucidating the inherent fractality and other fundamental properties of this gendered ritual cosmos. Finally, drawing upon Gell's (1998) analysis of sorcery, a range of comparative cases drawn from ancient and contemporary Hindu myth and ritual will be considered in an attempt to unravel the roots of the *jagya*'s ritual efficacy.

# 4

## On Hindu Divine Sexuality and Marriage

### INTRODUCTION

Within the rich historical body of Hindu ritual and myth, sexual imagery in general and divine marriage in particular have enjoyed a remarkable career, fraught with major fluctuations in meaning, significance, and function, and have evinced an outstanding flexibility and ability to assume different forms and manifestations in keeping with the shifting currents of the ideologies and beliefs of their time.[1]

My aim in this chapter is not to present a comprehensive or exhaustive account of this vast topic, but to mention a number of prominent landmarks in the development of these ideas in Hindu thought, particularly those pertaining to divine sexuality and marriage, in order to provide the necessary background for appreciating the divine matrimonial encompassment dominating the village *jagya* and *hom* discussed in chapters 5 and 6.

A presentation of the nature of contemporary mythic (*Purāṇic*) divine marriages and sex, with their strong ascetic overtones, will be followed by examination of their Vedic counterparts and a discussion of the ancient ritual "technology" of procreating immortality out of death, which seems to dominate the Vedic sacrificial paradigm.

A brief mention of two Vedic nonviolent ritual wombs, which appear to anticipate the subsequent move away from violence toward the "vegetarian," peaceful *pūjā* and the ideal of *mokṣa*, will be followed by a note on the multifarious roles of fire and *tapas* (internal heat) within the Vedic sacrificial cult. Next, I shall discuss the major metaphysical transformation that took place around the time of the Buddha, which included a radical change within the Hindu mind-set toward procreation and seems to be echoed by a major shift in the nature of divine marriages. This will be followed by a brief examination of the Upanayana (male initiation)—one of the cardinal Hindu *saṁskāras* ("rites of passage" or life cycle rituals), which all appear to be dominated by the ancient pan-Hindu concept of "second-birth."

Finally, the remarkable, age-long resilience of the Vedic bond between life and death, as epitomized by the Vedic sacrifice, and the later attempt to obviate this bond for the ideal of *mokṣa*, that is, the rejection of *saṁsāra* (the endless succession of life and death), will be examined. This will provide the necessary basis for appreciating the general point I wish to pursue in this section, namely, that it is the aforementioned ancient bond that Thamgharian Brāhmaṇs attempt to sever through the realization of divine matrimonial encompassment in the village *jagya*.

## ASCETIC DIVINE SEX

The main Hindu attitudes toward sexuality and procreation seem to oscillate between passionate adherence, in literal or abstract forms, and complete rejection, and, more often than not, conflicting views appear to flourish side by side. This seems to apply to divine and mortal marriages alike; both may be portrayed as ideal, desirous, and vital, yet are often laden with inherent dangers and referred to with great ambivalence. As stated in chapter 3, the position Thamgharian Brāhmaṇs adopt toward marriage (as epitomized by the notion of matrimonial encompassment) and sexuality provides a striking example of the ambivalence Hindus in general seem to feel toward these issues. Often, as argued by O'Flaherty (1980a: 47), this results in what appears as cognitive dissonance or a yawning gap between belief (or explicit statements) and practice. She writes,

> [T]hough most Hindus *say* that a *yogī* is doing the best thing (an expression of what they think they think), they act in a way that belies this belief, by procreating as much as possible (an expression of what they do think) (ibid.).[2]

The deep-seated Hindu ambivalence toward marriage is also reflected in mythical divine marriages as exemplified by south Indian Tamil shrines, where the goddess's wedding is portrayed as a *sacrifice*, the victim being none other than her divine groom (Shulman 1980: 90–294). Let us now proceed to take a closer look at the nature of divine marriages and the ambivalent feelings they evince toward sexuality.

Hindu thought conceptualizes divine marriage predominantly as a mechanism for controlling and channeling the goddess's ambivalent *śakti*, in order to enrich her divine partner and human worshippers.[3] Unlike human unions, where the energetic transaction between the sexes is primarily aimed at begetting offspring, divine marriages seem to remain barren even when sex is regularly performed since, as Fuller (1992: 44–48) notes, for Hindu gods and goddesses, sexuality and parenthood are usually divorced from procreation. Hence, although childlessness often vexes spinster goddesses and seems to be one reason for eliciting their anger, many married goddesses appear quite content to have no offspring (ibid.).

Moreover, divine sex seems to be redolent with strong ascetic overtones and divine procreation is often reminiscent of great *yogīs* who may shed their sexual fluids (such as sweat, tears, or semen) into any form of womb (a pot, the earth, a river, or someone's mouth) and thus engender an embryo (O'Flaherty 1980a: 39). An obvious example is Lord Śiva, who is forever torn between intense asceticism and passionate eroticism (O'Flaherty 1981). The god's prolonged lovemaking with his faithful consort Pārvati or Satī, never results in ordinary pregnancy and birth; his son Skanda (the lord of war) is created from Pārvati's spittle, while Gaṇeś (the remover of obstacles) is engendered from the water she uses to wash herself after having passionate intercourse with Śiva (O'Flaherty 1980a: 37–39; Kramrisch 1984: 208). Therefore, it is perhaps little wonder that the marriage of Śiva and the goddess is at times idealized as *non*-sexual (Shulman 1980: 141), and both Gaṇeś and Skanda remain strictly celibate (ibid. 103).

Chapter 10 of the *Swasthāni*, a well-known text among Thamgharian Brāhmaṇs, narrates the following tale about the birth of Śiva's son, Kumar (Skanda's other name):[4] Agni (the fire god) interrupted Śiva and Pārvati while the divine couple was engaged in wild love-making and had to accept some of Śiva's semen into his mouth, becoming pregnant. Unable to bear the sperm's intense heat, he spat it into a sacred *kuś* (*kuśa*) grass growing on the banks of the Ganges. The virile potency of Śiva's semen was so great that the *ṛṣis'* wives, who came to warm themselves in the heat of the burning semen, were impregnated. Suspected by their husbands of adultery, they vomited Śiva's semen into the Ganges, who thus conceived and finally gave birth to the six-headed baby Kumar. In striking contrast, Pārvati never conceived after drinking Śiva's seed (O'Flaherty 1980a: 49).

Therefore, unlike Westerners who, until recently, could only be born from their mother's womb, the Hindu deities appear to have been cloning themselves using surrogate wombs from time immemorial. The reason behind the Hindu gods' appeal to what is often (perhaps wrongly) termed *modern* reproductive technologies, lies in the view that procreation fuels *saṁsāra*—the wheel of recursive births and deaths characterizing a mortal state of being. In other words, in order not to lose their immortality and divine status, the Hindu gods must abstain from humanlike procreation (Fuller 1992: 47).

The ascetic nature of divine marriages is given a striking illustration by what is perhaps an almost ubiquitous representation of Śiva throughout the Hindu world—the Śiva-*liṅgam* (phallus)—where the *liṅgam* stands in the middle of a *yoni* (the goddess's sexual organ).[5] According to O'Flaherty (1981: 256–58), the Śiva-*liṅgam* represents sexual satiety and control (but *not* their denial), while Daniélou (1964: 224–25) views the *liṅgam* as fundamentally *procreative*, yet the fact that Śiva's *liṅgam* always emerges *out* of the *yoni* instead of penetrating *into* it (Kramrisch 1984: 242–49), suggests the contrary. To my mind, as has already been suggested by Babb (1975: 231–32),[6] the Śiva-*liṅgam* represents an inverted, *non*-procreative sexual union, where the semen of the erotic-ascetic Śiva is *retained*, making it an emblem of chastity and *contained* sexuality.

Another salient expression of the barrenness of divine unions is found in the *mithunas* (the state of being a couple, denoting marriage and sexual union) that often adorn the walls of Hindu temples, and in particular are carved onto the doorjambs of the *Garbhagṛha*[7]—literally, "the house which is the womb" or "the womb of the house" (Kramrisch 1976: 346–47, 157, 162). Viewing *mithuna* as a representation of the *mokṣa* ideal, a final release from the wheel of *saṁsāra*,[8] Kramrisch (1976: 156–65) speaks about the fundamental transformation that the devotee, entering the dark womb of the temple, undergoes in terms of rebirth: "The *Garbhagṛha* is not only the house of the Germ or embryo of the Temple as Puruṣa; it refers to man who comes to the Centre and attains his new birth in its darkness" (ibid. 163).

However, as the ethnographic literature on Hindu temple worship clearly suggests, this notion is not ordinarily found among contemporary Hindu temple-goers; the latter do not normally entertain the thought of attaining personal rebirth in the depths of the temple, but are instead predominantly preoccupied with the wish to attend to the corporeal needs of the deity residing in the *Garbhagṛha*, to please and identify with it, and seek the experience of *darśana* (vision, ocular exchange with a deity or a person), anticipating the reward of divine grace in exchange for devotion.[9]

Consequently, the ideally dark inner sanctuary of a Hindu temple is perhaps mysterious (Eck 1998a: 63) and may be a locus of "bounded creative power" (Good 1989: 193) and "secret growth" (Beck 1976: 235), but it certainly does not function as its title, *Garbhagṛha*, suggests. Rather than embodying a vibrant womb, it appears to be quite sterile. Furthermore, as Fuller (1984: 5) notes regarding one of the most popular south Indian temples—the Mīnākṣī temple in Madurai—nowadays the darkness of the *Garbhagṛha* is marred by the electric lighting that is often installed to illuminate it.

Divine sex, barren as it may seem, is nonetheless extremely potent and is in fact capable of preserving and maintaining the entire universe[10]—when

Śiva is separated from Pārvati the moon is said to wane (O'Flaherty 1981: 296). Moreover, as asceticism and sexuality are intimately connected in Hindu thought (Eliade 1978: 228), where power and particularly sexual power (*śakti*) are often viewed as exhaustible or in terms of an energy balance, ascetic practices are believed to augment one's power considerably,[11] hence abstinence from procreation does not necessarily reflect a lack of fertility. Indeed, a number of gods and particularly goddesses seem to be overflowing with fecundity, and it is the latter's ability to radiate their *śakti* onto a certain territory, kingdom, the soil,[12] or upon individuals[13] that makes both agricultural production and human reproduction so entrenched within and dependent upon the power of the goddess (Fuller 1992: 47).

The role of Aditi, the Vedic earth goddess, the prototypical mother and the ultimate source of progeny and growth (Miller 1985: 87, 138), was later usurped by Bhūme or Bhūdevī (Shulman 1980: 139), maintaining the earth's role as "the eternal womb of creatures" (Manu 9.37)[14] throughout Hindu tradition. Unlike most other Hindu divine consorts and mothers, Bhūme seems to be a rather "corporeal" goddess and is literally a terrestrial divinity; she is frequently likened to a cow,[15] and is often a mother in quite a humanlike sense. Similarly to mortal women and other worldly goddesses, such as the river Gaṅgā (Fuller 1992: 193), she is often said to menstruate,[16] though this may only take place once a year. Moreover, Bhūme actually conceives and gives birth to semidivine and earthbound entities, among them houses,[17] temples,[18] all agricultural crops, and vegetation (which are seen, at times, as the fruits of her union with the rain or with heaven),[19] as well as the goddess Sita, the furrow (Daniélou 1964: 261).

Although Bhūme is Viṣṇu's second, lower-status wife (Fuller 1992: 33),[20] the god himself is rarely (if at all) explicitly involved in his wife's (pro)creative activities, hence for example, when the earth gives birth to Sita (Rāma's devoted wife), no father is actually mentioned. Sita herself gives birth to two sons through ordinary procreation and thus has only an ambiguous claim to divine status. Unlike real gods and goddesses who reside in heaven, she becomes a true goddess only upon returning back into the earth (O'Flaherty 1980a: 79–80, Doniger 1999: 8–15).

The Hindu king is normally regarded as the reincarnation of Viṣṇu (Fuller 1992: 107) and, like him, he is the protector and preserver of the earth and the terrestrial order, is often regarded as her husband and lover, and is responsible for the rain as well as the prosperity and productivity of the soil of his kingdom.[21] In Nepal, this is reflected, for example, by one of the king's titles: Bhūpati,[22] the master or husband of the earth (Lecomte-Tilouine 1996: 33), as well as in the belief that the king (a reincarnation of Viṣṇu) and the queen engage in *hilo khelnu* (mud play) at their royal palace in Katmandu during the rice transplantation day to enhance the fertility of the soil (Rutter 1993: 364).

Nonetheless, the role of Viṣṇu in the production of rice in one Chetri village is said to be that of an *ascetic*; during the rice cultivation season the god descends into the earth where he becomes fully absorbed in intense sleepy meditation, and it is the *śakti* generated by the austerities he performs that facilitates the ripening of rice (Rutter 1993: 367–78).[23]

The deep-seated ambivalence toward procreation and sex exhibited by Hindu divinities to this day, mainly for preserving their fully-fledged heavenly status and immortality, seems to be at odds with and even to a large extent invert the general stance found amongst their Vedic predecessors, as will be outlined below.

## PROCREATING IMMORTALITY

Before embarking upon a brief examination of ancient Hindu myth and ritual, I wish to emphasize that the Vedas do not seem to provide one singular answer to the mystery of creation, death and rebirth, the nature of existence and the fate of man, and the cosmos. Instead, layers of messages, multiple meanings, and a (vast) number of parallel yet often disparate narratives may be found to run through the Vedic texts dealing with these issues. In addition, it must be borne in mind that, as notably highlighted by Heesterman (1993: ix), Vedic thought, which is speculative in itself, cannot but form the basis of speculative scholarly essays. In the following lines I shall focus on what is arguably one of the prominent Vedic modes of sacrificial creation, namely *sexual* (pro-)*creation*.

Vedic creation is largely predicated upon a violent process of sacrificial procreation and rebirth and, as Renou (1957: 78)[24] phrases it, "The creation is procreation, spoken of in biological terms. But the relationships mentioned are reversible: the gods also produce their parents . . . and no genealogy is permanent."

Malamoud (1989: 91–92) presents a concise and original translation of one of the well-studied Vedic cosmogonies, where the creator Prajā-pati—"Father god" (Hiriyanna 1993: 40), also known as "the lord of creatures" (Malamoud 1989: 91), the lord of Progeny (Daniélou 1964: 48), the lord of (biological) generation (Kaelber 1976: 367) or is simply thought of as a sexual procreator (O'Flaherty 1980a: 161), procreates, conceives, and fashions the world through the initial performance of the sacrifice in which he is the *victim*. Thus Prajāpati, who replaced the primordial Puruṣa (male, person) of earlier cosmogonies, invented and consolidated the link between violence, death, procreation, and rebirth, which later appears to dominate the entire institution of the Vedic *yajña* (sacrifice). I quote:

> In the beginning, Prajāpati is alone. He is overcome with a desire to procreate, which is also the desire to "become multiple." In order to realize this desire "he becomes [the] sacrifice,"[25] in which he is simultaneously the sacrificer, because it is his own project; the officiant; because he himself carries out all the necessary operations; and the victim, because the only oblatory material at this stage is his body. Thus, he performs the sacrifice and emits (rather than, strictly speaking, creates) beings, and disperses his person in the infinity of things which thenceforth form the cosmos.[26] There are several different accounts of the way he went about this. Some texts insist on the solitude of the god. Others teach that Prajāpati has [the female] Speech already within him, and that his first act was to bring it out of himself, so that it is, or thereby becomes, his "assistant," his feminine double, with whom he copulates;[27] there are, however, two competing narratives of what happened next. Speech, pregnant, separates from Prajāpati, gives birth to the creatures and then returns not to him but within him;[28] alternatively, the copulation occurs between Prajāpati's "mind" (*manas*) and Speech, but it is Prajāpati who becomes "pregnant."[29] (Malamoud 1989: 91–92)[30]

In this cosmogony, sacrifice and procreation, death and orgasm are flip-sides of the same coin, and it is here that the crucial apparent paradox, which later informed the entire Vedic sacrifice, seems to lie: new life may *only* issue out of *death*. This theme is more blatantly expressed in another famous Vedic cosmogony where Death unites with Hunger, and together they are able to procreate the world (Shulman 1993: 127–28).[31] This notion also appears to lie at the root of the "technology of enchantment" and "captivity" (Gell 1992, 1998: 68–72)[32] of the Vedic sacrifice, which may account for the spell it seems to have held over Vedic man and still holds over contemporary scholars alike.

The second apparent paradox that seems to be involved here is concerned with the enactment of a coitus between an enveloping male and a female inherent within his body, and the resultant *male* pregnancy. Particularly since this may be interpreted otherwise,[33] Malamoud (1989: 92) emphasizes that Prajāpati is *not* an androgyne but the totality of a man, who can either *encompass* a woman or keep her at a distance.

Commenting on another version of the above cosmogony where Prajāpati creates via the exercise of *tapas*, Kaelber (1976: 345–46) stresses that interpreting *tapas* here in terms of *ascetic heat* or austerities is inadequate and misleading, since it fails to express the clear *sexual* connotations implied by the text.

Prajāpati emerges out of his procreative self-sacrifice, having "emitted" the world and its creatures, empty and dislocated from his act of creation, and is inflicted by exhaustion spoken of in terms of sexual fatigue.[34] Concomitantly, the created cosmos is fundamentally blemished, "a nightmare of egalitarian chaos" or a "metaphysical mess" (Smith 1989: 59), equally problematic for the creator and his creatures (Smith 1989: 54–65; Malamoud 1996: 41). In his distress, and in order to rectify his faulty creation, Prajāpati asks the newly created gods to restore his "distributed personality," to borrow from Gell (1998), via another sacrificial and cosmogonic process, which has become the prototype for the *agnicāyana* and the Vedic sacrifice in general.

Hence "every sacrifice repeats the primordial act of creation and guarantees the continuity of the world for the following year" (Eliade 1978: 229),[35] and the *agnicāyana*, literally the piling up of fire, is a year-long sacrificial process explicitly conceptualized as gestation, in which the magnificent Vedic fire-altar embodying Prajāpati is built.[36]

## The Gendered Sacrifice

According to Malamoud (1989), the Vedic sacrifice is a personified, gendered, and sexual entity, whose body in its entirety is supremely *male*, and is in itself made of multiple objects, characters, gestures, and utterances of speech, which are all perceived as indefinitely arranged living bodies. All of these are *sexed* beings that normally form *mithunas*, that is, they are arranged as couples. What was true regarding the mythical Prajāpati also applies to the numerous *mithunas* that make up his reconstituted "personality," that is, the body of the sacrifice; the presence of scores of *mithunas* does not render the latter an androgyne, instead the sacrifice is perceived as an overall male body enveloping femaleness.

The picture of the Vedic sacrifice portrayed by Malamoud may be couched in the terms employed in the present work as a gendered, fractal sacrificial-body, where the basic structure of male encompassing or forming a union with female is recursively manifested on the sacrifice's numerous scales of existence. However, unlike the fertile notion of the Thamgharian matrimonial encompassment, Malamoud (1989: 81–82) stresses that "the *mithuna* is fertile in the sense that it sets in motion, acts as a catalyst for the general fecundity of the sacrifice, yet not because it may give birth to offspring of its own."

Correspondingly, the *mithuna* between the *jajamāna* (Vedic sacrificer) and his wife is a mere reminder that they are a permanent couple, but actual sexual union between them during the sacrifice is not even hinted at, and each is involved in symbolic copulation with various objects, performed through mere sight rather than touch.[37] According to Malamoud (1989: 80), the texts recursively mention that these *mithunas* are meant "for procreation," which he interprets as *general* fecundity and, in particular, each stage of the sacrifice procreates the next so to speak, and the sacrificial energy proceeds through these series of couplings (ibid. 96). In the present terms, what Malamoud (ibid.) describes is a fractalic chain reaction of self-generation.

Other scholars point toward additional gendered aspects of the sacrifice. The altar, for example, symbolizes the (female) earth and thus is called *vedi*,[38] since the gods obtained (*sam-ved*) the entire earth by encompassing the sacrifice on four sides (Kramrisch 1976: 17). According to Knipe (1975: 103), "[t]he Brāhmaṇas and Āraṇyakas do not miss the opportunity to identify feminine sexuality and fire [situated on the *vedi*] with the sacrifice," which he interprets as an elaborate *procreation* ritual.[39]

Heesterman (1993: 92) reflects upon the intimate association between the *vedi* and the Vedic earth goddess Aditi, noting that the former was arranged in the schematic shape of a woman with broad hips and narrow shoulders: one sacrificial fire (*gārhapatya*) situated between her hips and the other (*āhavanīya*) between her shoulders. That the straight line between these fires was thought to connect and lead from earth to heaven (Heesterman 1993: 51) implies that the *jajamāna*'s path to heaven and back during the sacrifice was imagined to pass through the earth goddess's womb, and indeed, according to Renou (1957: 84), the *jajamāna* engenders his future heavenly self through a mystical union that takes place in the *vedi*.[40]

The notion of the *vedi* embodying a ritual womb is a recursive one in the various textual sources on the Vedic sacrifice. One remarkable example is the secondary altar constructed for the goddess Niriti[41] as part of the *agnicāyana*. Here the *jajamāna* is supposed to follow and assume the place of his prototype, Prajāpati, and to be reborn, as the latter was after the gods poured him (in the form of seed) into a fire pot embodying a *yoni*—vagina or womb (Heesterman 1957: 17).

Elsewhere, it is noted that the sacrificial altar is the loins and the central fire is the vulva of the woman created by Prajāpati and with whom he copulates (Knipe 1975: 103).[42] Overall, the Vedic sacrifice seems to be taken as an operation upon a *ritual womb*, hence in the sacrifice "man is *born* [and not simply created][43] into a world made by himself" (Smith 1989: 103),[44] or "In truth man is unborn. It is through the sacrifice that he is *born*"[45] (Eliade 1958: 55) [my emphasis].[46]

The cornerstone of the Vedic sacrificial paradigm was a network of correspondences, equivalences, and identifications, which facilitated the intimate connection between micro- and macrocosms and the movement between apparently disparate planes of reality (Smith 1989; Shulman 1993: 16). As Heesterman puts it,

> In this world of floating forms there are no hard and fast lines; conceptually different entities and notions interchange with bewildering ease. All things, entities, notions, powers are connected with each other. (Heesterman 1957: 6)

Primarily, the *jajamāna* was identified with the cosmos, with Agni and with Prajāpati. Following the latter, he was the patron of the sacrifice but was also identified with the altar, the victim, and the sacrifice as a whole.[47]

Since being and knowing may come very close in Hindu thought—as Shulman (1993: 99–100) phrases it "[e]pistemology shades off into ontology, and vice versa: to know the truth is to become the truth, in a transformative and experiential manner"—this dense metaphysical system had to be *known* for the sacrifice to be effective. Therefore, the texts repeatedly state that "he who thus knows" is reborn through the sacrifice,[48] and this "knowledge of the sacrifice," which secures the entire world for the knower is, according to Knipe (1975: 103), primarily about *sexual intercourse*; that is, carnal knowledge.

## Anticipating Renunciation

I shall now turn to briefly consider two Vedic ritual wombs found in the rites of the *dīkṣā* and the Upanayana, which seem to deviate from the pattern of the violent sacrificial Vedic womb discussed hitherto. In fact, they appear to anticipate the notion of ascetic unilateral creation that was discussed in relation to the *Purāṇic* divine marriage and sex at the beginning of the present chapter. In both these events, gestation and rebirth have little to do with either death or sexual union.[49]

In the preliminary *dīkṣā* ("initiation"),[50] the *jajamāna* (often together with his wife) sits in a closed dark hut explicitly referred to as a womb, where he undergoes implicit death, returns to an embryonic state, is gestated, and finally reborn on a higher divine plane. This is facilitated by the *tapas* he himself generates through the practice of asceticism and by the fires present in the hut. Only now is the *jajamāna* able to enter the more violent procreative womb of the sacrifice.

Gestation and rebirth seem to adopt an even more abstract and unilateral form in the Upanayana, the "introduction" of a boy to a Vedic teacher, which was a compulsory initiation into Vedic society; the most important of all Vedic *saṁskāras*, it was conceptualized as a boy's "second-birth." The guru (teacher) was thought to conceive the boy by simply placing his hand on the latter's shoulder and after three days of pregnancy, in which the master carried the student in his belly, the boy was reborn as a high-caste. From that moment onward, the teacher was regarded to be the boy's true *father* and only as a "twice-born" could a young man embark on the study of the Vedas and the performance of the obligatory sacrifice.

Becoming "twice-born" enabled a Vedic man to marry, to produce offspring, and to sacrifice, that is, to go through an almost endless series of deaths and rebirths throughout his lifetime, and thus to periodically realize his identification with the "cosmic drama," oscillating between the poles of death and rebirth (Heesterman 1957: 6–7). Predominantly, a man wished to escape premature death via the sacrifice and attempted to follow his gods, who became divine and immortal by similar sacrificial means (Eliade 1978: 229; O'Flaherty 1985: 20).

Furthermore, not only did the sacrifice serve to make a Vedic man "complete" and "integral," but the maintenance, re-creation, and the vitality of the cosmos, as well as the preservation of its order (*ṛta*), were all contingent on the continuous sacrificial manipulation of the priests.[51]

## Fire and Tapas

Although virtually all aspects and elements of the Vedic sacrifice appear to induce scholarly debate, an almost unanimous general agreement exists regarding the pivotal role of fire, Agni (the fire god),[52] and *tapas*[53] within the actual and mythical sacrificial cult. In short, fire is where the oblations are placed and Agni is the messenger between heaven and earth; he carries the oblations to the gods and brings the latter down to the sacrificial feast, as well as leading the *jajamāna*'s *ātman* to heaven and back. Agni is the sacrificer *par excellence* but also the priest and the deity of the sacrifice. He is identified with the altar, with the sacrifice as a whole, and sometimes with the victim. Son of Prajāpati, he is also his father and is cosubstantial with him and, as was implied above, Agni often embodies the fiery sacrificial womb where the *jajamāna* is "cooked" and gestated.[54] More generally, all Vedic processes of cosmic and personal transformation, creation, procreation, and gestation are predicated upon *heat*, be it the combustive Agni or the inner fire of *tapas*.

*Tapas* is primarily an erotic, cosmic, and controlled heat which is also an internal mode of creation. It is often mentioned as the primordial, mysterious, cosmic source personified as (the female) Viraj. *Tapas* is strongly associated with procreation, pregnancy, incubation, desire, sex, and creativity within the sacrifice as well as with ascetic practices. It is a cosmic entity or substance, an inner capability and knowledge but also something to perform as well as something that can be given, acquired, or lost. It is closely associated with a number of deities, principally with Agni and other forms of intense energy and is also the heat of fermentation, ripening, digestion, cooking, ecstasy, mystical vision, and knowledge (Knipe 1975; Eliade 1978: 232–35, 442–44).

The above discussion supports the suggestion that at the heart of the Vedic sacrifice, quite like its mythic prototype, lies its somewhat enigmatic capacity to orchestrate man's (entire) life cycle stages, from death to rebirth, in a highly condensed, prompt ritual mode by means of heat (fire and *tapas*). The near simultaneity of death and birth, of killing and gestation, and the recursive, obstinate, and almost obsessive execution of the Vedic sacrifice, together with the Hindu tendency to connect what (most) Westerners usually wish to separate (the actor, the act, and the outcome, the present and the future, the sexes, nature and culture, ritualized and "secular/profane" activities, myth and reality), the propensity of Hindu categories to intertwine and the Vedic metaphysics pervading the sacrifice (where all things, notions, and powers are interconnected), all contribute to the amalgamation and fusion of the fibers of life and death, violence and procreation, and loss and regeneration, into a solid knot within ancient Hindu thought.

Thus in the Vedas, life (procreation, gestation, and rebirth) is imagined to be *born* out of death, forming an inseparable "couple" where one necessarily reflects and implies the other and, as Shulman (1980: 90) eloquently puts it, "life and death are two facets of a single, never-ending cycle."

## BIRTH INTO AGONY OR "THE NEMESIS OF REPRODUCTION"[55]

From the above discussion of the "shifting moods" in Hindu divine marriages through the ages, emerges a strikingly dichotomous picture that, as I see it, is but one aspect of the fundamental transformation of thought and practice that is often imagined by Western scholars as the ideological and metaphysical crisis, breakthrough, or even revolution that shook the Hindu world around the time of the Buddha (sixth century B.C.).

This transformation is largely referred to in terms of the transition from Vedism or Brahmanism to "classical" and later "matured" Hinduism. Its cardinal manifestations are usually thought to include the move from seeking rebirth and immortality by means of sacrifice to a number of un-iconic gods on an ephemeral altar in the open to the *Upaniṣadic* quest for *mokṣa* via the "internalization of the sacrifice" (or its abolishment) and renunciation, or through the householder's nonviolent worship of the embodied manifestations of honorable divine guests and the careful attendance to their bodily needs (mainly food) in a *pūjā*. The latter is characterized by the spirit of *bhakti* (devotion) and is performed with the primary objective of obtaining *darśana* and identifying with a divine entity, which often resides within a *murti* situated in a brick and stone temple. Moreover, the Hindu (higher) divinities and their (mainly high-caste) mortal counterparts are said to shift from a carnivorous diet to almost strict vegetarianism.

In essence, the cosmos may have remained a hungry one, as both the sacrifice and the *pūjā* are fundamentally oral affairs, yet the culinary etiquette had altered completely, with love replacing the violence that characterized the former.[56] The fundamental shift O'Flaherty notes in the employment and significance of the notion of *tapas* is highly pertinent to the present discussion. She writes,

> In the Vedas, *tapas* is able to accomplish the chief desideratum, fertility; in the *Upaniṣads*, *tapas* is the means to the new goal, Release [*mokṣa*]. Both are forms of immortality, both promising continuation of the soul without the body—Release giving complete freedom of the soul (or absorption into the Godhead), progeny giving a continuation of the soul's life in the bodies of one's children." (O'Flaherty 1981: 76)

I believe that by employing the perspective of *tapas*, O'Flaherty skillfully captures and brings to the fore the issue that arguably lies behind the great Hindu metaphysical and ideological transition discussed above, namely, the perennial, unrelenting Hindu quest for immortality.

In fact, the transformations in both the nature of divine marriages and the significance of *tapas*, highlight a single point, namely the radical change that took place in the Hindu mind-set toward *procreation*. If for Vedic man, progeny was of paramount importance and via the procreative sacrifice he attempted to intensify the cycle of death and rebirth, follow his gods, and thus delay death and become immortal, post-Vedic thought dictates that the most obvious way to achieve this is to relinquish procreation altogether—renouncing both the performance of the sacrifice and family life for the path of a *saṁnyāsin* (ascetic). The latter became the sole route for escaping the daunting *Upaniṣadic* metaphysics portraying the nightmarish *saṁsāric* trap, since "How can a man who is a householder find Release?"[57]

This general approach to procreation is given radical expression in the *Mahābhārata*, where the union of seed and blood (conception) is depicted as taking place in a fouled womb, moist with excrement, complemented by the note that "let man know that women are the continuers of the web of Saṁsāra" (Meyer 1995: 365). This seems to be a clear reversal of the positive female fecundity and general attitude toward women found in Vedic texts (Menski 1992: 49–51). Likewise, Tantric sex, whether literal or more sublime and symbolic was mainly based on the notion of "salvitic sexuality," that is, passion, desire, and sex had a soteriological value and were originally for generating, offering, and ingesting transformative sexual fluid, which was then employed for the attainment of salvation, not procreation (White 2000: 15–18).

It is perhaps little wonder that such ideas were easily embraced within the Hindu pantheon, since the gods and goddesses had already gained their immortality in immemorial times via the Vedic sacrifice and were now merely required to avoid procreation in order to preserve it, while they could still enjoy (safe) sex. However, for the Hindu masses who to this day passionately endorse the ancient Vedic notion that the father (or both parents) is reborn through his son (or children), descendants were and still are a matter of ultimate importance (O'Flaherty 1981: 68; Fuller 1992: 47; Jamison 1996: 16).

This is why the notions of *mokṣa* and renunciation were always "the object of extreme ambivalence within the mature Hindu tradition" (Shulman 1993: 117). Accordingly, the majority of present-day Hindus are "rebirth oriented" (O'Flaherty 1980a: 47) and the best the "man in the world" (Dumont 1960) can aspire to is most likely what Bennett (1983: 37–39) aptly termed "conditional immortality," that is, a temporary sojourn in *swarga* (heaven) before being reincarnated and born in a new body.

Given that divine procreation became obsolete, it was perhaps to be expected that when the brick and stone Hindu temples replaced the ancient ephemeral Vedic sacrificial arenas,[58] there was little sense in holding to and preserving the ancient sexual role of the Vedic ritual womb (which still seems to echo in the title *Garbhagṛha*), and by and large Hindu ritual wombs became sterile. Thus, it seems plain that what most Hindus felt (and still feel) they cannot practically endorse in personal life, namely a complete dismissal of (sexual) reproduction, they nonetheless strive to carry out within the realm of ritual and myth.

### The Upanayana

On the face of it, the Hindu *saṁskāras* and in particular the Upanayana (male initiation), marriage (for women), and the rites for the dead, often conceptualized as birth or rebirth,[59] may appear to constitute a major exception to the contemporary barrenness of Hindu ritual wombs mentioned above. For present purposes, it will suffice to briefly examine the Upanayana (*bartoon* in Thamghar or *bartaman* elsewhere in high-caste Nepal)[60] which, in its capacity as the most important male *saṁskāra*, associated with what is probably the pan-Hindu notion of "second-birth," seems to be an example of contemporary Hindu "ritual birth" *par excellence*.

Notwithstanding what may be implied by the notion of a "second-*birth*," I would like to argue that the present-day Upanayana reflects the general Hindu ambivalent stance toward procreation, and is an expression of the deep-seated desire to avoid ordinary birth from a feminine womb. Indeed, by and large, no "ritual womb" is operated upon in contemporary Upanayanas, and no reference is made to procreation or even to the asexual, unilateral ascetic reproduction process that characterizes the Vedic Upanayana.

Instead, contemporary Hindu initiation (Thamgharian *bartoon* included) is mainly concerned with investing the sacred thread (*janaï* in Nepal), hearing the *gāyatrī* mantra (verse from the *Ṛg Veda*) from a *guru* (teacher), and frequently includes a dramatic farcical performance, in which the young man supposedly becomes a *yogī* (ascetic), has to beg for alms, and sets out on the arduous way to Banaras, yet is soon called to return.[61]

Thus Babb (1975: 79) concludes his description of the Upanayana in Chhattisgarh (Madyha Pradesh, India) by noting that "[h]aving been invested with the sacred thread and heard the mantra, the boy is now complete as a twice-born man."[62] As Manu (2.144–49)[63] clearly put it, a birth in (from) a woman's womb is a mere coming into existence "but the *birth* that a teacher who has crossed to the far shore of the Veda produces for him [an initiate] through the verse to the sun god, is real, *free from old age* and *free from death*."[64]

Although Manu adds that the latter is a boy's *Vedic* birth—the verse to the sun god is his mother and the teacher is the father (Manu 2.150, 170)—as I see it, following Doniger and Smith (1991: xl–xliv), this is merely an appropriation of tradition and Manu employs Vedic rhetoric ("in truth man is only *born* out of the sacrifice" and "Vedic birth") for advocating the new goal of *mokṣa*. Similarly, I believe that Hindus usually evoke the notions of "twice-born" and "second-birth" mainly as statements about *social status* and the degree of transformation involved, *not* about its inherent nature.

Moreover, these notions are deeply rooted within a frame of thought that views procreation and birth as *evils* to be avoided. Therefore, it seems that within the ritual realm, gestation and birth, like *tapas*, are synonyms for (inner) transformation, which if taken literally are thought of in terms of ascetic, unilateral creation (as practiced among gods and goddesses)—*not* procreation.

I would like to suggest that at least part of the extraordinary career the Vedic notion of "second-birth" in the Upanayana has enjoyed in Hinduism is rooted in its nonsacrificial and asexual, unilateral ascetic nature. Unlike the Vedic sacrifice and "well ahead of its time," the Upanayana fitted with relative ease into the ideological and ritual move away from violence and procreation, and was adopted as an "old barrel for fresh wine." Ironically, unlike the Vedic Upanayana, which enabled a young man to set off on his lifelong sacrificial journey as well as establish a family, that is, embark on the main bifurcated path to *immortality*, present-day initiation clearly roots him even deeper within *saṁsāra* and away from *mokṣa*. This is since, in the main, its aim is to allow a boy to marry and thus consummate his procreative role as a householder, yet as we have already seen, the best the latter may aspire to is "*conditional* immortality."

## CONCLUSION

The above discussion suggests that one of the main elements lying behind the post-Vedic ideal of *mokṣa* (and the overall rejection of *saṁsāra*) is the immense resilience of the Vedic link between death and life (with the biological processes that lead to and nurture it), as epitomized by the sacrifice. Accordingly, death and violence could not be removed from the texture of experience, whether for gaining immortality, on ethical grounds (*ahiṁsā*—the ideology of nonviolence and nonharming) or both, without obviating life itself. Thus, rather than being seriously challenged, it seems that the Vedic bond of life and death, taken as a fundamental characteristic of an earthly existence, was imagined as impossible to dissolve, hence the only avenues open for man were either its acceptance (the householders' path), or total rejection (asceticism).

The remainder of the present section is devoted to a detailed description, examination, and analysis of the village *jagya*, which is the most elaborate and complex of all Thamgharian *pūjās* I attended, paradigmatic of all village *saṁskāras*.[65] The *jagya*, I would like to argue, presents a local attempt to break (but not to break with) the solid bond and mutual implication of life and death in order to eradicate death while advocating life. I will attempt to demonstrate how, through manipulation of the image of matrimonial encompassment on a divine scale and via a complex system of identifications and correspondences, Thamgharian Brāhmaṇs enact an erotic cosmic drama at whose climax divine coitus sets into motion a fiery ritual womb, in which the householder and his family are regenerated and reborn.

By employing a selection of ritual and conceptual means within the general framework of the *pūjā* and "borrowing" (with very little if any awareness of it) a number of notions and elements that may be found within the Vedic sacrifice, Thamgharians attempt to engender rebirth and thus temporarily avert death, pave a direct and safe path to *swarga* after death, and attain (conditional) immortality. As will be demonstrated below, the *jagya* presents a ritual arena that realizes a cosmic fractal embodying the notion of matrimonial encompassment, which is the ultimate Thamgharian *modus operandi* for procreating life out of *life* without the need of recourse to and in order to avoid *death*.

## NOTES

1. Fuller (1980, 1992, 1995); O'Flaherty (1980a, 1999); Shulman (1980); Good (1989); Leslie (1992); Pintchman (1994); Jamison (1996); and Handelman and Shulman (1997).

2. Original emphasis.

3. Babb (1970, 1975); Beck (1974); Fuller (1980, 1992); O'Flaherty (1980a); Shulman (1980); and Good (2000).

4. See the translations in Bennett (1983: 292–93) and Ranjan (1999: 69–70).

5. This can clearly be seen in the photograph of a Śiva-*liṅgam* being worshipped, on the cover of Fuller's *The Camphor Flame* (1992).

6. See also Meyer (1995: 338) who regards the erect *liṅgam* as an emblem of unbroken chastity.

7. This is the innermost part of a Hindu temple and the place where its principal deity resides.

8. Kramrisch (1976: 346).

9. Gonda (1970: 76–86); Stevenson (1971: 368–417); Babb (1975: 108); Appadurai and Breckenridge (1976); Fuller (1984: passim and particularly 14–22, 1988, 1992); Eck (1998a); and Gell (1998: 116–21).

10. Daniélou (1964: 201–3); Gonda (1970: 130); Fuller (1980: 326); and Good (1989: 193).

11. O'Flaherty (1980a: 120) and Fuller (1992: 36).

12. Fuller (1992: 44) and Dumont (2000: 426). A local Thamgharian example may be the earth goddess Bhūme, who presides over the rice fields and is worshipped there in miniature temples situated in large trees.

13. Kiran Neopane (50), one of the area's supreme priests who specializes in conducting Devī *Putiks* (complex prolonged *pūjās* dedicated to the goddess, which are often employed for *bir* exorcism), told me that it is only by virtue of the goddess's *śakti* he had managed to absorb (while meditating in the service hut near the regional Kāli temple where he spent the nights over a two-week period), that he was able to engage in such perilous *pūjās*.

14. Doniger and Smith (1991: 201).

15. O'Flaherty (1980a: 28).

16. Misra (1966: 240); Marglin (1981: 169); Moore (1983: 442); and Lecomte-Tilouine (1996: 31).

17. Daniel (1984: 119–29). The conception of the Thamgharian house was mentioned in chapter 2 and is further discussed in chapter 6.

18. Mallaya (1949: 134–35); Kramrisch (1976: 126–28, 165).

19. Derrett (1959: 110).

20. The first is Lakṣmī, the goddess of fortune and wealth.

21. Derrett (1959); Gonda (1969); Inden (1978: 43); Marglin (1981: 170–71); and Gupta and Gombrich (1986: 137, n2).

22. *Pati* is the Sanskrit word for lord, master, and husband.

23. This however, is far removed from the Thamgharian perception of rice cultivation, which will be analyzed in part III.

24. See also Renou (1953: 25) and O'Flaherty (1980a: 31).

25. *Maitrāyani-Samhitā* 1.9.3.

26. *Śatapatha-Brāhmaṇa* 2.1.8.2.

27. The other, often quoted version of this episode is of the lone Prajāpati who, moved by desire, generates a female entity, Vac (the word or speech), within himself

through the exercise of intense "heat" (*tapas*), brings her out of his body and copulates with her (Renou 1957: 84).

28. *Kāthaka-Samhitā* 12.5.

29. *Śatapatha-Brāhmaṇa* 4.1.2.6.ff. For other examples of pregnant men in Hindu mythology see Meyer (1995: 372ff.).

30. All notes apart from note 27 and the latter part of note 29 are original.

31. *Śatapatha Brāhmaṇa* 10.6.1–5.

32. Gell (ibid.) analyses the notion of captivation as a form of artistic agency, embodied within an object (index), which enchants or fascinates a spectator by virtue of the latter's inability to decipher the origin of the object.

33. e.g., O'Flaherty (1980a: 26–28).

34. Eliade (1978: 227–29, 233); Malamoud (1989: 92); Smith (1989: 58–60).

35. Emphasis removed.

36. Renou (1957: 85); Malamoud (1989); Smith (1989: 60–68).

37. This also forms part of the scene in other Vedic sacrifices, for example in the *aśvamedha* (royal horse sacrifice), the principal queen lies beside the dead stallion and performs simulated sexual intercourse with it (Eliade 1978: 218, 237). The stallion is identified with both the sacrificing king as well as with Prajāpati (Smith and Doniger 1989: 212).

38. *Vedi*, which is feminine in Sanskrit (Renou 1957: 84), is the grass bed where the offerings are put and the seat of the divine guests (Heesterman 1993: 104).

39. Knipe (ibid.) further notes that by the time of the early *Upaniṣads*, what once may have been an elaborate procreative ritual (the Vedic sacrifice), became little more than an esoteric speculation. I shall revert to this shift or decline in the sexuality of ancient Hindu ritual later.

40. What is further suggested here is the assimilation of the theme of sacrificial procreation (discussed above) with another major narrative of the Vedic sacrifice describing the temporary ascent of the *jajamāna* to heaven in order to be reborn there. About the *jajamāna*'s journey to heaven see, for example, Hubert and Mauss (1964: 19–49) and Smith (1989: 104–19).

41. Identified with and representing the earth's evil aspect (Heesterman ibid.).

42. *Bṛhadāraṇyaka Upaniṣad* 6.4.2–3.

43. My emphasis.

44. *Śatapatha Brāhmaṇa* 6.2.2.7.

45. *Maitrāyaṇī-Samhitā* 3.6.1.

46. For additional commentary on the Vedic sacrifice in terms of marriage, copulation, gestation, and rebirth see Heesterman (1957), Kaelber (1976: 348–49), and Smith (1989: 116).

47. Heesterman (1957); Eliade (1978: 229–31); and Malamoud (1989: 92–93, 1996: 41–42, 281 n36).

48. Heesterman (1957: 6–7); Smith (1989: 69–81); and Shulman (1993: 99–100).

49. Here I follow the general scholarly view (Heesterman 1957: 15, 218–19 and passim; Renou 1957: 114–16; Hubert and Mauss 1964: 20–21; Kaelber 1976: 357–67; Kramrisch 1976: 156–57; Gonda 1980: 377–85); and particularly Eliade (1958: 52–57). However, compare Smith (1989: 82–119, 207), who views the Upanayana as a sacrifice.

50. Performed mainly prior to the *soma* and *agnicāyana* sacrifices.

51. Eliade (1978: 229–30); Miller (1985: 220–26); Doniger and Smith (1991: xxxvii).

52. For example, Heesterman (1993: 86–110 and passim).

53. For example, Knipe (1975: 107–21 and passim).

54. Renou (1957: 68–69); Gonda (1980); and Malamoud (1989, 1996: 23–53).

55. Borrowed from Gell (1975: 252–53) who, albeit in a very distinct ethnographic context, discusses the dynamics of a strikingly similar association between active sexuality, reproduction, and death among the Umeda of Papua New Guinea.

56. Renou (1953); Eliade (1973: 111–14); Biardeau (1976); Inden (1978, 1992: 85–117); Dumont (1980: 146–51); Smith (1989: 193–225, 1990); Doniger and Smith (1991: xxx–xliv); and Fuller (1992: 57–105).

57. *Devībhāgvata* 4.13.32 and 35, in O'Flaherty (1981: 78).

58. Kramrisch (1976: 27, 145–49); Snodgrass (1994: 139).

59. See, for example, Inden and Nicholas (1977: 35–66), Knipe (1977), Rutter (1993: 264), and Parry (1994: 191–222).

60. See Gray (1979) and Bennett (1983: 59–70).

61. See for example, Campbell (1976: 74–77), Inden and Nicholas (1977: 69–61), Bennett (1983: 67–70), Pandey (1993: 116, 139) and Gray (1995: 34–39).

62. Gray (1979) argues that in the *bartaman* (Upanayana) a boy is reborn in the sacrificial fire, yet more recently he mentions that it is only *after* recieving his (*gā yatrī)* mantra that a boy is considered to be "twice-born" (Gray 1995: 37).

63. In Doniger and Smith (1991: 32).

64. My emphasis.

65. Enacted mainly for marriage and *bartoon* (male initiation).

# 5

# The *Jagya*—Ethnography

## INTRODUCTION

What is perhaps appropriate elsewhere, namely a detailed, comprehensive description and analysis of a single, complete ritual performance, has no place here for a number of reasons. First, such an endeavor, even if applied to a short *jagya* (one that "only" lasts for one or two days), may well merit an entire book since it involves numerous actions, ritual gestures, mantras, and complex performances, many of which bear little direct relevance to the particular theme advanced in the present work. Second, as will be discussed below, the *jagya* is a general title that may accommodate various different village *pūjās*, each with its own explicit narrative and distinct objectives.

My aim is to look beyond the particular themes and prolonged, idiosyncratic liturgy dominating each of these performances, which to a large extent obfuscate the essential *common* basis that seems to be shared by all Thamgharian *jagya*s. More specifically, I wish to expose the *jagya's* logic of design and operation and highlight the framework within which the different narratives and intentions of each specific event are dramatized, experienced, interpreted, and reflected upon.

Thus, the description herein will focus on the *jagya*'s "core" and will attempt to present, in as bold a manner as possible, the salient and significant features of its structure, the cardinal parts of its performance (in particular the climactic rite of *hom*), and the major local commentary, to the extent to which these may be employed to shed more light on the Thamgharian notion of matrimonial encompassment. I believe that this approach will facilitate a better understanding of the *jagya*'s meaning for Thamgharian Brāhmaṇs, even though the ethnographic picture may not be a comprehensive one.

Following a number of introductory remarks, I will examine the twofold meaning of the term *jagya* (as sacred microcosmic arena and ritual performance) and the main kinds of ritual functions it lends itself to. A short discussion about the predicament of "symbols" in the Hindu world in general and in Thamghar in particular, and a presentation of the alternative term suggested and employed here—"embodied representation"—will ensue.

Next, I will provide a detailed description of the *jagya*'s main stages, beginning with the preliminary rite of *purbāṅga* and moving to the ritual construction process of the sacred arena itself. This will include a short note on the identity of the Thamgharian Bāstu, the house god and the indubitable consort of the earth goddess, Bhūme (Bhūmī), who are the prominent divine cosmic players within the *jagya* and *hom*.

Finally, the climactic performance of *hom* (fire ceremony) and the *jagya*'s concluding rites of *snān* (the shower in its energetic, invigorating water) and the distribution of auspicious *prasād*, will be depicted. A detailed analysis of the *jagya* and *hom* will be provided in chapter 6.

## A Royal, Heavenly Microcosm

In Thamghar, the term *jagya* (also *yagya* or *yegya*)[1] denotes both the temporary, vegetal ritual enclosure normally constructed in the *āghan* (*āgan*)[2] for *ṭhulo kām* (lit. big, laborious ritual work), and the arduous and prolonged ritual performance enacted within or around it. The former, with its two basic forms (discussed below), serves as the ritual arena for a plethora of different *pūjās* and although village priests often tend to claim otherwise, no two village *jagyas* I attended were identical.[3]

Erudition of the Hindu "orthopraxis"—the allegedly unaltered tradition based on the Vedas—is one major source of priestly legitimacy, yet conformity is obviously hard to maintain, particularly since in Thamghar, and as Bennett finds regarding high-caste communal worship of the clan deity (Bennett 1983: 132), neither the elaborate and complex structure of a *jagya* nor most of the liturgical procedure involved is encoded in any textual form.[4]

Villagers explain that the *jagya* is a (temporary) *mandir* (temple), and that "one has to imagine the *jagya* as *sansār* (*sangsār*, the world)." Mahesh Uperti (50), one of the village's learned priests, told me that the *jagya* is a *pratimā* of the world, describing *pratimā* as a special sort of *murti* (a manifestation of the gods). According to Smith (1989: 72–80) *pratimā* is a Vedic concept denoting a sort of connection, a resemblance but not complete identity, as well as a relationship of representation and mutual participation between a whole, a prototype and one (or more) of its parts or counterparts. Smith's interpretation of *pratimā* seems to come very close to local Thamgharian perceptions, which I shall attempt to capture via the concept of "embodied representation" as will be discussed below. In addition, it also evokes the notion of a fractal reality, which, as will be discussed later and further argued in chapter 6, seems to correspond well to Thamgharians' perceptions of the microcosmic *jagya* and of the world as a whole.

Holding a *jagya* is a costly affair[5] since it often requires the collaborative effort of a number of village priests and entails the entertainment of many guests whom one is obliged to invite, as well as the neighbors and other *kul* members whose help is crucial in the construction of the ritual enclosure. Thus it is little wonder that a *jagya* is only performed for a limited number of the most exclusive, dramatic and important *pūjās*. Primarily, a *jagya* is held for enacting the principal Hindu *saṁskāras*, that is, *bartoon* and marriage, as well as a number of other special *pūjās* such as Bāstu *pūjā*, *caurāsī* (*chaurasi*, eighty-four) *godān*, and *purān*.

Bāstu *pūjā*, celebrating the major *saṁskāra* of the house, its marriage, is probably the most complex and picturesque of all village *jagyas*, and will be discussed in some detail in the next chapter. *Caurāsī godān* is an elaborate *pūjā* in which an elderly person (a man or widowed woman celebrating or approaching the age of eighty-four) donates eighty-four *godāns* (*dān* embodied in a cow) to priests and lay Upādhyāya Brāhmaṇs. Each *godān* is given in order to ensure the safe passage through one of the eighty-four *narak kuṇḍas*—ponds of hell—believed to be part of Yama-rāj's (the god of death) city, which any mortal may have to cross after death. Through the successful completion of *caurāsī godān* villagers hope to avoid Yama-rāj's ponds altogether and pave a smooth path into *swarga*.

*Caurāsī godān* is often held following a *purān*, a prolonged *pūjā* that, when performed in its own right, may last for up to eleven days and as implied by its name, is centered around the public reading of Hindu *Purāṇas* (mythology).

Village priests count eighteen different types of *purān* and each is performed for achieving a specific objective such as the household's protection and the longevity of its members, assisting an unredeemed deceased member of the family (or *kul*) to make the transition from *bhūt* (ghost) to *pitṛ* (*pitṛi*, ancestor) or to reach *swarga*, for begetting offspring (particularly sons), procuring wealth, and almost any other desire of the heart.

The majority of villagers I addressed about the meaning of *purān* quoted the following story about King Parikchit's (son of Arjun, the primary hero of the *Mahābhārta*)[6] hunting excursion to the jungle:

Prior to returning to his palace having only managed to kill a snake, King Parikchit placed the latter on the head of Angira *ṛṣi* (*rishi*, sage), who was sitting on the ground in the midst of the jungle absorbed in *tapasyā* (meditation),

A purān.

mistaking the pious *ṛṣi* for a tree. When Angira's sons came to visit their father and saw the dead snake on his motionless head, they cursed the offender, condemning him to die from snakebite within seven days. Having learned of this curse, the king held a lavish seven-day *purān* in the hope of averting it. His efforts were in vain, yet by virtue of the meritorious powerful words of the *purān* stories (the king was the first person to listen to a *purān* during the present degenerate Kāli *yug* [era]) he nonetheless managed to reach *baikuṇṭha*, Viṣṇu's heaven, following his death from the snake's venom.[7]

Thamgharians further explain that during a *purān* the *jagya* embodies both Parikchit's palace and Viṣṇu's *baikuṇṭha*, and the *kartā* (literally the actor)—the patron and the main performer of the ritual work (usually the householder who orders the *jagya* and bears its expenses)—personifies the king. This is realized in a rather dramaturgical manner: since the *kartā* is unable to focus on each sacrosanct word uttered throughout the ritual affair (as he must in order to reap its fullest *phal* (*phol*—lit. fruits, effects)), his role is assumed by one of the village Upādhyāya Brāhmaṇs who sits attentively inside the *jagya* until the *purān*'s conclusion. Other major players in the *purān*'s ritual drama embody the elephant god Gaṇeś, whose duty is to remove all potential obstacles from the *kartā*'s path toward his goals, two palace assistants or advisers, and four *ḍhoke* (gate keepers) who sit at the palace's four main entrances guarding it against *rāches* (demons) that may try to thwart the event.

*An eighty-four-year-old* kartā, *his wife, and other family members offer grains to the* kalas *(representing Viṣṇu), which is situated on the* bedi *at the center of a (simple)* jagya, *during a* purān *performed before the* chaurāsī godan. *The white mud pedestal to the right, where the* tulasi *(here missing) is normally planted, forms one of the* jagya*'s corners.*

The *śakti* (power) embedded in the *purān*'s words is great and merely listening to its stories is believed to enhance one's merit considerably and improve the odds of attaining *swarga*. This is why *purāns* often attract a wide audience, predominantly women, who arrive dressed in their best, for whom it is a rare occasion to gather and enjoy a short respite from the mundane household chores. Often, the *jagya* is large enough to accommodate the *kartā*'s entire nuclear family within its outer frame, thus allowing them to reap the greatest merit, while all other attendees sit around it.[8]

The present section, which aims to analyze the Thamgharian *jagya* and *hom*, may well be regarded as falling under the category of "symbolic anthropology." Thus, prior to embarking upon the ethnographic description, I ought first to clarify the way Thamgharians seem to view what anthropologists usually call *symbols* (or related terms such as representations, manifestations, and the like), as well as the manner in which these are employed here.

## Embodied Representations

Anthropologists studying Hindu culture have long been noting the failure of the Western concept of symbol and other idiomatic forms, such as metaphor or analogy,[9] to capture the flexible relationship between Hindu "symbols" and their referents, as well as the peculiarity of Hindu idioms in general.[10] A number of prominent examples include Inden and Nicholas (1977: xiv–xvi) who note that in Bengal, unlike Euro-American societies, symbols and metaphors are homologous to their referents and are often perceived as "things," be they words, laws, persons, etc. Moreover, in addition to "symbolizing," symbolic actions are also frequently thought to be capable of *affecting* their referents, and, if properly performed, they indeed seem to achieve their objectives.[11]

Often, the idiosyncrasy of Hindu symbols and idioms is expressed in terms of their "literalization," thus Daniel (1984: 106–7) argues that in Tamil and other south Asian languages "the line that divides the figurative from the literal is a thin and fragile one," and proceeds to describe a Tamil "figurative reality," stressing that it does not imply the denial of people's capacity for figurative speech and thought.

More abstractly, Trawick-Egnor (1978: 178–80, particularly note 2) observes the different status of metaphorical or analogical reasoning in Western dualistic thought vis-à-vis Hindu monism; where in Western logic analogy is never part of the syllogism,[12] it is an essential component of the formal logic of India as formulated in Nyāya[13] philosophy.

As will be recalled, very similar notions have already been encountered in Thamghar, notably that a coin given in *godān* is treated as a real cow and is also believed to be a manifestation of Lakṣmī, the goddess of wealth. Similarly, the bride is believed to become part of and be encompassed by her husband in marriage, and a householder maintains a particular analogous relationship or partial identity with his house, whose first floor is not merely taken to be *like* but *is* his inner self. In the terms employed here, all the aforementioned examples are, to various degrees, but different expressions of a fractal reality.

If so, perhaps we would fare better by renouncing the notion of symbolism or symbolic meaning altogether when dealing with a Hindu context, as more generally suggested by Gell (1998), and instead adopt his action-centered approach where emphasis is placed on agency, causation, result and transformation. The following chapter draws upon a number of Gell's valuable insights, yet to adopt his absolute objection to "symbolism," which serves well for advancing his theoretical stance, would be to deny Thamgharian Brāhmaṇs their capacity for figurative expression and thus cannot be fully embraced here. Hence, in order to capture the true spirit of the sometimes fluid boundary between a symbol and its referent or the tendency of local representations to shade off into embodiments and vice versa as well as the generally felt "literality" in Thamghar of what Westerners call idioms or figurative expressions, I propose to view these as *embodied representations*.

An embodied representation should be imagined as occupying one position (or more)[14] on a conceptual thread stretching between the ideal poles of total embodiment and "literalization" at one end, and pure representation (a symbol in the Western sense or an empty "skin") at the other, while being, potentially at least, in constant motion between these poles (never fully occupying either of them). However, the exact location of an embodied representation at any given moment is uncertain and may only be revealed in retrospect, through the effect its agency is able (or unable) to engender or through the role it assumes within a particular context.

I will now turn to describe the proceedings of the *jagya*'s main stages of construction and performance.

## *JAGYA LĀUNU*—THE CONSTRUCTION OF THE *JAGYA*

As villagers adamantly emphasize and as emerges from the various aspects of the edifice, its construction process, and from the ritual performance itself, the *jagya*'s leitmotif is that rather than being a mere ceremonial pavilion or ritual arena, it is an entity, a living body (or fractal of bodies as will be suggested below), a vibrating emblem of life, prosperity, and eternal growth. This notion is highlighted throughout the event by the formation of the creative unions of male and female entities that characterize much of the *jagya*'s structure, ritual procedure and rich accompanying symbolic behavior.

These are explicitly accounted for in terms of the utmost need to evince the vitality of this temporary temple and fashion it as an auspicious and *purā* (complete) locus of *sṛṣti* (*shristi*, creation). Concomitantly, even the mere insinuation of an association with death, no matter how indirect, must be banished from the *jagya*'s premises.

### The Preliminary Rite of *Purbāṅga*[15]

The performance of *purbāṅga*, up to seven days prior to the construction of the *jagya*'s sacred enclosure, is the modest opening of the entire ritual event. This *pūjā* aims to protect the house, its inhabitants, and the future *jagya* from the extreme pollution that a death or birth in one of the other *kul* families may cast, which would otherwise lead to the postponement or cancellation of the *jagya*. As I wish to recall, all *kul* members are viewed as *hād-nātā* ("bone relatives"), and are thus mutually affected by each others' moral substances.

At the heart of the *purbāṅga*, the householder's wife stands on the house's threshold in front of the door, facing in, with a large leaf plate full of *naibedde* (various offerings, mainly food stuffs made of rice flour and fried in *ghyu*, clarified butter) on her head, while her husband worships the *sapta dwār Mātṛkās* (seven unmarried mother goddesses, the house's fierce and powerful guardians)[16] and Gaṇeś, the Hindu remover of obstacles, located above the door's lintel. The householder family's priest then performs *bāsudhārā* by pouring a thin stream of *ghyu*, the highly pure and auspicious substance, on the *Mātṛkās* so that it flows down onto the plate held on the wife's head. Following the *purbāṅga*, the family members must comply with various restrictions, such as refraining from sexual congress and eating impure items of food, in order to guard their enhanced state of purity.

## Alignment in Time and Space

The preparations for the *jagya* often last for most of its first day and include purification of body[17] and house,[18]cooking and arranging *naibedde* (done by the *kartā*'s wife), putting the ritual paraphernalia in order and finally the construction of the *jagya* itself. Unlike most of the complex worship performed within or around its precincts, the *jagya*'s construction, particularly the ordinary one (figure 5.1) regularly employed for marriage or *bartoon* for example, is a rather festive and informal event, where the *kartā* and his close family members work in tandem with various other kinsmen and neighbors, discussing, arguing, and giving advice, often without any priestly supervision. The priest need only be consulted regarding the *jagya*'s auspicious location and the optimal moment (from a planetary point of view) at which the work should commence.

The construction is carried out in a clockwise movement likened to an auspicious *pradakṣiṇā* (*pardacchinā*, circumambulation), starting from the *īśān* (northeast) or the eastern side of the *jagya*;[19] thus it seems that the *jagya* evolves into existence while being simultaneously worshipped. Rather like in the Vedic sacrifice of the *agnicāyana*[20] discussed in chapter 4, so also in the *jagya* it is difficult if not impossible to draw a sharp line separating the establishment process from the ritual performance; the overall construction work, which often includes various *pūjās* is thought of in terms of a ritual process and does not seem lower in significance than the more formal liturgy that follows, on the contrary, it appears to be one of the momentous components of the entire event.

In the establishment of a *jagya*, both the celestial configurations (the "cosmic design") and the *kartā* serve as the reference points with which it must be made compatible in terms of time and space. Accordingly, the time at which the construction work commences must correspond to the position of the celestial configurations[21] but also to the *kartā*'s horoscope chart.[22] Similarly, the *jagya*'s square structure must be aligned with the cardinal compass

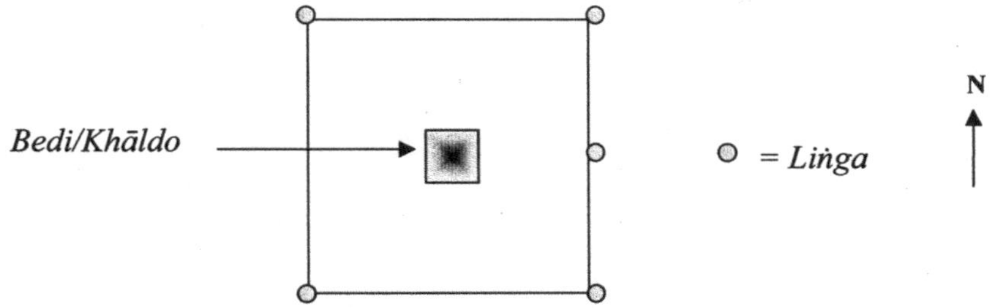


**Figure 5.1. A simple *jagya***

directions (and also often with the position of the house), and its dimensions should be consistent with the *kartā*'s bodily proportions.

The basic structure of an ordinary *jagya* (figure 5.1) is relatively simple: four *liṅgas* mark the corners of a square (whose sides normally measure approximately 1.5–2.0 meters, in accordance with the *kartā*'s height),[23] with a fifth *liṅga* located in the middle of its eastern side. Each *liṅga* is made of a number of tree trunks, long branches or poles of bamboo, sugar cane, banana, *sami*,[24] and *sāl*[25] trees,[26] and all are tied together and thrust into a hole dug in the ground. The prototypical *liṅga* is the one made of the *sāl* tree, which must

*A simple* jagya *viewed from the east. Early in the morning, the officiating priest (left) explains the procedures of the* jagya *to the* kartā *(right). At the center of the* jagya, *behind one of the* liṅgas, *is the* bedi. *In front of each* liṅga *a special* rekhi *for its* kalas *is already prepared.*

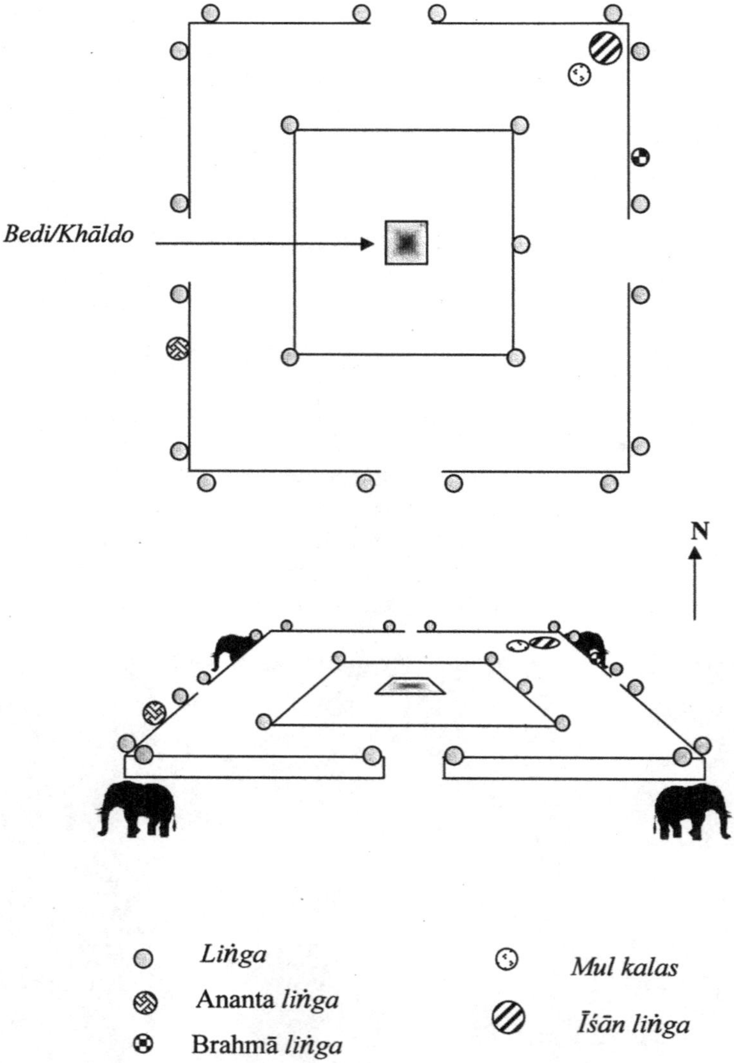


**Figure 5.2. An elaborated *jagya***

be a *syāulodar* (*seoldar*), that is, have a *syāulo* (*seola*)—a number of small leafy green branches—crowning its apex. As glossed by Baburam (42), the *syāulo* is the *liṅga*'s head and a sign that it is alive; "tell me" he said, "how can someone be alive without a head?" adding that there is no place for *dead* "things" in the *jagya*, a place of growth, vitality, and *sṛṣti*.

A number of *pūjās*, such as *purān*, *caurāsī godān* and Bāstu *pūjā*, entail the construction of an elaborated *jagya* (figure 5.2), which incorporates an ordinary one (figure 5.1) and is established around it. This larger *jagya*, likened to Parikchit's palace or Viṣṇu's *baikuṇṭha* (heaven), has four gates, one on each of its sides, corresponding to the cardinal compass directions and is bound by eighteen large *liṅgas*. Sixteen of these are arranged in eight pairs, four of which stand at the *jagya*'s four corners with the remaining four "guarding" each of its gates. Two additional elongated bamboo *liṅgas*, representing the god Brahmā and Ananta (the boundless cosmic king serpent), stand at its perimeter on the eastern and western sides respectively.

The most important of all *liṅgas* in an elaborated *jagya* is an additional six-meter-long *bamboo* pole, which stands within the *jagya* very close to its most auspicious *īśān* corner, with the *jagya*'s *mul* (chief) *kalas*[27] placed at its foot. Villagers often tie a banana trunk containing an almost ripened bunch to this *liṅga* in the hope that the *jagya* will similarly come to fruition or, as they put it, its *phal* (fruits, effects) *pākcha* (will ripen/become cooked).

To the top of this major *liṅga*, villagers attach a bell whose ringing is said to please the deities summoned to the *jagya*. Just below this, two ribbons—one red, the other white—denoting female and male entities respectively, are tied. As glossed by one householder, this is a sign that the *jagya* is alive and underway, and the white and red ribbons are tied together since they are married, thus making the *jagya* a locus of *sṛṣti* (creation and divine emanation), since "for *sṛṣti* two [male and female] are needed."

As the name may suggest, each *liṅga* is regarded as *puruṣ* (male) and villagers explain that the *liṅga* is the *mālik* (master, ruler, and guardian) of the direction in which it is located.[28] In addition, each *liṅga* also represents the entire cosmos and, as shown to me by Ganesh Neopane (45), villagers refer to the top of a *liṅga* as *akās* (heaven, the seven upper worlds), while *pāttāl* (the seven under worlds) are said to be buried in the ground, and the earth is believed to be located between these two extremes.

Notwithstanding their title, the *liṅgas* are not necessarily associated with Śiva (although a number of villagers claim that they do stand for this virile god) and most priests and lay villagers with whom I discussed this thought that each *liṅga* is a manifestation of Bāstu Puruṣa, the house god who is the usual partner of the earth goddess Bhūme.[29]

The sixteen *liṅgas* situated at the perimeter of an elaborated *jagya* are also called *dikagadges* (elephants)[30] that, according to a number of villagers, carry the universe on their backs.[31] One householder thought that the five *liṅgas* of a simple *jagya* stand for the *pañchāyan deutā* (*deotā*, the five house gods), while another, who often serves as a priest, told me he thinks that an elaborated *jagya*'s *liṅgas* are the sons of Brahmā; by and large however, most other Thamgharians were unfamiliar with these notions.

The *liṅgas* are normally cut from trees growing in the family's *bāri* (non-irrigated field) near the house, or in the forest just outside the village. When brought to the house and before they can be incorporated into the *jagya* or even simply placed on the ground, the *kartā*'s wife must perform *parsanu* (*parsine*). This is a modest rite where *achetā* (a mixture of *cāmal*, unbroken husked rice grains, and *abir*, auspicious vermilion powder)[32] is scattered over the *liṅgas*. This is regarded as a *liṅga pūjā* and is also performed in order to ward off the evil spirits that may have come along from the forest. In addition, *achetā* is also used for the *ṭikā* applied on the brow of each of the men involved in the carrying.[33] This *ṭikā*, as explained by Radika (39), is for *sait parnu* (*parne*)[34] that is, to confer auspiciousness and blessing over the men and thank them for their efforts.

Notwithstanding the difference in structure, villagers adamantly state that both an inner, ordinary *jagya* and an elaborated one are in fact *barābar*, that is, "equal" or homologous projections or manifestations of the universe.

## Bāstu is not Vastu

Prior to proceeding to the description of the next stage, in which the *jagya*'s modest "fire-altar" is established, I will briefly discuss the personality and identity of Bāstu Puruṣa, the Thamgharian *ghar deutā* (house god) who, as mentioned above, is also one of the *jagya*'s prominent characters.

On the face of it, the Thamgharian Bāstu may simply be the local manifestation of the Indian Vastu,[35] the demon god who occupies (in fact literally "fills") the *Vāstupuruṣamaṇḍala*, "the plan of all architectural form of the Hindus" (Kramrisch (1976: 22 and passim).[36] As the vast research conducted on Hindu mythology, architectural manuals and ethnographic accounts clearly demonstrates,[37] Vastu, as he usually appears in the *Vāstupuruṣamaṇḍala*, is a demonic figure; his head is turned downward, pointing toward the east or northeast (*īśān*) direction and the entire Hindu pantheon resides on his back, holding him in place. As such he is mainly described as a nameless "thing," a corpse, an impotent and inactive goat-headed *asura* ("anti-god"), a timid-looking old man, an ugly, hunchbacked and crippled creature.

However, Vastu (often also called *Vastavya*, *Vāstupa*, or *Vastopati*) is associated with a plethora of quite distinct figures and mythological traditions, which not only narrate his descent to earth as a demon, but also often portray him as the personification of "the sacrificial victim" or its impure remainder. At times he is closely associated with Paśupati-Rudra, and thus with Śiva. Moreover, Vastu also appears as *the* guardian of the cosmic order or one of eight such guardians standing around the *Vāstupuruṣamaṇḍala*; he is also frequently depicted as the (Vedic) cosmic Puruṣa (primordial man)—the origin and source of existence—or is simply viewed as the lord of the dwelling.[38]

Out of this confusing profusion, the Thamgharian Bāstu with his dual main manifestations (residing in and embodying the house, as well as surrounding the universe of the *jagya*), seems to be in accord with only the last four representations of Vastu, while all the others are totally unknown to villagers. The striking discrepancy between the Thamgharian Bāstu and his demonic counterpart is couched in visual terms: Bāstu, quite unlike the demon Vastu, does not exist *under* the house (held down by the mighty load of the Hindu pantheon), but enters it triumphantly in his *pūjā*, manifested as a large *liṅga* in an upright position. Alternatively, he is found *standing* guard at the perimeter of both types of the village *jagya*. Unlike the impotent Vastu, the Thamgharian Bāstu is a virile god *par excellence*. He is the partner of Bhūme and is explicitly engaged in prolonged coitus with the goddess in Bāstu *pūjā*, as will be discussed in the following chapter.[39]

## A Ritual Cosmos Brought to Life

Returning now to the process of the *jagya*'s construction, after the men (the *kartā* and his assistants) place and consolidate the *liṅgas* in the holes dug for them in the ground, the *kartā*'s wife purifies the square area they envelop by plastering the ground with fresh cow dung paste mixed with mud. Then, at the center of the *jagya* she fashions the *bedi*—where the fire of *hom* will burn in due course—using extremely pure mud dug from the family's *khet* (rice field),[40] located outside the village. This is performed by two of the family's or *kul*'s (auspicious, highly pure, and divine) *kanyās*.

Similarly, in order to transform the plain village water (used to fill the *jagya*'s *kalases*) into a heavenly pure and potent substance, it is imperative to mix it with a few drops of Gaṅgā *jal* ("the water of the Ganges"), brought from the nearby river, which like all streams surrounding the village is viewed as a rivulet of the holy Ganges.

The *bedi* is a small, square mud platform (no higher than ten cm), whose size must be proportional to the householder's arm. Before molding the mud, the *kartā*'s wife covers a *supāri* (Areca nut, often employed in village *pūjās* as a *pratimā* for a living *kanyā*, a virgin goddess) with a *janaï* (the symbol of high-caste status, the sacred thread worn by initiated males), perceived as female and male divinities respectively, and performs *bheṭi garnu* (*garne*) by placing a coin over the couple before covering them with mud.

The union of the *janaï* and *supāri* in the *bedi* is accounted for in terms similar to those used for other such *mithunas* found throughout the *jagya*—the need to turn the *jagya* in general and the *bedi* in particular into a "complete" locus of *sṛṣti*. *Bheṭi garnu* is an ordinary act of honoring a deity, person, temple or a sacred book, as well as a gift to a higher entity, and is *not* viewed as *dān*; as Madhu Lamichane (40) told me in private, the coin is thought to serve as a witness to the *janaï* and *supāri*'s marriage.

Resembling its Vedic predecessor (the *vedi*), the Thamgharian *bedi* is thought to be a feminine locus of *śakti* and a manifestation of the earth goddess Bhūme. One village priest, Surya Prashad Ghimire (65), added that the *bedi* also represents Prajāpati *deutā* (god), but could not provide much in the way of exegesis for this identification.

Often, particularly when a large *hom* is planned, a *khāldo* replaces the *bedi* at the hub of the *jagya*. The *khāldo* (lit. a pit) is a square hole dug in the ground by the *kartā*, which serves as an Agni *kuṇḍa* (Agni's pond of fire). Like the *bedi*, its dimensions must be proportional to those of the *kartā*'s arm or palm. The *bedi* and *khāldo* are interchangeable and, for example, a *jagya* may start with a *purān* where *hom* is performed on a *bedi*, and be followed by a *caurāsī godān* where the *bedi* is removed and replaced by a *khāldo*.

When the digging of the *khāldo* is complete, the priest utters the following mantra: "you are Pṛthivī, you are Bhūme, you are the fourteen *bhuwan* [worlds], you also carry the world. Today I am doing your *pūjā*."[41] In private, householders often noted that the *khāldo* is the *yoni* (vulva) of the goddess (Bhūme).

Unlike an open-sided, five-*liṅga jagya*, an elaborated one is at times furnished with bamboo walls and a conical roof topped with a *gajur*,[42] but like its simple counterpart, this *jagya* is also short-lived and is demolished when the *pūjā* is over. Even when no such roof is built, the tall curved *liṅga* tips, inclined toward the center, often seem to converge to an imaginary point high above the *jagya*, but only rarely connect in reality. Ordinarily, a square piece of red cloth called *canduwā* is tied between the *liṅgas*, approximately 1.8 meters above the *jagya*, and various *naibedde* are placed upon it. These are distributed as auspicious *prasād* (the blessed "leftovers" of the gods imbued with their grace and *śakti*) at the *jagya*'s conclusion.

The circuitous, constructive movement around the *jagya* continues with two additional series of threefold clockwise *pradakṣiṇās*. In the first, villagers tie the *toran* (an elongated rope with green leaves interwoven with fresh flowers) on the *liṅgas* surrounding the *jagya*, 1.5 meters above the ground.[43] A number of the edifice's most powerful guardian deities are situated on some of the *toran*'s leaves, among them the *das* (ten) *toran Mātṛkās*, manifested in the form of red dots (made of *abir*) on a large mango leaf. Next, the *jagya* is *bādhnu* (*bādhne*, tied) with a long white cord (*dhāgo*), carefully fashioned as an extended *janaï*, which surrounds it and is suspended near the *toran*. According to Prakash Khanal (46), a well-off village householder, the *toran* and *dhāgo* are tied for protection like other cords (*rakṣā-bandhan*), ropes, pieces of cloth, or mantras that are believed to be capable of keeping evil spirits and impurity at bay.[44]

Having enveloped the organic *jagya* with potent protective threads and mantras, the *kartā*, accompanied by his family and priest, commences a colorful clockwise worship of the ritual *mesocosm* thus fashioned. In one of the *jagyas* I attended—a sumptuous *purān* held in Ramesh Bastakoti's household—the present stage took the form of a joyful parade: the entire family, carrying various offerings in their hands, walked around the *jagya* dozens of times in a clockwise direction, keeping their auspicious right side toward the latter. Ramesh walked pouring water out of a *kalas*, creating a line of auspiciousness around the *jagya*, while two of his adolescent boys blew a *saṅkha* (a large conch taken as a manifestation of Viṣṇu) and played a ritual bell for the elimination of *pāp* (sin) and bad effects. During another circuit, the family's young *kanyā* placed small *dunus* (*duna*, ritual leaf plate) on top of the modest *rekhis* (ritual diagrams) drawn by the family priest at the foot of each *liṅga*, which were then filled with *dhān* (unhusked rice) by her mother. Ramesh then placed another *dunu* on top, filled it with water and performed a minor *phul-achetā pūjā*[45] to the leaf *kalas* thus created, while the family priest continuously chanted sacred mantras. The prolonged *pradakṣiṇā* included the performance of numerous *pūjās* and various ritual gestures for the abundance of *liṅgas*, deities, and various other beings that make up the *jagya*, as well as the different aspects of the cosmos and humanity (including the four Vedas, castes [*jāts*], elements, and colors), believed to be found there. At the conclusion of this stage the *jagya* was replete with copious sacred leaves, multicolored threads, ribbons, fruits, and decorations—evidence of the numerous *pūjās* that had been performed around it.

Following this, the officiating priest purifies the *jagya* by sprinkling *tilāni pāni*[46] over it and toward each of its sides, and turns to worship the *bedi* and the surrounding *liṅgas* by the performance of *phul-achetā pūjā*. As was the case during *parsanu*, the *achetā* is believed to have a double role: it is food for the deities as well as a substance capable of engendering an apotropaic effect against evil. The priest now begins to draw the main *rekhi*—a complex geometrical contrivance aligned with the major compass directions (see figure 5.3)—using a dual line: red *abir* powder (female) alongside a white line made of rice flower (male).

A number of villagers explained that the *rekhi* is *deutā basne thāũ*, the place where the gods sit during the *pūjā*, others thought it is a *mandir* (temple) in itself, and most explained that, like the *jagya* as a whole, it is *sansārko*

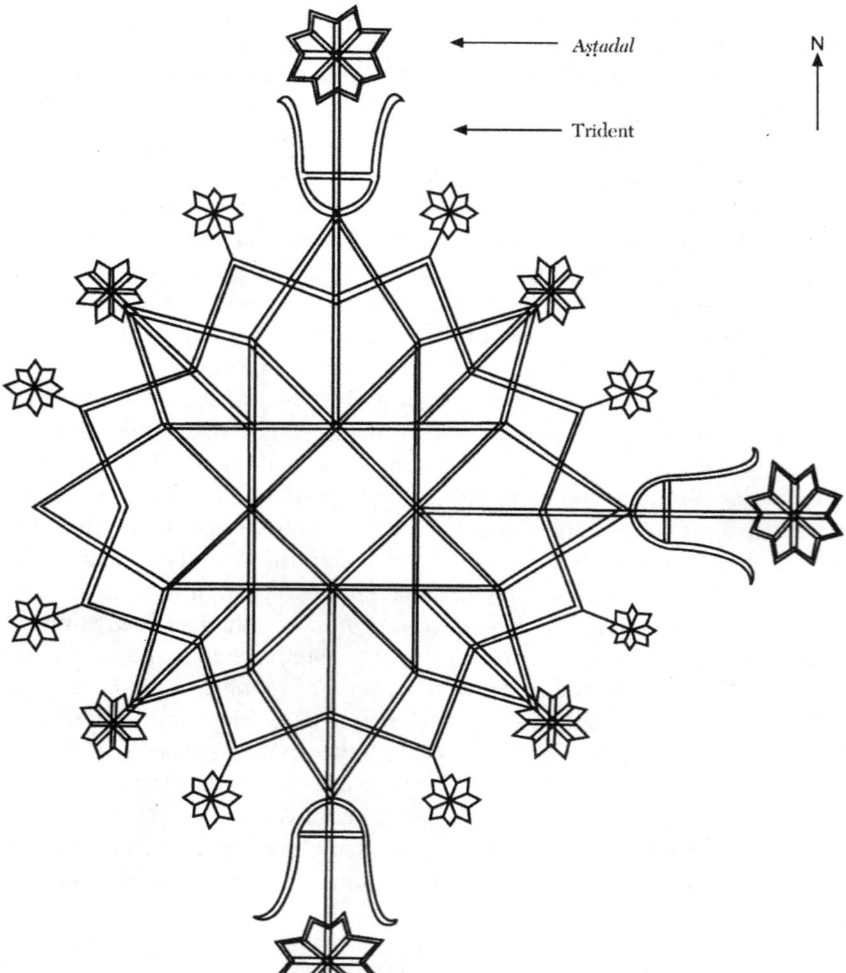


**Figure 5.3. A *rekhi***

*rup*, built in the form of or embodying the universe.[47] Furthermore, villagers often relate to the *rekhi* in "personlike" terms noting that it has the ability to *bādhnu* (tie) the honorable deities to their place, until the completion of the *pūjā*.

The priest draws the *rekhi* from the top of the *bedi*, situated at the navel of the inner *jagya*, toward the perimeter by trickling the powder through the fingers of his auspicious right hand. Notwithstanding its complexity, every village priest can skillfully draw a *rekhi* from memory within a number of minutes.[48] The use of red and white lines in the *rekhi* is often explained in terms of the substances involved: the extremely pure rice flour and the red *abir* (a marker of auspiciousness and good fortune) but, being gendered entities (male and female respectively), Thamgharians add that their combination is a means of fashioning the *rekhi* as a "complete" entity.

Lay villagers and priests alike provide very little by way of exegesis regarding the various elements within the *rekhi*, with the exception of a number of common graphic symbols such as the trident, *aṣṭadal* (lotus) and a triangle, which are ordinarily associated with or represent Śiva, Viṣṇu, and Devī (the goddess, usually Durgā, Kāli, or Śakti) respectively, yet are not necessarily employed in accordance with this "code."

Next, the priest covers the area immediately surrounding the *bedi* (or *khāldo*) with green *kuś* grass[49]—a sign of life and growth—and via a prolonged liturgical process, places the diverse paraphernalia in designated places on the *rekhi*. This includes numerous leaf plates overflowing with various offerings,[50] and a number of ribbons and threads as well as the un-iconic manifestations of the gods.

Gaṇeś is manifested by a lump of fresh cow dung with a sacred *dubo* grass[51] thrust in it, while Śiva is represented by a *diyo*, a small oil lamp made of a burning wick coming out of a copper saucer or a larger elaborated oil container with a saucer for the burning flame. One villager commented that while the saucer is thought to be Śiva, the wick is a manifestation of his (consort) *śakti*. The *kalas* is the embodiment of Viṣṇu, and Devī is manifested by *achetā* placed in a small leaf plate.

As the gods must not sit on the bare ground, all their manifestations are placed on an *āsan* (throne) in the form of a large leaf plate full of *dhān*, *cāmal*, or both, one on top of the other. A small *kalas* is placed at the foot of each of the *jagya*'s five *liṅgas*, and the *mul kalas* is located under the long bamboo *liṅga* at the most auspicious *īśān* corner.

Apart from a single *kalas* representing Viṣṇu, each *kalas* is worshipped as the divinity associated with its *liṅga*. Lay Thamgharians and priests alike explain that the *jagya*'s *mul kalas* is (an embodied manifestation of) the entire cosmos: it contains all the seas, the terrestrial and cosmic *tīrthas*,[52] the holy Ganges, the entire Hindu pantheon,[53] numerous entities, and various cosmic elements. Among many others, the *mul kalas* is said to contain *sae auṣadh (auṣadhi)*, literally one hundred medicines. This prototypical remedy for every possible illness is represented by a plethora of herbs, grains, fruit, and substances that, being interwoven with potent mantras and imbued with all the powerful forces that are at play during the ritual performance, are thought to embody a particularly potent and charged (cosmic) liquid that is put into effect in the final shower (*snān*) during the *jagya*'s closing stages. The *mul kalas*'s "neck" is garlanded with red and white ribbons, which once again are referred to as a sign of it being "complete."

The structure of a Thamgharian *jagya*, particularly the *kalases* surrounding the *bedi*, may be illuminated by comparison with another type of *jagya*[54] held at the royal palace in Katmandu.[55] The latter commemorates one of the classical Hindu cosmogonies, where Viṣṇu in his Varāha (Barāha, wild boar) avatār[56]or Śeṣa (*Shesh*) *nāg* (the mythical serpent god) form, is worshipped at the *jagya*'s navel after saving Pṛthivī (the earth goddess) by bringing her out of the primordial cosmic waters. The five *jala ghaṛā* (*ghadas*, water vessels, *kalases* in Thamghar) surrounding the central *bedi* (in the royal *jagya* there are two additional *bedis*, one for Bāstu and the other for the nine planets), are imagined as the ancient *samudra* (*samundra*, seas) surrounding the earth. The *jagya* as a whole is taken as Bhūme, but it is also regarded as the *āsan* (throne) of Viṣṇu, and is based on the back of eight *dikagadjas* (elephants) manifested in the form of *liṅgas*.[57]

The remainder of the present chapter is dedicated to the presentation of the backbone of the *hom*, the elaborate dramatic fire worship, which is often viewed in Thamghar as the *jagya*'s climax and seems to epitomize its significance.

## HOM

*Hom* is the indispensable part of all village *jagyas* and is also found at the core of all important village *saṁskāras*[58] or other *pūjās* that do not necessarily involve the construction of a full-scale *jagya*.[59] Occasionally, a *jagya* may require the recursive performance of *hom*, and a *purān* often ends with an additional *lākh* (100,000) *batti* (oil lamps)—the nocturnal worship of a large, square, burning ritual lamp (1.5 $m^2$) containing 100,000 wicks.

The three main preliminary rites of *hom* include Brāhmaṇ *baran*, *saṅkalpa*, and *prāyaścitta* (*payachista*) *godān*.[60] In Brāhmaṇ *baran* (or as villagers often simply call it, *baran garnu*—to do *baran*, that is, to choose or nominate), the *kartā* officially appoints his priest to officiate over the *hom*, or as Rutter (1993: 410) eloquently puts it, he "inaugurates in him [in the priest] the power of mediation between men and gods." This is done by simply placing a (*achetā*) *ṭikā* on the priest's forehead, followed by the performance of *ḍhognu*.

The main purpose of *baran* as explained by Mukunda Raj Sharma (35), who was the *kartā* in one of the village *purāns* I saw, is twofold: to honor and worship the priest who is taken here as a *deutā* (god), but also to "bind" him (in conceptual chains) to the *jagya* until the successful completion of the *hom*.

Unlike most other village *ṭikās*, which normally convey a form of blessing and can be described as "top-down," that is, where the donor has a higher ritual status than the recipient, the *baran ṭikā* is a "bottom-up" one and therefore, similarly to other "donations" or "gifts" a *jajmān* gives to his priest, it is also believed to contain an ominous load of *dān*.

Next comes the *saṅkalpa* where the priest "informs the god(s) of the purpose of the *pūjā*" and utters the *kartā*'s earnest intentions and expectations from the ritual performance. The following declaration formed the main part of the *saṅkalpa* uttered prior to the performance of *hom* in Mukunda's *purān*:

> I [name] am doing this *hom* together with my wife [name] and children [names] in order to get and preserve Lakṣmī [wealth and good fortune] and bountiful crops, achieve all wishes of the heart, purification of the body, get all the returns mentioned in the *smṛti*[61] and *śruti*,[62] and all the fruits of the *jagya* and *hom* in *martya-lok* [in this life, world], to beget a son and a grandson, for general success, for being prestigious, for *sukha* [happiness] and prosperity, to eliminate *pāp* [sin] I have done in this or in previous life, to avoid the harm of *bhūt-pret*, *picās*, *ched* [evil spirits and specters of the dead], to enjoy the fruits of this *pūjā* along with my whole family, to eliminate illness and pacify the *nāg-nāgini* [male and female serpents deities] residing under the house and in the *khet* [rice field], and for the protection of the house, this is why I do this *pūjā* [*jagya*] and *hom* as one of its *aṅgas* [organs, parts].

Prior to this, the priest should "locate" the *kartā* and the *hom* in their right context, that is, he explicitly specifies the time[63] and place[64] of the performance, the age and astrological situation of the *kartā*, his kinship details,[65] and the names of the fasting members of his family, who perform the *hom* (and the *jagya*) with him.

It is important to realize that the notion of the *kul* members sharing a single body or substance (a bond that appears to be temporarily suspended by the *purbāṅga*) is even more deep-seated regarding the nuclear family. Accordingly, since the father (*kartā*) is perceived as the defining and unifying element of the entire family, all its members are thought to perform and be affected by the ritual process as he is. This is evinced and reinforced at certain moments during a *jagya*, particularly in the *saṅkalpa* for its *hom*, by every family member touching the father with their right hand (supported by the left).

In *prāyaścitta godān*,[66] the *kartā* gives *dān*, embodied in the form of a coin representing a cow (*go*), in order to enhance his ritual state and the effects, success, and general fortunes of the ritual procedure. The priest must then drag the small leaf plate, in which the coin sits atop *cāmal* grains, as if it were a living cow, before purifying the *kartā* by spraying *tilāni pāni* on his body, using a sacred *kuś* grass in lieu of the cow's tail.

*A* kartā *and his wife (touching his leg) during* saṅkalpa.

Following the performance of "*diyo*, *kalas*, Gaṇeś *pūjā*," that is, worship of Śiva, Viṣṇu, and Gaṇeś, the priest arranges the *samidhā* (firewood twigs)[67] for the *hom* on top of the *bedi* or on the floor of the *khāldo*. Different priests prescribe alternative ways of organizing the *samidhā*, which are stacked on top of each other to form a hollow tower consisting of five to nine levels,[68] but most agree that the uppermost part should be fashioned as a triangle using three twigs. The purport of the *bedi* and tower of *samidhā* is evinced by the mantra the priest utters while he arranges the wood: "*He* [hello] Agni, this is your *garbha* [womb] please come and stay in it, I am about to perform your *pūjā*."

In the following stage, Agni, manifested in the form of a burning coal, is placed in a small clay or iron pot near the *bedi*, where the priest performs its ten *saṁskāras* in a highly condensed and brief form, starting with *garbhadhāna* (conception) and culminating in *bie* (*bihā, bibaha*, marriage). Ideally, Agni should be "virginal" and *cokho* (pure), that is, fire that has not been used before, by virtue of being "churned," produced anew in a wooden drill. I have only seen villagers employ such a drill once, for lighting a fire for a *lākh batti* at the end of *purān*.[69]

Karananda Raj Upadhya (86), the highest priest in the area, told me that Agni's pot should have a lid since it is Agni's *garbha* (womb) and the growing baby should be *invisible*, but, he complained, the priests nowadays are too busy

*A* rekhi *with the offering placed at its edges and the* samidhā *piled in the center, forming a triangular shape at the top. As the* hom *is performed within the house for Bastu* pūjā, *no* bedi *is constructed and the* samidhā *are piled upon some soil taken from the* khāldo.

and hasten their work, thus they neglect the cover and perform the *saṁskāras* in the open. Following his marriage, Agni is ready for the next stage of *biu rākhnu* (*rākhne*), literally "to put the seed," in which the priest places a burning ember (Agni) into the hollow construction of the *samidhā* (*yoni*, *garbha*).

As Karananda added, although apparently it is Agni who is born and undergoes its *saṁskāra* (*sanskār*), the general belief is that it is in fact the *kartā* who attains his rebirth in the fire, and the *saṁskāras* performed are *his saṁskāras*.[70] The *kartā* (through his priest) can thus be seen to engender his own ritual rebirth in the *jagya* via the *hom*. This is a most significant detail, as it clearly demonstrates that in Thamghar, most *saṁskāras* and other *pūjās* performed in a *jagya* are ultimately imagined as transformative processes that culminate in fiery gestation and rebirth.

When the *samidhā* are ignited, the priest begins to offer *ghyu*, *cāmal*, and *dhān* into the flames of Agni, "the mouth of the gods," which is regarded as the ultimate manner of feeding them.[71] The grains are thrown into the fire by hand while the *ghyu* is offered using a *suro* (carved wooden stick) that is rapidly dipped in a copper saucer full of *ghyu*. The priest counts the number of times the *suro* touches the saucer, as each deity demands a different and specific number of such offerings. In addition, the *kartā* and his family may have made *bhākals* (vows) to make a certain number of offerings to a particular deity or an ancestor, and the priest should offer *hom* accordingly. The number of these offerings, which should be a multiple of the propitious number 108, may reach several thousand, and when *hom* is performed indoors, the entire ground floor fills with thick smoke, sweat pours off all present, and the house becomes redolent with the auspicious odor of burning *ghyu*. *Hom*, village priests testify, is the hardest of all the ritual procedures they must endure.

In *śeṣ* (*shesh*) *caru* (*charu*) *homnu* (*homne*),[72] performed just before the *hom* draws to a close, the *kartā* leads his family in a *pradakṣiṇā* around the *jagya* and, according to the priest's orders, they offer the remaining *caru* into the fire and clasp their hands together in an honorary gesture of *namaskār* (*namaste*, the Nepalese greeting, salutation). At this moment the priest utters the following mantra "I offer this *caru* and *samidhā* to the master of the *jagya* [Agni], who is full of brightness, and for Pṛthivī [Bhūme]. I am doing this for the purification of the world. You are Agni in the form of flame, please fulfill my wishes." This stage is accompanied by the loud sounds of the *saṅkha* and continuous bell ringing, believed to be highly effective in eliminating *pāp* and deterring evil.

## FINAL RITES

The *jagya* reaches its end in the performance of *snān* and the distribution of *prasād*. *Snān* (bath in Sanskrit)[73] is the title ascribed to the communal shower of the *kartā* and his close family in the potent water of the *mul kalas*, which is poured from the roof of their house through a metal strainer. As Pramod Banjara (28) explained, villagers believe that the potency of the sacred water, imbued with the *śakti* of all the substances mixed in the *kalas*, is considerably intensified by the prolonged ritual performance and is thus highly efficacious in removing *pāp*, *dos*, and various impurities. In order to attain maximum results, it is imperative that one is completely soaked with water.

Prior to the *bhāter* (lavish public banquet) that awaits the honorable priest(s), the family and the large number of other invitees,[74] the family priest distributes *prasād* to all those present. This is conducted hierarchically, beginning with the *kartā* and his close family and moving downward in accordance with the ladder of age and *kul* seniority.

*Prasād*, villagers explain, is the auspicious blessings, the "grace" and particularly the *śakti* of the deities, imbued within the offerings left by them at the end of the *pūjā*. Of the various colorful forms of *prasād* (among them a variety of foodstuffs and other edible substances, ribbons, protective threads (*rakṣā-bandhan*), flowers, ashes, and so on), the *achetā ṭikā* is probably the most prominent one and seems to be of the greatest significance. Like all

Snān *at the end of a* jagya.

other *ṭikās*, it is attached to a person's *nidhār* (forehead), the most sacred and pure part of the body where the goddess Babi inscribes one's biography and fortune in life on the sixth night after birth.

While discussing the *ṭikā*, Sharmananda Subedi (71) testified that years ago he participated in one man's cremation as a *malāmi* (a man who helps to

*Eknath Lamichane's two younger sisters with the large Dasaī* achetā ṭikā *on their foreheads. An identical but smaller* ṭikā *is given in ordinary* pūjās.

carry the bier to the cremation grounds), and just before the skull was burnt he suddenly saw numerous, miniature compressed letters written all over the brow, which clearly were inscribed by Babi. Prakash Ram Khanal (61) added that our body, like the universe, is divided into fourteen *bhuwan* (worlds), the highest of which is Brahma *lok* (heaven, where the god dwells). This is the brow, the premier and most *cokho* (pure) part of the body, and thus a *ṭikā*, which is the gods' auspicious and sacred *prasād*, should be placed there.

The general purpose of the *ṭikā* is captured by the mantra uttered by the priest attaching it: "May Brahmā, Viṣṇu and Maheś war (Mahesor, Śiva) give the *kartā* [or any other recipient of the *ṭikā*] long life, may all the gods and goddesses protect him/her, may the life within these gods be embodied within you."

Intentional removal of the *ṭikā* from the forehead is highly inauspicious and villagers stress that it should fall off by itself. The paramount importance Thamgharians ascribe to the *ṭikā* may be discerned from the words of Prabha Dhakal (42) at the end of *caurāsī godān* held for her father-in-law. Prabha inquired whether in my country (Israel) there is also a *ṭikā calan* (*chalan*, custom or tradition). On hearing my negative response she was bewildered: "*Ho-ra, ṭikā nalagāikana, kasori* (*kasari*) *tapāīharu bācnu* (*bachnu*) *huncha ta*?"—really (unbelievingly) if so, how do you (people) survive without a *ṭikā*?

Finally, the *kartā* gives the priest(s) the *dakṣiṇā* (*dacchinā*, in cash) atop a leaf plate full of *cāmal*, performs *ḍhognu*, and honors him by attaching a *ṭikā* to his forehead. Later on, the householder's wife cleans the entire area of the *jagya* and meticulously collects all *grains* that have been offered, packing them in a special cloth that the priest takes with him to be used in his house. These (mainly) rice grains are believed to embody the heavy load of the family's *dān* and villagers are extremely careful not to leave even a single grain containing evil influence behind. The *jagya*'s paraphernalia is now tidied away, the flowers and remaining offerings are taken for *selāunu* (*selāune*)—to be "cooled" in the village *dhārā* (spring), and the remaining *achetā tulasi* in the *āghan*.

## NOTES

1. Turner (1996: 522) notes that *jagya* or *jagge* is the popular form of the Sanskrit *yajña*, which literally means sacrifice, with the *bedi* as its altar, and is the place where the bride and groom sit during the marriage ceremony.

2. The yard in front of a house, where the *tulasi*, the basil manifestation of Viṣṇu, is planted on top of a decorated mud pedestal.

3. The stylistic variations are mainly, if not entirely, accidental.

4. With the exception of particular ritual segments, texts are solely used as sources of mantras.

5. A particularly lavish *jagya* may cost up to 60,000 NRs (in 1997 prices), equivalent to the costs incurred in the construction of a medium-sized house.

6. Renou (1953: 6, 1957: 127) notes that King Pariksit (Parikchit) is the first notable king mentioned in the *Atharvaveda* and that the *Mahābhārta* was first recited in front of his son.

7. Otherwise, Thamgharians explain, following such an inauspicious death he would normally be condemned to *narka* (hell).

8. Most village *purāns* I attended were dedicated to Viṣṇu, yet occasionally the principal deity may be either Devī (the goddess) or Śiva.

9. All of which denote that in one way or another, something stands for, represents, refers to, is understood or experienced as or in terms of something else (Lakoff and Johnson 1980: 5; Lewis 1980: 111; and Wagner 1986: 6).

10. This is concurrent with the similar growing unease regarding "symbolism," found in general anthropology, as may be exemplified by Gell (1975, 1998), Lewis (1980 passim and particularly chapter 6), and Godlier (1999: 121–25).

11. Thus Inden (1978: 32–33) coined the term "effective symbols," referring to a part that contains the whole and is thus able to affect it.

12. Discussing this point, Copi (1972: 352) eloquently states that an argument by *analogy* cannot be demonstratively valid and has no "logical necessity."

13. One of India's six classical philosophical systems whose "distinguishing feature is its belief in the utility of analysis and in the reliability of reason" (Hiriyanna 1996: 84).

14. This is since a "symbol" may represent an idea and/or embody several entities on various scales of existence simultaneously (e.g., the aforementioned *godān*), where each may differ in its level of "literalization." Concomitantly, a notion/entity may be represented in several forms and levels of embodiment at one and the same time (as will be demonstrated by the various manifestations of matrimonial encompassment in the *jagya*).

15. From Sanskrit, literally meaning the first part or limb (Turner 1996: 187).

16. According to Slusser (1982: 344–45), these mother guardians of encompassed space usurped the traditional role of their Vedic predecessors, the *dikapālas*, which still guard (albeit in the local form of Bāstu) and play a prominent role in a Thamgharian *jagya* as will be demonstrated below.

17. Rather like the Vedic *dīkṣā* mentioned in the previous chapter (Hubert and Mauss 1964: 10, 13 n56), in Thamghar this mainly includes a purifying bath in the village *dhārā* (spring) and the cutting of fingernails. Men wear a pure *dhoti* (a white cotton loincloth) and shave their head, except for the *ṭupi* or *śikhā*—the auspicious top strand often tied to a knot at the crown of the skull. In addition, all participating members of the family must avoid eating cooked rice (as it is considered *biṭulo*, impure) until the end of the ritual performances of each day.

18. The householder's wife covers the walls and floor (usually only the ground floor) with a paste of fresh mud and cow dung.

19. The auspicious direction of the rising sun, "where the gods arrive from."

20. Snodgrass (1992: 48–49); Malamoud (1996: 41).

21. This is calculated by an astrologer, usually a knowledgeable priest, who has mastered the complexities of the Nepalese *pātro* (astrological calendar and almanac).

22. Every Thamgharian (man or woman) owns a *cinu* (or locally *chinā*), an astrological birth chart prepared by an accomplished astrologer during infancy.

23. Or other dimension of his body.

24. *Prosopis Spicigera*.

25. *Shorea Robusta*.

26. The term *liṅga* is used for each individual trunk as well as for all of them tied together.

27. Ordinarily, a *kalas* is a small copper ritual flask, with the exception of the *jagya*'s *mul* (main) *kalas*, for which one of the household's heavily decorated *gāgris* (large water vessels) is used.

28. Indeed, the names of some of the *liṅgas* surrounding an elaborated *jagya* (hardly known to lay villagers) are in accord with those of the ancient Hindu cosmic protectors and the regents of the compass directions, often named *Lokapālas*, *Aṣṭadikapālas*, or the *Vāstupuruṣas*. The latter may number four, five, eight, or ten and stand around ritual arenas, *yantra*, and various *maṇḍalas* found throughout (and beyond) the Hindu world (Daniélou 1964: 129–32; Eliade 1973: 219–20; Beck 1976; Kramrisch 1976: 91–94; and Snodgrass 1994: 149–54).

29. The clear marital and sexual bond between these divinities is evident in different contexts, among them Bāstu *pūjā* (discussed in the following chapter) and rice agriculture (as will be elaborated in part III).

30. According to Turner (1996: 311), *diggaj* (from Sanskrit *diggaja*) is one of the four or eight elephants that support the world.

31. This cosmic image resembles the magnificent Kailasa Śiva rock temple in Ellora (eighth century), which is "carried" on the back of numerous giant carved elephants as shown in Brown (1971: LVII).

32. *Achetā* is an extremely auspicious, pure substance, the *primary* type of offering given in *all* village *pūjās*, and is thought of as a highly desirable food for the gods.

33. In order to enhance the *achetā*'s adhesion to the brow, villagers often mix it with some *dahi* (*dai*, yoghurt). Ṭhe giving of a *ṭikā* is complemented by auspicious placing of fresh flowers on the head or behind the ears.

34. *Sait* is a highly desirable yet rather elusive state of affairs, characterized by high auspiciousness and good fortune, and is mainly associated with cosmic compatibility in time and action. *Parnu* is to cause, to become, to fall into.

35. As is often the case in Nepali, Thamgharian Brāhmaṇs pronounce the Sanskrit V as in Viṣṇu, *vedi*, and Vāstu as B (Biṣṇu, *bedi*, and Bāstu).

36. Yet, as Kramrisch herself was well aware, the *Vāstupuruṣamaṇḍala* does not always constitute the Hindu temple's actual plan but is "a prognostication," a forecast of its content or part of an "architectural rite" performed during its construction (1976: 228). Regarding this issue see also the recent discussion in Meister (2003).

37. Volwahsen (1969: 43–46); Beck (1976: 226–28); Kramrisch (1976: 21–97, 1984: 51–70 and passim); O'Flaherty (1980b: 276); Shulman (1980: 90–91); Moore (1983: 227, 1989: 179–82); Meister (1991: 270–71); and Eck (1998a: 59–60).

38. Kramrisch further suggests that the confusion regarding the identity of *Vāstupuruṣa* in the architectural *maṇḍala* stems from a break in the architectural tradition (1976: 79). This is also probably the reason behind the general lack of interpretation and reasoning in the ancient Indian geometric and architectural manuals.

39. Fragmentary evidence may suggest that the Thamgharian Bāstu usurped the place of Viṣṇu as the god of the house and the husband of Bhūme, yet villagers refused to embrace this possibility.

40. Thamgharians often refer to their rice fields as *akās kheti*—heavenly (or "of the sky") fields.

41. Translated from Sanskrit by a village priest.

42. This is the (ideally) golden, uppermost metal part of a Nepalese (Newāri style pagoda) temple (Slusser 1982: 148), which corresponds to the *stūpikā*, the pinnacle of a Hindu temple (Kramrisch 1976: 348–56).

43. Binding a *jagya*, a temple or a house with a *toran* is an auspicious act of worship, believed to have a strong protective effect on the bounded entity.

44. The *jagya*'s *toran* and *dhāgo* bring to mind the ropes that hold the worlds together in Vedic cosmology (Renou 1957: 78).

45. The offering of *achetā* and *phul* (flowers) is the most prevalent basic act of worship known in Thamghar, performed numerous times throughout a *jagya*.

46. This mixture of Gaṅgā *jal*, *til* (sesame seeds), sacred *kuś* (*kuśa*) grass and *jāo* (*jau*, barley) is considered extremely potent.

47. In other contexts, namely, during the harvest of rice, a different type of *rekhi*, made of ashes, is used as an apotropaic figure intended to repel ghosts and demons from the harvest.

48. The *rekhi*'s size, shape, and relative complexity depend on the *pūjā* to be performed and every priest has his own plethora of patterns passed down from his father.

49. *Kuśa*, *Eragrostis Cynosuroides*.

50. Mainly a variety of grains (particularly *dhān* and *cāmal*), *dahi* (curd), various foodstuffs (made of rice flour and fried in *ghyu*), as well as fruits and many flowers.

51. *Cynodon Dactylon*.

52. A *tīrtha* is a Hindu pilgrimage site conceptualized as "a ford," and is usually an actual crossing place over a river or a confluence, believed to originate in heaven and connect it to earth as epitomized by the holy Ganges (Eck 1981). All rivers and springs surrounding Thamghar are seen as rivulets of the Ganges.

53. For example, Viṣṇu is said to "live" in the *kalas*'s mouth, Śiva in its throat, Brahmā in its body, while all other Hindu deities simply reside within the water.

54. Also called *maṇḍap*.

55. *Rāj purohit* (the Nepalese royal priest) Mr. Ramesh Pandey (p.c.).

56. Manifestation or incarnation, particularly of Viṣṇu.

57. Cf. O'Flaherty (1981: 40–41) about the erotic version of this myth where Viṣṇu saves the earth from the depths of the cosmic flood and then repetitively makes love to the heated, passionate earth who becomes his wife.

58. As found elsewhere in high-caste Nepal (Gray 1979: 88; Bennett 1983: 52–123; Rutter 1993: 254, 339). This echoes Gonda (1980: 164), who notes that fire-worship was the most permanent element in the ancient Vedic *saṁskāras*.

59. There, *hom* is simply performed on a *rekhi*.

60. These stages pertain to almost any form of village *pūjā* officiated by a priest.

61. "What is remembered," the teachings of the ancient sages as compiled in the formal books of law and conduct.

62. "What is heard," the Vedas and other texts, which are regarded as direct revelation.

63. Starting with the creation of the world by Brahma and his age, the present *kalpa* and *yug* and the planetary situation, and ending with the exact time when the performance commences.

64. This starts from the universe, the geographical location (e.g., north of Bhārat, India, "from where our forefathers fled the Muslims") and ends with "here in the village of Thamghar at the foot of Kālika *mandir*."

65. This includes the *kartā*'s *gotra*, *kul*, and his ancestors going back three generations.

66. According to Turner (1996: 398), *prāyaścitta* is propitiation and expiation, particularly for impurity or loss of caste, yet as I wish to recall, in Thamghar (though not without exception) *dān* does not include impurities.

67. These are usually made of *sāl* tree, but may vary in accordance with the principal deity being worshipped.

68. Again, the shape and number of levels often depend on the identity of the principal deity worshipped.

69. Thamgharians are completely unaware of the sexual connotations that "drilling a fire" had in the Vedas (Heesterman 1993: 94).

70. For a similar Chetri notion see Rutter (1993: 264).

71. In addition to the *jagya*'s principal deity, a long list of other gods and goddesses are explicitly fed here; in effect, villagers note that the entire Hindu pantheon is nourished through the *hom*.

72. *Śeṣ* means the remaining or remainder, and *caru* means the mixture of *dhān*, *cāmal*, sacred *kuś* grass and *ghyu*.

73. Turner (1996: 625).

74. As may be expected and in accordance with chapter 2, these public meals are said to be prone to the evil manipulation of the village *boksis*, and my impression is that for most participants it is rarely a convivial event of geniality. Instead, people constantly complain about the social imperative that compels them to participate.

# 6

# The *Jagya*—an Analysis

## INTRODUCTION

This chapter begins with a number of extended introductory remarks about the meaning of the titles *jagya* and *hom*, as well as the *jagya*'s obviation of the boundary between context and content. This is followed by an analysis of the Thamgharian *jagya* and *hom* in accordance with the main stages of construction and development.

Following a discussion of the *jagya*'s preliminary rites, I will turn to examine the manner in which it is aligned according to the *kartā* and the macrocosm. An analysis of the gendered fractal cosmology evinced by the *jagya* and Bāstu *pūjā*, the most elaborate and picturesque of all village *jagyas* I attended, will ensue in an attempt to illuminate the meaning and implications of the divine matrimonial encompassment dominating them.

A brief note about the inherent capabilities of the *rekhi* will be followed by an examination of the two primary modes by which the *jagya*'s divine *śakti* and grace are incorporated by its participants, namely, through the *snān* and *prasād*, and finally, the efficacy of the entire ritual operation will be considered via a comparison with other relevant Hindu events. Throughout, particular attention will be paid to the import of gender attribution to nonhuman entities and the sexual significance of specific actions during the *jagya*'s construction and operation.

The analysis herein does not attempt to exhaust every possible interpretative avenue or meaning, nor does it cover the implications of all its ritual acts. Instead, it is intentionally focused on the *jagya*'s major and most significant aspects, which are relevant to the main theme of the present work, namely the place in which the notion of fractalic matrimonial encompassment occupies in people's perceptions of themselves, their world, action, and interaction.

## A Matter of Terminology

Prior to analyzing the *jagya* and *hom*, the vernacular meanings of these terms need to be considered. This is crucial for understanding what follows, particularly since these meanings seem to differ considerably from their ancient imports and those attributed to them by many other contemporary Hindus, mainly Indian Brāhmaṇs. In the course of discussion, local notions and perceptions regarding the significance and violent or peaceful (*hiṁsā/ahiṁsā*) implications of the concepts of *pūjā*, *bali*, and the various manifestations and aspects of fire will be explicated and glossed upon.

In Thamghar, the terms *jagya*, *hom*, *bedi*, and *jajmān*[1] seem to belong to what Smith (1989: 217) calls "canonical categories," which are Vedic categories employed as markers for tradition in order to make present-day practices or concepts seem "Vedic," that is, canonical and orthodox.[2] As most villagers have little knowledge of Sanskrit and are unfamiliar with the majority of ancient Hindu texts, they are oblivious to the fact that, as will be demonstrated below, the present-day village *jagya* differs considerably from the ancient *yajña*. Thus, when mentioned in myths read during *purāns*, the *yajña* (and in particular the *aśvamedha*—ancient horse sacrifice) is always referred to by village priests using the local term *jagya*, in order to bestow it with its ultimate source of reputation and authenticity.

Thamgharian Brāhmaṇs, for whom terminology is of little concern, do not draw profound distinctions between grain offerings (as well as various cooked and mainly uncooked food-stuffs, vegetables, water, flowers, incense, and the like) given to the deities in a *pūjā*, and the offering of *ghyu* and grains (mainly rice) into the fire in *hom*. Likewise, the term *bali* is used to denote not only animal (blood) sacrifice, but also the "slaughter" of vegetables[3] and vegetal offering to demons, in lieu of meat. All Thamgharian Brāhmaṇs with whom I discussed this, considered the above to be different yet equally pure and auspicious modes of worship that fall under the general category of *pūjā*. Almost every village *pūjā* includes *bali* (in the form of the offering of grains in lieu of meat to demons and ghosts) as one of its *aṅgas* (limbs, organs).

In line with other high-caste Nepalese communities (Bista 1972: 95–100; Bennett 1983: 140), the blood of a slaughtered animal is considered pure and auspicious and is consumed as a highly beneficial *prasād*; in particular it is deemed efficacious for alleviating women's fertility problems. The most auspicious, extremely pure and effective Dasaī *ṭikā* is *always* mixed with some drops of the sacrificed animal's blood. Concomitantly, men who believe that they become drained of their *śakti* via ejaculation, are very fond of eating *rakti* (*rāuto*, goat's blood cooked with salt), which is regarded as an extremely *tāgatilo* (potent) substance.

According to the general Thamgharian view, the blood of a sacrificed animal is consumed by the divinity that demands it, and is utterly pure. Only one sophisticated village priest, who spent a number of years studying in Banaras, thought that the sacrificial blood is somewhat *biṭulo* (mildly impure) and explained, in agreement with the view of many Indian Hindus particularly in the south (Fuller 1992: 87–96), that it is not the gods who consume it but their *senā* (army) of demons and ghosts.

As was already discussed in chapter 2 and as also found by Babb (1975: 54) and Shulman (1993: 35) in India, a Thamgharian *pūjā* is thought of primarily as an *oral* affair, centered on the careful attendance to the "cry of (mainly) divine bellies." Thamgharian Brāhmaṇs clearly state that, as Bennet (1983: 48) finds elsewhere in high-caste Nepal, a *pūjā*'s menu depends on a particular deity's taste or whim, but in contrast with what seems to be the case in many places in India, especially in the south (Biardeau 1976: 139–40; Fuller 1987), divine preferences do not seem to have any moral significance or implications regarding purity or pollution and do not reflect or constitute the basis for a "divine hierarchy" (Babb 1975).

Similarly, village priests do not hesitate before sacrificing a dove during an exorcism session for example, and other village Upādhyāya Brāhmaṇs (not to mention Jaisis or Chetris) do not think twice before killing a goat (for meat or offering alike). Likewise, the honorable *pūjāris* (special appointed priests) who stay at the Kāli temple near Thamghar throughout the ten days of the Dasaī festival, during which time they engage in various austerities and must avoid entering the village in order to maintain their elevated state of purity, do not refrain from sacrificing animals to the blood-craving goddess.

These Thamgharian attitudes are in accordance with the generally indifferent local approach to vegetarianism; quite similarly to that of other high-caste Himalayan Hindus (Parry 1979: 89; Fuller 1992: 93; Berreman 1993: 51), Thamgharians are very fond of mutton and would eat it regularly if they could only afford to do so. Vegetarian *bhāters* are unheard of and if a family cannot afford to include meat in a festive meal, they will postpone the religious function that this celebrates.

Present day village animal sacrifices have very little in common with their Vedic predecessors. Most remarkably, one of the central notions that was present in the *yajña*, namely the identification of the *jajamāna* with the victim who served as his *substitute* (Hubert and Mauss 1964; Smith and Doniger 1989), an idea that as Gray (1979) implies may still be found among some

Nepalese Chetris and is also prevalent in India (Fuller 1992: 84–85), is totally unheard of in Thamghar.

I wish to stress that similarly to the village (animal and "vegetarian") sacrifices, the Thamgharian *jagya* and the performance of *hom* are also far removed from the Vedic *yajña* and *homa*, despite acquiring their titles from the latter. Thus, death, which played a major role within the Vedic sacrifice, has no place within the Thamgharian *jagya* as villagers adamantly emphasize. Furthermore, even the sacrifice of a male goat in Bāstu *pūjā*—the single occasion in which a *jagya* does involve the killing of an animal—is merely done to satisfy Bāstu's culinary demands and, from a Thamgharian perspective, does not seem to introduce an element of death or impurity into the *jagya* and does not taint its essential vitality. What appears to be true for the Thamgharian *jagya* applies equally to the offering of *ghyu* and grains into the fire in *hom*. In clear contrast to the Vedic sacrificial cult, the fire of *hom* is clearly detached from any ritual slaughter in present-day Thamghar, although they may be performed in tandem.

Therefore, quite similarly to the attitudes of other high-caste Nepalese (Bennett 1983) or those found in central India (Babb 1975: 55, 250), but unlike the Katmandu valley Chetris (Gray 1979), their neighboring Newārs (Hoek 1992; Gellner 1996: 158), present day Tamils or nineteenth century Indian Mewars (Fuller 1987: 29, 1992: 113), who view *hom* in terms of (or as actually including) animal *sacrifice*, Thamgharian priests and lay householders alike view *hom* as a highly auspicious and pure mode of *pūjā* and insist that it does not ordinarily include *bali*.[4]

From a Western perspective, oblation into a fire may appear to involve, even epitomize a violent act of destruction and possibly killing and death, but Thamgharian Brāhmaṇs do not share this perspective. As Netra Lal Paudel (33) told me, the general belief in the village is that the destructive aspects of Agni are only (rarely) manifested when the deity, infuriated with rage, appears in an uncontrolled form as fire burning in the forest or flames that raze a hay barn.[5] Otherwise, well-controlled within the human body, *hom* or *culo* (*chulo*, hearth, cooking place), fire is associated with the non-violent processes of eating and cooking, redolent with strong sexual undertones.

As discussed in chapter 4, during the chaotic drama of violence and creation that was the Vedic sacrifice, the colorful personality of Agni was not merely associated with destruction and killing, sacrificing and devouring life, but also with eroticism, procreation, gestation, cooking, and rebirth. It is the latter aspects that still seem to resonate in the Thamgharian notions regarding the above three cardinal fires as will be demonstrated in their brief examination below.

As mentioned in chapter 5, in Thamghar, the disappearance of grains and *ghyu* within the flames of *hom* is thought to be an auspicious sign of their actual consumption by the Hindu divinities, since "all the gods eat through Agni" who is thus believed to embody a cosmic mouth situated within Bhūme's womb on/in the *bedi/khāldo*. Moreover, *hom* is often portrayed as the ultimate *pūjā*, the finest manner in which the gods are fed with the purest, most delectable food, and may also be entertained by its appealing smell. As Krishna Prasad Sharma (46) remarked before the performance of *hom* in his house, "*hom* is like a meal where, instead of doing *peṭ* (stomach) *pūjā* to the inner digestive, cooking fire, we feed the gods through Agni (*pūjā*)." A short explication of the significance of fire in the context of the *culo* and the body's inner-fire will render this point lucid. It is thus necessary to refer back to the discussion in chapter 2 about the construction of the house. The latter is infused with *life* when the householder's wife lights a fire and cooks rice on the *culo* for the first time. As I wish to emphasize, this fire is explicitly equated with the digestive-cooking fire in the stomach, burning within the body (likened to a *mandir*) as long as a person is *alive*, and as villagers put it, "when *āgo marcha* (the fire dies), the person in whose body it burns dies with it."

The verb *pāknu* (to cook) is not merely employed to describe the literal process of cooking, but as often found elsewhere in the Hindu world,[6] it also denotes the ripening of crops, sexual maturation[7] and a woman's pregnancy. The inherent fecundity of the fire is clearly evinced by the prevalent local custom of letting a newlywed couple spends their nuptial night near the *culo*, in an attempt to ensure that a son will be conceived. Accordingly, it may be suggested that when Thamgharian women describe the wondrous augmentation in the quantity of rice within the cooking pot (as a result of its inherent *saha*) in terms of *pāknu*, this is a euphemism for the process of pregnancy thought to take place there. This may further aid in explaining the notion that the rice cooking process and particularly the final product (*bhāt*), are regarded as *biṭulo* (mildly polluted),[8] similarly to a pregnant woman, which is the main reason why it cannot be offered in a *pūjā*. Obviously, the mild pollution of *bhāt* cannot be associated with the rice (supposedly) being *sacrificed* in the cooking pot, primarily since, as discussed above, Thamgharians do not view sacrifice as a polluting process. The notion of cooking as a process of pregnancy offers an additional reason why only pure and sexually mature persons (post-*bartoon* males or married women) may engage in cooking rice.

What all this amounts to, in my view, is the conclusion that according to Thamgharian perception, fire is saliently associated with and is an agent of life, not death; it is completely devoid of any connotation of violence and destruction and is endowed with strong sexual overtones. Moreover, it becomes apparent that the notion of *pūjā* (the offering of food to the deities) in general, and *hom* (offering into the fire) in particular, are closely related to and interwoven with the processes of cooking, eating, digestion, and gestation.

In short, the main point I wish to advance here is that the Thamgharian *jagya* is not a "sacrifice," nor is the *hom* a "*fire*-sacrifice," and thus to refer to or examine it in the terms generally employed for the Vedic sacrifice (e.g., Hubert and Mauss 1964), present-day Indian sacrifices (Fuller 1987), the Nepalese fire sacrifice as analyzed by Gray (1979) or along the lines of Biardeau (1976),[9] would be misleading and out of place.

As will become clearer below, a number of the *jagya*'s aspects certainly seem to be in agreement with several principles that dominated the Vedic *yajña*, yet the chaos, darkness, and annihilation of life, which characterized the latter (Shulman 1980: 90), are wholly absent.

It is important to note that none of the notions discussed hitherto seem to create any problem in villagers' minds; if at times these appear inconsistent or even paradoxical to us (Westerners), it is only due to the inapplicability of Western modes of thought, with their tendency to seek crystal-clear coherence and obsessive employment of absolute categories and values, to the Hindu relative and somewhat fluid perceptions of reality and intertwined categories, as notably argued by Marriott (1989). Put differently, what we may see as indigenous incongruities are mainly an artifact of *our* habits of thought, perceptions, and manner of description and expression.

## On Frame, Context, and Content

Worship and ritual, so enmeshed within and completely inseparable from other realms of Hindu life, have occupied a significant place within the body of scholarly work dedicated to Brāhmaṇ-Chetris. Nevertheless, the *jagya*, commonly found throughout high-caste Hindu Nepal, has received only little scholarly attention, and most accounts seem to imply that it is a special sort of sacred pavilion or ritual arena (established mainly for celebrating the major Hindu *saṁskāras*, that is, marriage and the Upanayana), a complex yet rather "mute" context or a relic of forgotten autochthonous metaphysics. Accordingly, the ritual enclosure of the *jagya* is often portrayed as being of little or no relevance to the various rituals enacted within.[10] This seems, however, to be rather exceptional in the Hindu world, where "contextual sensitivity" or the pervasive emphasis on context in all realms of life, is normally the rule (Ramanujan 1989b; Hess 1993).

As will be demonstrated in the present chapter, the Thamgharian *jagya*, unlike its counterparts, displays very close rapport between stage and performance, context and content, or, to use Bateson's (1987: 177–93) and Goffman's (1959, 1974) terms, it is the frame employed by people in their attempt to make sense of the reality projected and experienced within. Moreover, as implied by the dual denotation of the term *jagya*, the distinction between stage and performance is obviated in a manner that brings to mind the Vedic *agnicāyana*; the *jagya*'s ritual performance is intertwined with its construction process and the latter seems to come to an end, that is, the "body" of the *jagya* appears to fully assume its life (similarly to a newly constructed Thamgharian house), only when a fire is ignited on the *bedi* in the climactic yet concluding stage of *hom*, a short while before the whole edifice is dismantled.

This may appear incongruous with Western habits of thinking, yet it seems quite prevalent elsewhere, as suggested by Ingold (1993), who draws upon the field of art to point out that unlike Western thought, in non-Western societies process has privilege over form and priority lies with the "act of painting," while the final product is often short-lived.

## INITIAL PERFORMANCES

### Time Enmeshed in Space

The preliminary rite of *purbāṅga*, although performed "off-stage" so to speak, as it takes place prior to the construction of the *jagya*, and notwithstanding the modest procedure involved, is clearly crucial to the success of the entire event. That villagers provide little exegesis regarding the exact meaning of the rites enacted or the significance of the symbolism involved in *bāsudhārā*, leaves much scope for extrapolation.[11]

Although Thamgharians have not openly made this connection, I think that the aim of the *purbāṅga* may be elucidated by comparison with a similar procedure that takes place during Bhai Ṭikā in Tihār, in which the general objective of the performances (which take a slightly different form in every household) is to enable sisters to protect and ensure the life of their brothers. It is particularly the sister in her role as a divine, powerful virgin goddess, as Bennett (1983: 246–52) notes, who is capable of protecting her brothers from Yama-rāj—the mighty god of death. In the main, this is achieved by worshipping the brothers and fashioning an auspicious seven-colored *ṭikā* on their forehead, as well as by creating a number of potent barriers to prevent Yama-rāj from reaching and gaining a grip on their lives. These barriers, made of water and oil marked with sacred *dubo* grass (believed in this context to be a manifestation of Viṣṇu), are made around each brother as well as in front of the door of the house. Finally, one of the sisters crushes a large nut on the house's threshold (on the door's lower lintel in exactly the same place where the wife stands during *bāsudhārā*), which is thought to "seal" the house from the invasion of Yama-rāj.

In addition, *bāsudhārā* brings to mind an ancient Hindu rite of similar title: *vasodhārā*, which consisted of a very slow trickling of *ghee* (clarified butter, *ghyu*) onto a consecrated fire, kindled at the bottom of a pit in the middle of a house, perceived as a potent "shower" of wealth and blessing, but which was also capable of destroying fierce enemies (Derrett 1959: 120). Since *dhārā* is the local pronunciation for a village spring, marked by the free flow of pure water, and *Vāsudeva* is one of Viṣṇu's names (V is often locally pronounced as B), *bāsudhārā* may be interpreted as Viṣṇu's *dhārā* of *ghyu*. Being one of the purest and most potent liquids known in Thamghar, the *ghyu* creates a strong barrier at the entrance of the house, which is identified with the *jagya*. In this manner, *bāsudhārā* seals the house and the future *jagya* and thus prevents the invasion of (mainly) death and birth pollution emanating from other *kul* families, which would otherwise destroy or delay the *jagya*.

Discharged over the seven fierce, single mother goddesses (*Mātṛkās*) situated above the door, the *ghyu* becomes imbued with their enormous force, thus it is particularly the wife by virtue of her *śakti*, who is able to accept it onto her head. Acceptance of an overly forceful flowing substance onto the head is a known Hindu contrivance against its potentially harmful aspects, as exemplified by the Ganges's forceful flow from heaven; being too potent for the earth to handle, it must first flow over Śiva's head (O'Flaherty 1981: 230), a theme that is often depicted in oleographs found above the doors of many village houses.

Viewed from an additional perspective, *bāsudhārā* may indeed seem to connect heaven, represented by the *Mātṛkās*, and earth via the *kartā*'s wife (the mother of the house), yet apparently it does not merely connect these separate domains, but in effect it draws the latter into the realm of the former; being purified by the effective stream of *ghyu* in the *purbāṅga*, the household's members seem to ascend to a higher realm (in a manner reminiscent of the *jajamāna* following the *dīkṣā* which preceded some of the Vedic sacrifices), and like the Hindu divinities, at least as these are perceived in Thamghar, move beyond the reach of terrestrial pollution.

Furthermore, by sealing off the household (the house and its *ātmā*—the family residing within) against the invasion of inauspiciousness and impurity, the *purbāṅga* temporarily severs the organic connection between the family members and the social body of its *kul*, disconnecting the household from the passage of ordinary time for the duration of the *jagya*. In effect, the *purbāṅga* encapsulates time within the space of the house in order to create a particular time frame in which the future *jagya* and its main actors would be protected from the uncontrolled and often violent effects associated with the ordinary flow of time.

As with other entities (or ideas) that Hindus in general and Thamgharians in particular deem too potent or complex to be confronted directly, time is conquered by being incorporated, co-opted, and encompassed within the house. In Thamghar, as found elsewhere in the Hindu world,[12] time is often personified as *kāla* (*kāl*, death) thus the *purbāṅga* appears to contain death as the first step in the larger project of the *jagya*, whose primary objective lies in the complete elimination of death in order to induce rebirth into (conditional) immortality.

As a consequence, once under control, time is no longer deadly and may be further manipulated and operated upon. Thus death is eliminated within the future *jagya*, but from the vantage point of a higher scale that ensures domination and authority. From within this dense spatio-temporal capsule fashioned in the *purbāṅga*, Brāhmaṇs later begin to engender the procreative cosmic fractal of the *jagya*, which appears to rise into existence as though erupting out of a fracture in the fabric of the apparent reality.

## Displaying Connections

Creation myths are often particularly revealing about the building blocks of people's perceptions of the world and the way they organize their cosmology.[13] Likewise, I believe that the coming into being of the microcosmic *jagya*, reenacting the cosmogony that appears to dominate Thamgharians' imagination, is highly illuminating regarding the premises underlining the local construction of reality and perception of the world.

Rather like its fully-fledged, brick and stone counterpart—the Hindu temple—which is both "terrestrial and extra-territorial" (Kramrisch 1976: 7), a Thamgharian *jagya*, being in command of time, operates within its own time frame, allowing Agni and the *kartā* to undergo their *saṁskāras* within a number of minutes and be reborn in the *hom*, yet it is not completely beyond the horizons of ordinary time flow. Indeed, the main concern in the *jagya*'s incipient stages of construction is its alignment in time and space with both the *kartā* and the celestial bodies, and this theme is underscored once more in the *saṅkalpa*.

Alignment with the universal design is an obvious measure for ensuring the harmony of the established *jagya* and its main actor(s) with the universe. Similarly to the execution of any other major action, such as building a house or setting off on a journey, it is carried out in a manner that guarantees that the cosmic forces will be harnessed in the most beneficial manner (Beck 1976: 214) and is clearly part of Thamgharians' general effort to maintain, like many Hindus elsewhere, an auspicious compatibility with their cosmos. To my mind, this recursive need to realign oneself with the universe provides further evidence for Hindus' acknowledgement that the general flow of everyday life constantly draws the world into a degenerative process of disintegration, propelling it deeper into an entropic, disharmonious vortex, which must be counteracted.

An additional reason I would like to suggest for the alignment of the *jagya* with the macrocosm is to ensure that its main actor, the (at times "deflected") person or entity for whose benefit it is performed, is replaced into an auspicious orbit of life (or afterlife) and action. Similarly, capturing time within the space of the *jagya* may be viewed as an attempt to resynchronize the entire cosmos into a tuned parallel trajectory as part of a general effort to reconstitute the ideal pattern of relationships in the universe.

As already implied and as will be further demonstrated below, from a Thamgharian perspective, the latter is a perpetual, cosmic sexual relationship within a divine matrimonial encompassment, where procreation and regeneration of life are ever present and death is completely obviated. Given that alignment has powerful hierarchical implications (as highlighted in marriage where the bride is aligned according to the groom), the *jagya* clearly diverts from and in fact inverts the ordinary relationship between the cosmos and the *kartā*. While the latter ordinarily aligns *himself* according to the former, the *jagya* takes both the universe and the *kartā* as its macrocosmic reference points, with which it must become attuned.

Compatibility in time is achieved by aligning the *jagya* with the celestial configuration as well as with the *kartā*'s horoscope, while spatial compatibility is achieved by the dimensional proportionality between the *jagya*'s square shape (the general manner in which the universe is imagined in Hindu cosmology)[14] and the *kartā*'s body. This suggests that the *jagya* does not merely equate but actually attempts to fuse micro- (*kartā*) and macrocosms into one unitary whole—the cosmic embodiment of the *kartā*, enveloping the universe he creates within his (ritual) body. Similarly to the relationship between a man and his house, the *kartā* identifies with his act as well as with the outcome; cause and effect are merged into one and, like the great Hindu cosmic gods, the *kartā* is simultaneously transcendent and immanent within the terrestrial heaven he has fashioned for himself as his cosmic self-portrait. As I shall argue later, the *jagya*'s alignment with both the *kartā* and the cosmos provides the basis for the efficacy of the entire ritual event.

I shall now turn to discuss Thamgharian Brāhmaṇs' "gender-sensitivity," with particular reference to the way it is evinced in the *jagya*, in order to provide an additional angle for stressing the identification of actors with their act and what they act upon, and facilitate the examination of the particular manner in which this microcosmic ritual arena comes into being and is operated upon.

## ORCHESTRATING A MICRO-MACROCOSMIC MATRIMONY

### On "Gender-Sensitivity" and "Sexual Labor"

The ethnographic description presented in the previous chapter provides the grounds for arguing that the Thamgharian *jagya* manifests a high degree of "gender-sensitivity," in a manner that will be illustrated below.[15] By this I do not merely mean that gender is attributed to different parts of the edifice or to its entirety, but that rather than being random, these attributions serve to fashion a certain *harmony* (locally described as a state of *pūrā* or completeness) and, moreover, that gender matters and makes a difference.

This characteristic further distinguishes the Thamgharian *jagya* from its Nepalese counterparts (including the royal *jagya*), as well as from most other Hindu ritual arenas. With the notable exception of the Vedic sacrifice discussed in chapter 4, most other Hindu ritual arenas, be they the south Indian ritual enclosures and ceremonial courts (Beck 1969: 359, 1976), the sacred pavilions erected for royal rituals in medieval India (Inden 1978), or the Hindu temple itself (Volwahsen 1969; Kramrisch 1976; Michell 1977),[16] seem to evince little gender-sensitivity.[17] Chapter 4 has provided some background, which may partially account for this apparent diminishing concern for the gender of the edifice and its various parts in Hindu ritual arenas.

Furthermore, as may be already apparent, in the Thamgharian world in general and within the *jagya* in particular, the ascription of gender to nonhuman entities is not simply an aesthetic or grammatical attribution made in passing, but a meaningful statement about the inherent *śakti* and agency of an entity, and about cause and effect. For example, as will be recalled from the discussion of *dān* transference in chapter 3, the attribution of female gender to certain divine or semidivine entities[18] is an explicit reference to their inner capabilities to carry the heavy ominous burden of *dān*.

This notion seems to be epitomized by the local belief that it is particularly the *śakti* of the goddess associated with the nearby river, deemed as a rivulet of the holy Ganges (a potent manifestation of the goddess Gaṅgā), which makes it such an efficient expiatory medium for stripping bathing believers of their sins and impurities. That this appears to be a pan-Indian notion suggests that the Hindu organic and divine cosmos[19] in general is also a fundamentally gendered one, as already implied by Trawick-Egnor (1978: 127, 138, and passim). Moreover, it suggests that the gender of nonhuman entities often plays a crucial role in the indigenous perceptions of their role and significance.[20]

In line with Thamgharian women, the gender of nonhuman female entities does not merely denote their intense *śakti*, but also their high susceptibility or "permeability" to external bio-moral influences such as pollution. Hence, unlike *dhān* (the grain while still enveloped within its male husk), Brāhmaṇs will never accept the feminine *cāmal* (husked grain) from the hands of the village "Untouchables." Likewise, it is mainly *cāmal* that is deemed suitable for *dān* transferences, and I will revert to the subject of the gender of *dhān* and *cāmal* in more detail in chapter 7.

As demonstrated by the *jagya* and as will be clarified below, gender ascription also expresses an anticipation of a particular process and its outcome, and reveals the inherent potencies of the relationship between different entities or, as is the case in the *jagya*, between the structure's different "limbs" (*aṅga*), to use local terminology.

Taken together, these points suggest that the attribution of gender to nonhuman entities in Thamghar is primarily a sign of *life*, that is, of the gendered entity being alive, which is mainly thought of in terms of the potential to reproduce, and more often than not is a token of divine or semidivine status. In contrast, other rather lifeless, passive and negligible beings in the Thamgharian universe are ascribed a neutral gender and often villagers are unable to decide what is the sex of such entities. More generally, the attribution of gender implies having personlike properties as Gell (1998: 96) sees these, that is, being the source of and the target for social *agency*.

It is significant to note that in sharp contrast with the relative gender ascriptions of south Indian Tamils (Trawick-Egnor 1978: 164) or what is at times a rather superficial and "fluid" Hindu "divine" gender (i.e., of deities and divine entities) as analyzed by Doniger (1999), in Thamghar gender is (not without exception) intrinsic, unalterable, and of great significance.

Thamgharian Brāhmaṇs' "gender sensitivity" does not end with the attribution of gender to nonhuman beings, but is also given a salient expression regarding human action and interaction with such entities. Stating that many village activities, including the construction of the *jagya*, are "gender sensitive" does not merely mean that they are characterized by considerable attention to a strict gendered division of labor, but also that they are often thought of in sexual terms and consist of what I prefer to call "sexual labor." I would like to emphasize that the latter denotes neither involvement in what we may consider as explicit ordinary sex, nor the mere metaphoric use of sexual imagery, but instead refers to actions that, from a Thamgharian perspective, are construed as being equal to a sexual act or having an intrinsic sexual significance.

A prominent example of sexual labor is the annual replanting of the *tulasi* (basil), the sacred manifestation of (the male god) Viṣṇu in the *āghan* (*āgan*).

As villagers clearly emphasize, only widows or men may perform it because the engagement of a married woman in this task is perceived as entailing an intimate involvement with the god and must thus be avoided. According to Brabha Bastakoti (42), adhering to this rule enables married women to escape the destiny of the mythical Brinda, Jalandhar's faithful wife, who committed adultery with the disguised scoundrel Viṣṇu.[21]

This example highlights the import behind the attribution of gender to non-human entities; clearly, the action of planting the *tulasi* in itself is not simply "feminine" or "masculine" and no ordinary gendered division of labor is involved, but the action itself, depending on the actor's gender, may potentially be perceived as being equal to a sexual act. Since replanting of the *tulasi* by a married woman is viewed as forming a union with the god it embodies, this may be an additional example of the local imagining of the identification of actor with action.

As may perhaps be expected, most forms of labor in Thamghar, particularly those actions that do not imply production or creation, are not perceived in sexual terms and may not even be gendered at all, while a number of tasks may transcend gender boundaries altogether. For example, reverting once more to the discussion of *dān* in chapter 3, I wish to recall that collecting *dān* is primarily the role of both (female) *kanyās* (regarded as the ultimate embodiment of *dān* in *kanyādān*) and (male) village priests, while the entire transference is facilitated by the high status of the receivers and the particular *female* qualities of its various mediums of conveyance. Thus in addition to the reasons mentioned in chapter 3, and in sharp contrast to the in-married *kanyādān buhāris*, it is perhaps also the gender incongruity between the male priests and the ominous female "gifts" they collect that enables them to avoid being identified with their action and what they act upon, namely the female entities imbued with *dān* and its harmful effects.

Hence, depending on the specific context, the gender of the object acted upon, the form of agency it is able to exert and whether the action is perceived in sexual terms or not (beyond the presence of a gendered division of labor) may imply *correspondence* or *incongruity* between an actor's gender, the act performed and the entity acted upon. This often has serious implications regarding the identification of an actor with his or her action, its significance and consequences.

Another relevant example is Thamghar's rice cultivation process where, as in many other places in the Hindu world,[22] sowing and plowing are viewed as virile actions, reserved for men in accordance with the division of sexual labor, while the transplantation (locally imagined as merely a change of location within Bhūme's womb) is an exclusively female task. Here the gender of the actor is in line with the nature of his or her labor and the state and gender of the objects acted upon, which facilitates their salient identification. These issues will be discussed in more detail in chapter 7.

Similarly, in the *jagya*'s construction, there is a clear association, even merger between the actor, his or her action and the "object" or divine entity acted upon; the *kartā* thrusts the (virile) Bāstu *liṅgas* via his male assistants, while his wife fashions Bhūme (the feminine *bedi*) at the *jagya*'s core. This is further highlighted by the *kartā* becoming explicitly *embodied* by the *jagya*, built in accordance with his bodily proportions.

Attentiveness to Thamgharians' gender sensitivity is of particular significance since, as may already be apparent (and will become clearer below), gender imagery provides the ultimate frame through which the *jagya* is locally perceived and is the primary medium for apprehending its dynamics and operation.

However, these issues are rarely expressed directly or overtly in these terms, and villagers evince extreme shyness (*lāj mānnu*) about the *jagya*'s procreative potential becoming explicit, displayed or realized. This is because Thamgharian Brāhmaṇs are ordinarily highly reserved regarding any sexual matter; in fact there seems to be a general taboo regarding any public reference to sex and most people are extremely shy about it. To illustrate this, suffice it to mention that a husband and wife hardly ever touch each other in public, let alone share the same bed,[23] and any public expression of affection between them is viewed as implying eroticism and is thus eschewed.[24] While for young men a premarital sexual affair is nothing to boast about and is often referred to (at least in public) with considerable contempt, a girl's prestige is forever marred by a mere reference to the potential violation of her strict chastity, and such incidents often tend to end in tragic circumstances.

Concomitantly, reference to sex in the presence of one's parents is unthinkable, even when the "children" are themselves (elderly) parents, and sexual matters never form part of the village public discourse. Thus for example, Ram Krishna (33), an open-minded young householder, once asked me to accompany him to his remote *bāri* (nonirrigated field), fifteen minutes walk away from the nearest house, before daring to whisper in my ear his wish and hope that I would be able to provide him with a medicine to ease his pregnant wife's delivery.

As a result, although Thamgharians emphasize that the *jagya* is alive and (like all other entities in the world) must become *purā*, a "complete" and auspicious locus of *sṛṣti*, and while this is realized via the frequent joining of male and female elements or entities, only rarely (or in private) would a householder be willing to admit that this, in effect, denotes marriage, let alone coitus. However, to my mind, villagers' statements and actions clearly suggest that for them, being alive, the state of *purā* and the essence of *sṛṣti* mean the potential for sexual reproduction, and thus these terms are in effect employed as a euphemism for procreation in marriage.

The main argument I wish to advance here is that within the gender-sensitive context of the *jagya*, where the major actors (the *kartā* and his wife) identify with their sexual actions, which involve fashioning *mithunas*, that is, the joining together of male and female entities to ensure *sṛṣti*, it is the salient presence of the notions of marriage, sex, and reproduction that prevent villagers from providing a more precise exegesis or discussing this openly.

Hence elsewhere in the Hindu world and mythology,[25] divine marriages are perhaps referred to in a much more explicit manner than in the fertile Thamgharian *jagya*, despite the former's largely barren and nonreproductive nature. Likewise (with the notable exception of Bāstu *pūjā* discussed below) of all the *jagya*'s *mithunas*, Bāstu and Bhūme often appear to be the least explicitly "wedded" couple, although they embody the most active and reproductive divine couple in Thamghar.[26]

Nevertheless, the employment of the terms *liṅga* and *yoni* for the encompassing Bāstus and the central *bedi/khāldo* respectively, and the identification of the *kartā* and his wife with the results of their "sexual labor" during the *jagya*'s construction, clearly suggests that the *jagya* realizes the ideal state of matrimonial encompassment on a cosmic divine scale, culminating in luminous fiery coitus, pregnancy, and rebirth in the *hom*.[27]

Thus, divine matrimonial encompassment emerges as the process through which the universe of the *jagya* comes into being. Indeed, each of the twenty-four *liṅgas*[28] brought to the *jagya* is greeted, similarly to other high-caste grooms upon arrival to the marriage *jagya* held at the bride's *māiti*,[29] by the performance of *parsanu* (*parsine*). The *liṅgas* are brought from the nearest possible place in order to minimize the work involved, but the mud representing Bhūme must be carried from the distant rice fields, even though there is obviously no shortage of earth within the village.[30] This seems to suggest that not only are the rice fields purer but that Thamgharians also imagine them to be the place where the earth goddess actually resides, where she is most active and where her temples are situated. What may also be suggested here is that in Thamghar, the fields are gendered female (quite similarly to what Cameron (1998: 176) finds in a low-caste Nepalese settlement) vis-à-vis the "masculine" village. Indeed, Bhūme's apparent absence from the village's landscape is accounted for in a story narrating the commencement of rice agriculture and this will be discussed in some detail in chapter 7.

In contrast with her mortal counterparts, Bhūme is neither tainted by impurity nor inauspiciousness, and she obviously does not carry with her or embody any portentous *dān*. These characteristics, belonging to human terrestrial existence, do not normally affect Thamgharian divinities. Like the *kanyās* who dig up the mud and bring her to the village, Bhūme is perceived as a particularly pure, potent goddess, yet like other single goddesses she is also renowned for being rather capricious and at times even potentially harmful. Consequently, she needs to be encompassed through marriage in order for her *śakti* to become benevolent and accessible prior to its employment within the *jagya*. Like mortal women, the goddess seems to be redefined according to her husband, the cosmic *kartā*-Bāstu, and whether she is manifested in the form of a *bedi* or *khāldo*, these must be constructed according to the dimensions of the *kartā*, thus confirming his identity as the *jagya*'s principal deity. In this regard, Heesterman's comment regarding the supremacy of the Vedic king in the *rājasūya*, the ancient royal consecration, is pertinent:

> The stature of the king completely overshadows the gods who are only concerned with limited spheres of action, and who are on several occasions integrated into the person of the king. (Heesterman 1957: 225–26)

Similarly, in a Thamgharian *jagya*, the *kartā* incorporates the cosmos and the Hindu pantheon within his ritual body and thus assumes the position of a "cosmic householder."

## Imagining a Gendered Fractal Cosmology

I believe that the *jagya* should be viewed as an exercise in fractal imagery on a cosmology dominated by the meta-design and logic of matrimonial encompassment, forming the fourth fractal dimension or image that dominates the way Thamgharian Brāhmaṇs perceive themselves and the world around them.[31] Moreover, I would like to argue that it is not merely by chance that the *jagya* resembles a fractal but that this is intentional since, as will be demonstrated later, it is this fractality of the cosmos displayed by the *jagya* that forms the basis for the latter's efficacy. Villagers are obviously not aware of this nomenclature, yet the fractal concept certainly emerges from their statements.

As will be recalled, Thamgharians recursively emphasize that the *jagya* (the entire edifice as well as each of its cardinal elements or *aṅgas*—limbs) is a microcosmic manifestation, which should comply with and embody the ideal modality of existence, where the sexes are united in a state of marriage, and indeed they endeavor to construct it accordingly. Implicit within their statements and actions is therefore the idea that the *jagya*'s various elements are (intended to be) *self-similar*, which is a primary characteristic of a fractal reality. On the face of it however, the *jagya* does not necessarily appear to maintain visual self-similarity, and the meta-image of matrimonial encompassment is displayed in varying forms and degrees of explicitness. While matrimonial encompassment can be saliently seen on the "macro-level" of simple (five-*liṅga*) or elaborated *jagyas* with the *bedi* at their core, elsewhere, within the *bedi* itself, the *rekhi* or the red and white ribbons that are tied to the *mul kalas* or flutter from the pinnacle of the *īśān liṅga*, it is evinced in a rather abstract form. Moreover, this image is not apparent in the individual *liṅgas* found in both *jagyas*, yet the latter provide the only clear manifestation of the overall unmistakable portrait of the cosmos as an erect male body, an image that remains rather implicit regarding the edifice as a whole.

In my view, this situation is not merely due to villagers' limited success in objectifying the "fractal" they have in mind, but is an inherent *property* of a fractal reality. As I wish to recall, a fractal reality possesses a subjective aspect where its observable details (and "effective dimensions") depend on both the scale of the object observed and the distance of observation. The latter is obviously not at play here since all observers, with the possible exception of the *kartā*, view the *jagya* from a similar vantage point. However, the different details observed in each of the *jagya*'s elements, seem to result from their existence on different *scales*.

Furthermore, it is imperative to bear in mind Bourdieu's comment regarding the logic we often ascribe to practice. He writes:

> Practice has a logic which is not that of the logician. This has to be acknowledged in order to avoid asking of it more logic than it can give, thereby condemning oneself either to wring incoherencies out of it or to thrust a forced coherence upon it. (Bourdieu 1992: 176)

This statement captures well the obvious limitations embedded in the ascription of an *etic* logic to an *emic* practice, and in particular the employment of Western metaphors, such as the fractal of matrimonial encompassment discussed here, in order to analyze and comprehend indigenous tangible or mental constructs. Nevertheless, to my mind, the way Thamgharian Brāhmaṇs imagine and strive to construct the *jagya*, conceived as a reflection and manifestation of the macrocosm, does indeed come very close to the Western construct of the fractal.[32]

Following Heesterman's (1957: 114–22) and Inden's (1978: 46) analysis of the rotary movement around Hindu royal microcosmic ritual enclosures, I tend to view the recursive *pradakṣiṇā*, the clockwise auspicious movement around the *jagya* which includes tying a *dhāgo* (thread) around it, adorning it with a *toran*, and various other complex performances, as a concerted effort to imbue the *jagya* with vitality and life for setting this ritual universe in motion.

In addition, the *dhāgo* seems to bind all the *liṇgas* together, uniting them into one body and thus consolidating the marital bond between Bāstu and Bhūme, which may indeed be necessary in view of the rather porous and fragmented nature of the *jagya*'s male skin, in contrast with the ordinary perception of human male bodies. Moreover, through their dynamic circular movement around the temporal shrine of the *jagya*, villagers appear to worship it as "the pivot of the cosmic rotation as well as the way between heaven and earth" (Heesterman 1957: 122), activating it and spiraling it as an *axis mundi*, piercing the center of the universe (Shulman 1980: 40–41).

As I imagine it, through Thamgharian eyes the *jagya* appears as an enshrined, vibrating fractal universe, pregnant with potentiality. Like other fractals, the body of the *jagya* "obviates the contrast between one and many" (Gell 1998: 139) and like an onion or a Russian doll it has a number of skins and multiple inwardly pointing heads. The latter converge at a single point above the edifice, granting it the figure of an upright male, in accordance with the ancient Hindu perceptions of the cosmos[33] and the Hindu temple[34] as having the shape of the primordial man—Puruṣa-Prajāpati. More specifically, I think that for Thamgharians, the *jagya* is the cosmic manifestation of the *kartā* as Bāstu, enveloping the earth goddess, Bhūme, in an erotic matrimonial encompassment, likened to Viṣṇu's *baikuṇṭha* (heaven) or a royal palace and firmly supported on the wide shoulders of sixteen colossal elephants.

### Bāstu *Pūjā*: Encompassing Immortality via Divine Matrimony

At this point, the question to be considered pertains to the meaning, significance and implications of the divine matrimonial encompassment embodied by a village *jagya*. These issues may be best explored in light of the following brief examination of a number of the crucial moments in Bāstu *pūjā*.

Although other village *jagyas* may last longer, Bāstu *pūjā* is certainly the most elaborate, complex, dramatic, and picturesque of all the *jagyas* I observed. As Prakash Paudel (41), a wealthy young householder, remarked during the Bāstu *pūjā* held in his house, Thamgharians often say that Bāstu *pūjā* enacts the house's marriage, but as will be demonstrated below, more specifically it celebrates the divine marriage and sexual union of Bāstu and Bhūme.

Given that matrimony is explicitly and recursively proclaimed to be the ideal "complete" state for all living beings, emblematic of unlimited fertile *sṛṣti*, safety, and the epitome of auspiciousness, it is little wonder that the house, which embodies the most significant and crucial enshrined entity for Thamgharians, must attain this supreme idyllic state lest it become a disastrous locus of misery, *alacchin* (bad luck, portentous sign) and eventually death. Hence, Bāstu *pūjā* is deemed vital for the well-being of the household, and its performance is of utmost importance; householders, particularly if afflicted by various difficulties or suffering from recursive misfortune, often attempt to perform Bāstu *pūjā* as soon as possible, sometimes immediately following the *praveś* (entrance) *pūjā*, marking the completion of the construction process. Nevertheless, due to the large expenses and effort involved, as well as the relative scarcity of days in which a suitable planetary arrangement exists, Bāstu *pūjā* is often postponed and may be performed years later, as was the case in both my host households. Without fully accounting for it, villagers note that unlike humans who normally hold a series of *jagyas* throughout their lives, a single (successful) performance of Bāstu *pūjā* suffices for turning a house into a permanent locus of safety, dynamic (pro)creativity and harmony.

As explained by Shiva Ram Banjara (32), a dynamic Thamgharian householder, during the Bāstu *pūjā* held for his new house, it is advisable to invite all village priests to officiate jointly over the complex performances involved, which are regularly preceded by various preliminary smaller *pūjās*. The latter, villagers note, are designed for manipulating the *kartā*'s *graha daśā* (unfavorable celestial situation and malignant effects, plight), in order to prepare the "cosmic atmosphere" and induce an auspicious "energetic milieu" so to speak, to ensure the success of the ensuing major event. Like *saṅgat* (proximity and contact among human beings), which is thought to engender a significant mutual influence, so also when two different *pūjās* are *bādhna* (*bandha*) *gareko*, "bound together" as villagers put it, their *śakti* and overall impact are considerably amplified.

Two of the prime features of an ordinary village *jagya*, namely the merger of micro- and macrocosms and the identification of the main actor with his action and the outcome, are given a remarkable expression in Bāstu *pūjā*; unlike the former, which must simply correspond to the dimensions of the *kartā*, the elaborated *jagya* of Bāstu *pūjā* is established alongside and surrounding the house walls. Thus, the house itself (normally identified with the householder and with the god of the house—Bāstu) seems to assume a cosmic stance and literally embody the entire ritual universe of the *jagya*. An additional, relatively small five-*liṅga jagya* is assembled inside the house on the ground floor, with Bhūme in the form of a square *khāldo* at its core.[35] The *khāldo*, a number of householders revealed in private, is thought of as the goddess's *yoni* (vulva).

The climax of Bāstu *pūjā* is neither the sacrifice of a white male goat (as is otherwise the case in rituals involving the killing of animals),[36] nor is it the performance of *hom* (as is the case in ordinary village *jagyas*). Instead, the high point of this spectacular *pūjā* is the dramatic entrance of Bāstu, who until this moment may seem almost aloof and is kept in a transcendental passive position, into action within the *jagya*'s core.

*The* khāldo *in the center of Bastu* pūjā*'s inner* jagya. *A* kalas *can be seen under each* liṅga.

This occurs when the householder takes the god (the subject and the main actor of the entire event, performing through his human counterpart),[37] manifested in the form of a large *liṅga*,[38] and thrusts him into the depths of the *khāldo*. The Bāstu *liṅga* is then tied tightly to an adjacent supporting pillar (one of the two poles that support the ceiling) and remains in this erotic position until the house is finally demolished or deserted years later.

Thereafter, *hom*, which as villagers explain should actually take place *inside* the *khāldo*, is performed on a *rekhi* set nearby, due to the obvious spatial constraints within the *khāldo* and the indoor *jagya*.[39]

The success or disastrous failure of Bāstu *pūjā* comes to light in the ensuing stage, when the *kartā* (re)fills the *khāldo* with the soil dug out of it earlier. A clear surplus of land, that is, a nicely rounded hill fashioned above the *khāldo*, is considered as a clear token of the success of the entire ritual operation and even a flat surface, though not the preferred option, is not considered portentous. However, if the land above the *khāldo* is concave in shape, the *jagya* is considered a total failure and a new one must be performed lest the house be considered doomed, forcing the family to abandon it in order to avoid future calamity.

Notwithstanding the reference to the *khāldo* and Bāstu *liṅga* as *yoni* and *liṅga* respectively, most villagers with whom I discussed this simply noted that Bāstu *pūjā* enacts the marriage of the house, and appeared too embarrassed to

*Bringing Bastu* liṅga *from the forest just prior to the performance of* parsinu.

*The* kartā*, helped by his priest (in white), placing Bastu* liṅga *in the* khāldo *at the center of the inner* jagya*, alongside one of the house's (black) beams. The colorful cloth seen above the* kartā*'s head is the* canduwā*, which is tied between the* jagya*'s* liṅgas*. Various* naibbede *are usually placed on the* canduwā *and distributed as* prasād *at the end of the* jagya.

provide detail regarding the exact process that takes place at the heart of the internal *jagya*, within the dark ground floor, redolent with fecundity and potentiality. Yet it seems plain that, as glossed by Muktinat Pokharil (46) who is one of Thamghar's epigrammatic priests, the convex mound of land emerging above the *khāldo* is evidence of Bhūme becoming *garbhabati* (pregnant), which is thus the main objective of the entire *pūjā*.

Hence, it appears that Thamgharians' reference to "marriage" is but a euphemism for its consummation. I am thus further inclined to surmise that Bāstu *pūjā*, whose focal point is the divine matrimonial encompassment and coitus of Bāstu with the earth goddess, reenacts the incipient union between the *kartā* as Bāstu and Bhūme (who was inseminated by the foundation stone (*jag*) placed by the householder during the construction of the house), which engendered the house in the first place.

If so, in Bāstu *pūjā* the house may be seen as reproducing itself from within; it is gestated and finally reborn from the vibrating ritual womb of the *khāldo*, encapsulated within its own body. Not only does this scenario echo the creation of the world by Prajāpati in the Vedic cosmogony discussed in chapter 4, but it is also implied in the Thamgharian notions of the wife's encompassment in marriage ensued by the father being born *in*, or *as*, his son.

I believe the above lends sufficient support to the idea that unlike Hindu (*Purāṇic*) mythology and much of contemporary ritual practice, where divine sexuality seems to be divorced from procreation, Bāstu *pūjā*, and as I would further like to suggest, the Thamgharian *jagya* in general, celebrate divine marriage, sexuality and procreation and merge these notions into one.

The disparity between ordinary, nonprocreative Hindu divine marriages and Bāstu and Bhūme's marriage in Bāstu *pūjā*, may best be couched in visual terms: while as I wish to recall, a Śiva-*liṅgam* ordinarily emerges *out* of the *yoni* in a denial of sexual reproduction, procreation is accentuated in Bāstu *pūjā* by the dramatic penetration of Bāstu *liṅga into* Bhūme's *yoni*.

Furthermore, Bāstu *pūjā* implies that in Thamghar transcendence may also mean immanence and the state of matrimonial encompassment obviously also suggests its inversion or, stated differently, encompassing a female through marriage also involves (and not merely implies) being encompassed by her via sexual intercourse. Likewise, human marriage and coitus mutually implicate each other, and the absence of one element (a sexless marriage or intercourse outside of marriage) is intolerable. Yet as will be recalled, in contrast with village men who tend to imagine sexual intercourse as a state of great peril, drainage of *śakti*, and loss of control, the house god does not appear to be affected by such mortal concerns—on the contrary, he seems inclined to remain in the above erotic posture indefinitely.

The main point I wish to advance here is that at the heart of the Thamgharian notion of matrimonial encompassment lies the intimate notional conflation of marriage, sexual relations, gestation, and rebirth. I am thus inclined to argue that what is perhaps the ideal of permanent matrimonial encompassment and coitus, which Bāstu is able to realize in his *pūjā*, represents a broader Thamgharian attempt (pertaining to all village *jagyas*) to break the ancient Hindu solid bond between birth and death in an effort to eliminate the latter without extinguishing the former.

The avenue chosen by Thamgharian Brāhmaṇs for keeping death at bay may be likened to cutting open and straightening the *saṁsāric* cycle and fashioning a perpetual linear process where procreative matrimonial encompassment equals ongoing eternal gestation. This becomes apparent if we consider that in Bāstu *pūjā* the earth goddess's conception does not come to fruition in explicit rebirth. This is not a result of ascetic informed ideas that attempt to preclude rebirth, but as I wish to maintain, is in order to gain the infinite capacity to live by establishing a process where life is forever created and the potentiality of rebirth is *unlimited*.

An alternative interpretation, informed by the Vedic *rājasūya*, where the central moment of the Hindu king's birth is not represented in the ritual and where entering the womb itself implies rebirth (Heesterman 1957: 117–20), may suggest that the goddess's eternal pregnancy, set into motion in Bāstu *pūjā*, means that the house is kept alive via a process of continuous rebirth, lasting as long as Bāstu and Bhūme's union remains uninterrupted.

Both the *rājasūya* and Bāstu *pūjā* are striking illustrations of the fine line that divides potentiality and actuality, cause and effect in Hindu thought, and thus the ongoing potentiality of the creation of life also implies, and in fact entails, its perpetuation.[40] The union of Bāstu and Bhūme commemorates the relationship of marriage and coitus by which the house came into being and symbolizes the house's perpetual, enlivening state of becoming. In this way, I would like to argue, Bāstu *pūjā* enunciates what Thamgharian Brāhmaṇs mean when they portray matrimonial encompassment in terms of a harmonious state, "complete" within itself, overflowing with fecundity and life potency, a state where eternal *sṛṣti* (procreation) is ensured by the perpetual union of the sexes, in which death is altogether annihilated.

Returning to the ordinary village *jagya*, we are now in a position to appreciate the manner in which it embodies a particular arena where the *kartā*, in his cosmic manifestation as Bāstu, sets into motion a process of divine matrimonial encompassment, where he is not only capable of encompassing the earth goddess, but also penetrates her *yoni* in the form of a burning ember, that is, as Agni, the fire god. It is through this dynamic, ephemeral and erotic process that the *kartā* is able to ignite and activate the earth goddess's divine ritual womb, where he himself is gestated (or cooked) by Agni's *tapas*, and from which he is finally reborn at the climax of the entire event.

The brief and condensed *saṁskāras* the *kartā* (identified with Agni) undergoes in this earthly yet divine fiery womb, restore him back to his previous stage in life, that of a married householder, yet he is now regenerated and invigorated having absolved any past sin and the remnants of all potentially disabling events accumulated throughout his life.

In sharp contrast with Bāstu *pūjā*, human marriages and ordinary *jagyas* are characterized by (a state and process of) matrimonial encompassment that is either beyond direct perception or else is rather temporal, involving only an occasional or a single and short-lived occurrence of sexual congress, gestation, and rebirth respectively. It thus becomes obvious why human marriages and ordinary *jagyas* fall short of the ideal realized in the marriage of the house.

Only within the dense, eternally erotic dynamism established in Bāstu *pūjā*, does death have no foothold, and so the house may come to enjoy a particular state of "immortality," conditioned solely on the wishes of its *ātmā*—the family residing within and the householder with whom it is identified. This, I would like to suggest, is the reason why for the house, a single performance of a *jagya* (Bāstu *pūjā*) is perceived as sufficient, while humans must repeatedly perform *jagyas* throughout their lives.[41] Furthermore, Bāstu *pūjā* emerges as the ideal *jagya*, an unfinished, endless *pūjā* or perhaps even an anti-*pūjā* or *antiritual*, as it engenders a situation where any further ritual action is made redundant.

## A Captivating *Mithuna*

I shall now turn to briefly examine an additional and "nonreproductive" (in the strict sense of the word) aspect of the union of the sexes as it is employed by Thamgharian Brāhmaṇs in the *rekhi*—a two-dimensional projection of the *jagya* and a sacred living map of the cosmos—made of two lines of rice flour, one white and the other red. Villagers' comments make it clear that the *rekhi* is a form of *yantra*, literally an "object serving to hold," an "instrument" or an "engine" as Eliade (1973: 219) has put it. Rather than merely forming a table or the hall for the forthcoming banquet held therein, I would like to suggest, following Gell (1998: 66–95), that it mainly serves as a cognitive spider's *trap* for the divine, royal dining guests.

Like the Trobriand Islanders' canoes, which function as weapons in psychological warfare (Gell 1992: 44–46); Tamil threshold designs (Beck 1976: 220; Gell 1998: 84–86); or Sri Lankan *yantras* (Kapferer 1997: 103), which are able to generate an apotropaic effect, the Thamgharian *rekhi* appears to captivate and thus seize the deities descending to the *jagya*, ensnaring them in its complex pattern. Thus it may seem that (divine) perception is imagined here as having an actual grip on reality.

Moreover, as I would further like to surmise, the capacity of the *rekhi* to bind and hold the great Hindu gods is not solely a corollary of "the fact that once one submits to the allure of the pattern, one is liable to become hooked, or stuck, in it" (Gell 1998: 82), but it also stems from the life breathed into its red and white chalky "ropes" by the nameless union of the male and female it embodies. Hence if like other places of worship in monotheistic religions, many Hindu temples are often employed as tools for captivating the *worshipper* and instilling a sense of divine grandeur in his or her heart, in a Thamgharian *jagya* and *rekhi*, it is the *gods* who are made captive.

## INCORPORATING DIVINE *ŚAKTI*

### *Snān*

Ostensibly, the powerful rite of *snān* is part of the Hindu tendency to imagine a ritual's tripartite progression being registered in chromatic and tactile terms as notably discussed by Beck (1969). She analyses how a *pūjā*'s initial and concluding stages, thought of as cool and white and characterized by inactivity, calmness, and stability, encompass an intermediate stage of red-hot transformative *tapas*. The latter is evidence of an active divine presence, intense *śakti*, fertility, and life, but may also imply danger.

Accordingly, it may be argued that after being heated by the *tapas* involved in the processes of the divine coitus, gestation, and finally rebirth from the burning womb on the *bedi*, the *kartā* and his family, like the *jagya*'s ritual paraphernalia or the manifestations of the gods in various other Hindu *pūjās* (Beck 1969; Babb 1975: 23–24; Eck 1998a: 57) must be cooled prior to the conclusion of the event.[42] This obviation of "thermic excess" (Moreno and Marriott 1989: 156) would thus facilitate the *kartā*'s reintegration into the social body of his *kul* without exposing the other members to uncontrolled, divine, transformative and potentially dangerous heat outside the protective ritual framework.

However, the water of the *snān* is not the cooling water of the village spring but the highly charged, efficacious water of the cosmic *mul kalas*, and it is mainly for this reason that I tend to construe the *snān* as completing and complementing the family's rebirth with a purifying, regenerative bathe in a potent cosmic flow. Although I cannot recall an explicit mention of the *mul kalas*'s water temperature, Thamgharian Brāhmaṇs' notion that ritual is a heat-generating process clearly suggests that this water (like all other paraphernalia involved) cannot remain cold. Moreover, hot water in general is thought to be an effective *auṣadh (auṣadhi,* medicine), and that the *mul kalas* contains not one but *sae* (100) *auṣadh* further supports the present line of interpretation.

Moreover, engendering rebirth seems to be the explicit role of water in a number of Hindu rites, for example, in the Vedic *rājasūya* (Heesterman 1957: 117–22) and the installation ceremony of medieval Hindu kings (Inden 1978: 48–51).[43] Furthermore, in Hindu thought water is usually viewed as a particularly fertile fluid, and like a *kalas*, it is often perceived as embodying a womb (O'Flaherty 1980a: passim, particularly 55; Slusser 1982: 350–53; Good 1983: 234).

The *snān* appears to be yet another expression of the *jagya*'s ritual fractality, since the *mul kalas*, like the *jagya* as a whole and its cardinal elements, is a microcosmic manifestation of the cosmos, albeit of a rather liquid, fluid nature. Put differently, the *snān* is a segment that may be imagined to encapsulate the whole, by sharing the fundamental principals of the latter, that is, maintaining a relationship of *self-similarity* with it, thus in effect, the *snān* reenacts the *jagya* in fluid terms.

## *Prasād* and *Ṭikā*

The hierarchical distribution of the auspicious and highly potent *prasād*, "the indispensable sequel to all acts of worship in popular Hinduism" (Fuller 1992: 74), marks the *jagya*'s conclusion and is one of its most significant stages, a prominent part of what in Marriott's (1976) terms is the ritual transaction of substance-code.[44]

Like the *guru*'s leavings, which are cherished by his followers in Hindu devotional movements (Babb 1991: 72, 185), or the *rebbe*'s leftovers consumed at a Hassidic *tisch* or the wine drunk from his cup (Daryn 1998), Thamgharian Brāhmaṇs feel that *prasād*, a source of "overflowing auspiciousness," is not only imbued with the *śakti* of, but also actually embodies the presence and agency of the divinities with whom it came into direct contact, and thus its distribution is one of the significant moments in all ritual events.

It may come as a surprise that of all the important forms of *prasād* (most of which are internalized by way of consumption), it is particularly the *ṭikā*[45] (attached to the forehead) that is singled out by villagers as possessing the utmost potency and significance; embodying the epitome and ultimate form of *prasād*, it is referred to as *indispensable* for one's existence. This is particularly perplexing since, as discussed in chapter 2, "a Hindu is what he eats" to paraphrase Parry (1985: 613–14), hence one would expect that the best manner for achieving a bio-moral transformation would involve direct internalization of the divine auspicious *śakti* and grace permeating the *prasād*, via consumption.

However, I would like to suggest that the main reason for the special status accorded to the *ṭikā* becomes apparent if we consider its application as an attempt to engender a more profound and holistic effect on a person, extending beyond a relatively short-lived gastro-transformation[46] whose merit may be reduced or obviated by the ensuing consumption of impure or cursed food. Since the brow is not only the most auspicious part of the body but is also likened to Brahmā's heaven, and is the place where a person's whole biography is inscribed in advance, it is there that villagers aspire to make an impact and engender a "global" constructive transformation, that is, to affect a person's entire fortune and life trajectory.[47]

Thamghar is not unique among high-caste Nepalese communities in cherishing the *ṭikā*. Indeed, interpretations of Brāhmaṇ-Chetri life have never failed to recognize the significance and cardinal social implications of the *prasād ṭikā* made of *achetā*.[48] Perhaps the hallmark of Nepalese rituals in general, the *ṭikā* is said to be a prime vehicle for the transfer of divine power and blessing (Bennett 1983: 133), the essential aspect of all Nepalese rituals embodying "the glue of social structure" (Rutter 1993: 409), as well as a principal tool for emphasizing and exposing status differences and epitomizing hierarchy and political authority.[49]

While scholarly discussion focused mainly on the hierarchical aspects of the *ṭikā*'s distribution, which saliently reflect the paramount importance of, even obsession with, social hierarchy in the Brāhmaṇ-Chetri milieu, present concern lies with what may explain or facilitate the *ṭikā*'s personal and social efficacy, allowing it to occupy such a prominent place in the life of high-caste Nepalese. To my mind, the main mystery associated with the Nepalese *ṭikā* is what lies behind the inherent and rather incredible capacity of these tiny red (mixed with *abir*), unbroken rice grains to carry and convey the enormous burden of divine *śakti* and grace discharged during a ritual performance, to its community of worshipers.

In Thamghar, the answer to this question, does not seem to lie with the "*symbolic meaning*' of the *ṭikā* or *achetā*, as this issue appears to be of minor relevance to villagers. Therefore, the range of suggested symbolic interpretation found in a number of studies of Nepalese high-caste communities[50] is of little assistance here, and the Thamgharian perception of the *ṭikā* cannot be illuminated by the meanings they evoke. The *ṭikā* discussed herein is an entity that seems to be situated close to the embodied pole of an embodied representation,[51] hence local concerns lie not with the "symbolic" or aesthetic value it may have, but with the *ṭikā*'s inherent qualities and mainly with its *identity*; as I wish to recall, according to Thamgharian Brāhmaṇs, the *achetā* is the food of the gods, it is *prasād* as well as (an embodied manifestation of) Devī—the general term denoting the goddess.

This is significant as it reveals that the *achetā*, like all the preferred mediums for the transference of *dān* discussed in chapter 3,[52] is a divine female, and thus it is its inherent *śakti* that may account for the *achetā*'s ability to attract, absorb, and carry much of the *jagya*'s massive divine load in a remarkably condensed form. This further illuminates an additional manner by which *achetā* is able to exert its agency, namely the apotropaic effect it generates during certain *pūjās* as mentioned above. Likewise, according to Ramirez (2000a: 110–11), at the height of the Dasaī celebrations in Argakot (central-west Nepal), the (goddess's) victory over the demons comes about when the *kartā* uses his left hand[53] to forcefully throw a handful of *aksata* (*achetā*), believed to embody a substantial force, toward the "locus of the demons." So great is the *aksata*'s power that crossing its trajectory must be avoided, lest one face instant death.

Moreover, sharing the *ṭikā* is in effect the distribution of the goddess's personality and agency, or the dispersal of her multiple embodied manifestations. Accordingly, by adorning their foreheads with a *ṭikā*, villagers literally place the *goddess* (embodied as *achetā*) on their brow and perform *ḍhognu* to her. Instead of the otherwise brief contact with the goddess during a *pūjā*, the *ṭikā* provides an avenue for a prolonged auspicious process of what Gell (1998: 116-21) views as the intersubjective experience of *darśan*, and this far exceeds the "official" conclusion of the *pūjā*. This highly auspicious act of devotion—the exchange of reverence with *śakti* and divine grace—ends only when the goddess's heavenly load is successfully delivered, draining her of all the divine qualities and *śakti* that held her in place, that is, when the *achetā* falls to the ground by itself.

Clearly, for the deities *achetā* is offering (food), and as such it may turn into their auspicious leavings, the epitome of *prasād*. For villagers, *achetā* is a powerful manifestation of the goddess, whose *śakti* is considerably augmented by the *jagya*'s ritual process, thus enabling it to carry her omnipotent cosmic grace. Needless to say, the perceptual resolution between the divine and mortal perspectives that seem to be involved here does not perplex Thamgharian Brāhmaṇs, who do not dare to think of their *pūjās* as (divine) cannibalistic affairs. The goddess's manifestation in the *achetā* may also account for the taboo on the *ṭikā's* consumption by humans (along with the fact that *abir* is inedible).

I would further like to suggest that the inherent capabilities of the *achetā* and particularly its ability to embody a highly efficient "magnet" for merit and divine grace, are related to the particular life stage of the goddess and are profoundly linked to the unbroken rice grains' "biography," a subject that will be further discussed in chapter 7.

## ON RITUAL EFFICACY

The *jagya* highlights two interconnected themes, central not only to contemporary Thamgharian *pūjās* but also to many past and a number of more recent Hindu rites elsewhere. The first is the identification, in effect the merger of the *kartā* (the main actor and subject of a ritual operation) with both the microcosmic enclosure (be it the *jagya* or any other temporary enshrined pavilion or edifice) he fashions as his own cosmic "self-portrait," and with one or several of the major Hindu gods. Among the latter, one may find the primordial Prajāpati-Puruṣa, Agni, Viṣṇu, or Bāstu in Thamghar. More abstractly, these mergers seem to be remarkable manifestations of the general notion in which an actor is identified with his action and the outcome, a cause with its effect.

The second theme, which derives from the former and may ostensibly appear paradoxical, is the notion that the main actor or patron of a ritual performance is acted upon and is finally reborn, or at least is fundamentally transformed and regenerated by the ritual universe he himself creates, embodies, and manipulates.

At this stage, it is necessary to refer back to the prominent Vedic cosmogony and ancient sacrifice discussed in chapter 4, where a number of the earliest expressions of the above dominant themes in Hindu thought and practice can be found. As I wish to recall, "father" Prajāpati was identified with the blemished world he emitted from within his own body via his procreative self-sacrifice, as well as with the sacrifice and with time (the year). Subsequently, the gods fashioned an additional sacrificial process (which became the prototype of the *agnicāyana*), through which they "restored" Prajāpati back to his initial state, and saved and recreated the cosmos in an optimal form. The Vedic sacrifice reenacted Prajāpati's primal sacrifice but in reverse, and was governed by similar identifications: the *jajamāna* was likened to the cosmic Prajāpati, and like him was also identified with the sacrifice, with Agni and with the animal victim.

As ancient Hindu texts overtly note, these fusions and equivalences were made explicit and realized through the correspondence of the sacrificial ground, the altar, and the sacrificial post's proportions to the *jajamāna*'s bodily dimensions.[54] The ultimate aim of the sacrificial process, according to the *Śatapatha Brāhmaṇa*,[55] was the rebirth of the *jajamāna* "into a world made by himself" (Smith 1989: 103).

A similar apparatus of identifications and logic of ritual operation also dominated the ancient year-long *rājasūya*, glossed as the "king-engendering" or "bringing forth the king," which was a Vedic royal sacrificial consecration ceremony (Heesterman 1957: 86, 116–122, 126 and passim). At the *rājasūya*'s heart, stretching his arms to the sky and standing as though he was at the center of the universe, the king impersonated the cosmic pillar and became the pivot of the cosmic rotation. Like Prajāpati who encompassed all beings, the king was thought to incorporate the entire universe and kingdom within his body. The *abhiṣeka* (sacred bath) he underwent was explicitly connected with copulation, reentering the womb, death, and rebirth at the throne—imagined as a cosmic womb located at the earth's navel. Through the complex ritual procedures, both the king and the cosmos were believed to be regenerated and reborn (ibid.).

In a similar vein, the Hindu king in early medieval India (A.D. 700–1200) was viewed as a human *axis mundi*, a microcosmic form of Puruṣa, and like the latter he was taken to be cosubstantial with his world (i.e., his kingdom, land, and people) and embraced the eight divine cosmic guardians within his body. In the *abhiṣeka*, the royal installation ceremony archetypal of all royal rituals of that time, the king along with his entire kingdom were regenerated in a process that reenacted the creation of the Hindu cosmos out of the body of the cosmic Puruṣa. The bathing pavilion and the king's throne were both seen as miniature replicas of the world, and the basic unit employed in their construction was based on the king's forearm length (Inden 1978).[56]

Another relevant example is the monumental temple established by a powerful, successful medieval royal conqueror, which served as a living sign of the "cosmo-moral" order he fashioned, an icon of the new Hindu "chain of being" established by the king and a major index of his power (Inden 1985b). Displaying his ability to construct an imposing temple (or better, a mighty temple complex), the king was thought to complete his *birth* as a new paramount ruler.[57]

The above twin themes dominating the Thamgharian *jagya* are also at the center of Tantra and particularly the Tantric employment of the *maṇḍala* (a meditation diagram and a mesocosm, an energy grid representing the universe), and can be found in Tantric (White 2000: 3–13) and other contemporary Hindu rituals. One apt example of the latter is the south Indian ritual discussed by Beck (1976: 223–26). In this ritual, performed every twelve years, the priest "establishes" Śiva's five faces as the god's cosmic body within the sacred ritual space, identifies with it and identifies Śiva with the original cosmic Brahma in order to effect a cosmic recharging of Śiva's temple images.[58]

On the edge of the hazily demarcated Hindu "ritual realm" and beyond it, Hindu (temple) architecture and town planning provide some of the most salient and literally monumental attempts to realize a merger of body, space, time, and cosmos.[59] Likewise, similarly to Thamgharian Brāhmaṇs, many other Hindus[60] evince considerable concern, at times even preoccupation, with personal and cosmic compatibility, and attempt to align themselves and arrange the world around them according to the "cosmic design."

## Micro- and Macrocosms

Following (mainly) ancient Hindu texts, scholars note that through the (re)establishment of a firm connection between man (micro) and the universe (macrocosm), Hindus aim to transform and channel what are otherwise perhaps completely unpredictable cosmic influences on an individual's well-being into more ordered avenues to achieve particular objectives. More specifically, it is generally claimed that the identifications, correspondences, and connections established within the ritual realm are at the root of its efficacy.

For example, according to Renou (1953: 18, 27–33, 1957: 84), the aim of the entire Vedic thought is to formulate *upaniṣads* (literally equations, correspondences) that can be manipulated within the sacrifice in order to gain purely material blessing and prolong the *jajamāna*'s present life.

Smith (1989: passim and particularly 30–38, 53–54, 68–69, 72–81) provides a comprehensive comparative discussion of much of the available scholarly commentary regarding Vedic ritualism and its system of equations, equivalences, homologies, and connections (or *bandhus*, *upaniṣads*, and *nidānas* in Sanskrit). He concludes that these serve as the basis for ritual action, that is, they are manipulated within the sacrifice in order to effect a change in the world, transform its entities and engender the rebirth of the *jajamāna*.

Heesterman (1957: 76, 116–22) views the manipulation of the *bandhu* connections between micro- and macrocosms as the source of the Vedic *rājasūya*'s efficacy, where the king is (re)born out of a marriagelike bond he himself establishes at his *abhiṣekah*, considered to be his womb. A similar line of argumentation, though via a somewhat different conceptual approach, is taken by Handelman and Shulman (1997: passim and particularly 61–111), who interpret a number of ritual (Vedic) and mythical (epic) dice games as simulations or "non-isomorphic limited models" of the cosmos, which are capable of affecting it, as well as the human players (e.g., the king), via the manipulation of the *bandhu* connections that *tie* them together.[61]

From a slightly different angle, following Hocart,[62] Inden (1978: 32–33) views the efficacy of Hindu royal rituals as a direct consequence of Hindu cosmogony where the cosmic man generated the phenomenal elements of the world out of his *own* body and thus united the cosmos within one single order of being. Accordingly, "a variety of beings and phenomena, considered elsewhere to belong to discrete realms, repeatedly and often dramatically influence others," as may be exemplified by the ability of man to influence the course of celestial events and vice versa. Consequently the king, who is a microcosmic part of the universe yet contains the whole within himself, could control and influence the entire cosmos, as its "effective symbol" (ibid.).

In another relevant example mentioned above, Inden (1985b) discusses the successful display of a medieval royal conqueror's ability to construct an imposing temple, serving as an index of his power, which appears to engender his *rebirth*.

With at least one notable exception, namely the aforementioned medieval temple (Inden 1985b), it may be difficult to relate explicit "ritual efficacy" to most Hindu architecture and town planning.[63] Nevertheless, it seems legitimate to surmise that the impressive endeavor to gain and display the connection and alignment of the Hindu urban landscape with the cosmos stems from the notion that since everyday life draws man's mundane world into an entropic vortex, it must be resynchronized with the divine ideal cosmos in order to maintain the mutually affective, primordial divine harmony.

Clearly this is but a higher scale reflection of similar aspirations still prevalent among contemporary Hindus, who maintain that alignment with the cosmos will enable them to "make the most beneficial use of shifting lines of cosmic force" (Beck 1976: 214), or that restoring the lost primordial cosmic "intersubstantial" equilibrium is the key for health and well-being (Daniel 1984: 6–7 and passim). Thamgharian Brāhmaṇs often express similar notions in terms of the need to avoid *alacchin*, glossed as inauspiciousness, *pāp*, *dos*, and ill fortune, acting instead with solid "unbroken" *sāit*, which may not only influence a man's destiny in the present life, but also improve the chances of arriving in *swarga* after death.

All the aforementioned valid lines of explanation may be applied, in varying degrees, to explain the efficacy of the Thamgharian *jagya*. Yet to a large extent, the question of *why* the connections, partial identity, or resemblance established between certain manifestations of micro- and macrocosms are imagined to *enable* their mutual affectivity, remains open. I would like to propose that at least one answer, perhaps the key to the issue at stake, is encapsulated within Gell's (1998) discussion of "the distributed person" and his insightful analysis of image sorcery,[64] to which I shall briefly refer below.

## The *Jagya* as the *Kartā*'s Distributed Personality

Gell (1998: passim, particularly 96–109, 221–58)[65] argues that the efficacy of image sorcery is based on the formation of a circuitous causal pathway in which the victim is ultimately made the victim of his *own* agency. This stems from the fact that the process by which a sorcerer makes a victim's representation inevitably involves the *impression of the latter's agency* on the image.

The core of Gell's argument is that agency may be mediated via images of a person or the objects created by him, that for social persons, existence is not limited to one's corporeal body but is present in everything that bears witness to one's agency, and that both persons and objects may be part of an effectual interaction within the realm of social relations. Accordingly:

> The "magical" aspect of volt [image] sorcery is only an epiphenomenon of our failure to identify sufficiently with sorcerers and their victims, our estrangement from them, not the result of their enslavement by superstitious beliefs entirely different from our own. (Gell 1998: 102–3)

Cause and effect are brought so closely together in image sorcery that the causal nexus linking the image to the person is made *reversible*; in Gell's own terms, the victim becomes both an agent and a patient (a recipient, the one acted upon), that is, the victim of his own agency (as shown in figure 6.1, which is adapted from Gell 1998: 103). This, he contends, is the major basis for the general tendency to attribute personlike characteristics to objects and images, and for regarding them as the sources of and targets for social agency.[66]

More generally, Gell argues that a person and a person's mind and personality should not be viewed as confined to particular spatio-temporal coordinates such as those of the body, but that these consist of the range of biographical events, material objects, various traces, and leavings, which can be

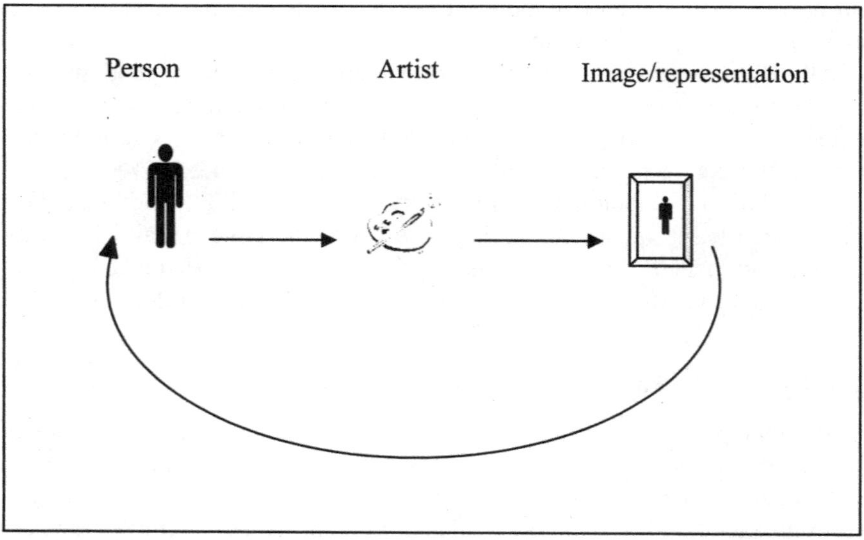


**Figure 6.1. Sorcery's reversible causal nexus of agency according to Gell**

attributed to that person; throughout his lifetime, a person, like an artist, distributes his personality and disperses his agency in the milieu beyond his physical bodily boundaries.

To my mind, the relevance of Gell's theory to Hindu life, and particularly to perceptions of ritual and general causality is so striking that it may appear as though he wrote it with the Vedic cosmogonies and sacrifice in mind. To demonstrate this we must recall that Vedic texts often narrate cosmic creation as originating from the primordial Puruṣa-Prajāpati, who *distributed* his body and personality in the universe by a procreative and (self) sacrificial process. Indeed, the authors of the Vedas often likened the primordial creator to *an artist*: Prajāpati is the demiurge Visvakarman, "he who makes everything," and creation is explicitly spoken of in terms of the procreative act of an artificer or an *artist* (Renou 1953: 25, 1957: 73–74).[67]

Moreover, unlike the Judeo-Christian tradition where God creates the world "*ex nihilo* or by breathing into formless matter" (Inden 1978: 32), in Hindu cosmology the primordial Prajāpati-Puruṣa does not simply distribute his personality through production (as one of Gell's artists) or (pro)creation, but dismembers his *own* being, emitting the world and its numerous creatures out of his *own* body, which thus becomes empty, drained and completely exhausted. Consequently, all the elements of the universe are not merely *representations* impressed with the cosmic artist's agency but are in fact fragments or segments of the creator himself, his body, personality, and agency, clearly suggesting the presence, or at least the potential of an even stronger, mutual, causal connection than the one envisaged by Gell.

This brief discussion demonstrates that Hindus have always evinced sensitivity to considerations of agency and its diffusion in the milieu, similarly to what Gell (1998) identifies in many other places the world over. This may be epitomized by the ancient declarations that "that which is I is he, that which is he is I," or in the words of the *Bhagavad Gītā*:[68] "They are in me and I in them" (Daniélou 1964: 44). Furthermore, I would like to suggest that this Hindu sensitivity gave rise to and influenced a plethora of other ideas, such as the Vedic notion that the gods produce (procreate) their parents (Renou 1957: 78),[69] the deep-seated Hindu belief that a father is not simply *like* his son but *is* his son (Shulman 1993: 92) or that he is (re)born in his son (Trawick 1990: 158; Malamoud 1996: 104), as well as the doctrine of *karma* itself, where "actors' particular natures are thought to be *results* as well as *causes* of their particular actions" (Marriott 1976: 109).[70] In addition, these notions of personality and agency distribution seem to be but different expressions of the Hindu view that *dān* embodies part of the person who hands it over (Mauss 1967: 54), and the fundamental cognition that Marriott and Inden (1977)[71] aptly term *dividualism*.[72]

My basic argument is that these are all salient examples of the entrenched Hindu perception of the roundabout causal pathway eloquently exposed by Gell, where cause and effect become *one*. Hence, it is perhaps little wonder that many Hindus seem to imagine that the intense causal nexus and the mutual influence and remarkable efficacy found within the *ritual* realm are also found to varying and often more modest degrees, *throughout* their world.

A consistent picture begins to emerge: following Prajāpati, the Vedic *jajamāna*, the Hindu king and the Thamgharian *kartā* are all *artists* distributing their personality and agency in the world, in the form of ritual arenas or temporary (and perhaps also permanent) temples, which may thus exercise agency, in reverse, on their creators. The *kartā* (pro)creates the *jagya* as his cosmic self-portrait by the enactment of a divine matrimonial encompassment and is reborn in its hub by his own agency via a circuitous causal effect; put differently, via the *jagya* a Thamgharian *kartā* may skillfully perform an auspicious act of self or "DIY sorcery."

Yet as may have been implied above, the *jagya*'s (and other Hindu ritual arenas discussed above) dynamics of causality and its agency trajectories are more complex than the rather simple two-way mutual effect Gell (ibid.) exposes in image sorcery.[73] As shown in figure 6.2, the *jagya* obviously has *two* artists and thus two prototypical[74] sources of agency. The *kartā* fashions the *jagya* as his self-image, proportional to his dimensions, but he embodies only

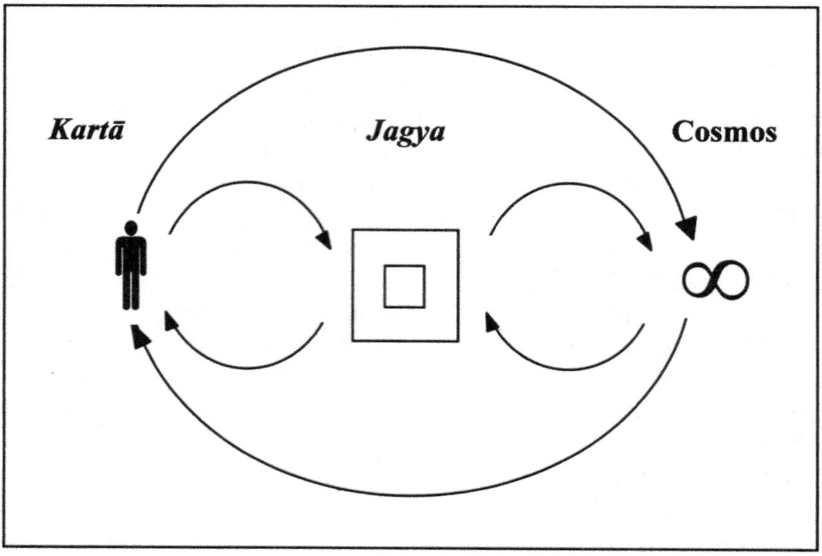


**Figure 6.2. The *jagya*'s causal nexus**

one of its primary "causes," the other being the macrocosm, which is the main source of the *jagya*'s square shape. Therefore, the *jagya* can be thought of as the embodied representation or distributed agency of both the *kartā* and the cosmos alike.

Three main circuitous causal pathways can be identified within the *jagya*, compared with the single, reversible causal nexus of image sorcery: the first and the second connect *kartā* and *jagya*, and cosmos and *jagya* respectively, as discussed above. The third and, to my mind, also the crucial causal nexus exists between the *kartā* (via the *mesocosm* of the *jagya*) and the macrocosm. Here the *kartā* operates upon and manipulates the *jagya* as a sorcerer upon a representation of his victim, and thus "bewitches" the macrocosm in order to ultimately become the true "*victim*" of his actions, engender a fundamental effect, and transform his *own* being. The regeneration of the cosmos, which seems an important by-product in a number of ancient Hindu rituals, is clearly not very high on Thamgharian Brāhmaṇs' list of the *jagya*'s *phal* or "fruits."

More generally, I would like to suggest that in order for the *jagya* and most other Hindu rituals discussed thus far to be efficacious, the Hindu cosmos *must* be imagined as a fractal and *maintain self-similarity across its scales*, manifested as micro-, meso-, and macrocosms. It is this effectiveness across scale, inherent within a fractal cosmology, that Thamgharian Brāhmaṇs, like Hindus in general, deem so auspicious, seek, and are preoccupied with when they build their ritual arenas, temples, houses, and cities; perform complex *pūjās*; or simply keep their bodies aligned with the cosmic design. I further believe that a "fractal efficacy" similar to the one discussed above may be found in other rituals and cosmologies elsewhere, mainly in "premodern" societies where people retain a fractal cosmology. Not only does it form the *key* for understanding ritual efficacy, but it may also suggest why so many people in various cultures the world over tend to imagine themselves and the universe they inhabit as parts of a cosmic fractal.

The *jagya* provides an actual, ritual, and mental locus, as well as a time frame in which a Thamgharian Brāhmaṇs may actively bring their imagined cosmic fractal to life; it seems to serve as an interactive medium in which a whole set of multidimensional images are dramatically displayed and manipulated, and thus a certain outcome is anticipated, crystallized in people's minds, and finally imagined to actually take place.

## A CONCLUDING REMARK

The marriage of Bāstu and Bhūme in the fractalic gendered universe of the *jagya* is a striking diversion from the general nonreproductive mode of contemporary Hindu divine marriages. Via a complex operation upon what is, ideally, a perennially procreative cosmic matrimonial encompassment and a fertile, fiery ritual womb, Thamgharian Brāhmaṇs attempt to eliminate death from the texture of experience, prolong their life, and procreate (conditional) immortality within their terrestrial plane. Thus the *jagya* proves to be an exception to the common view that:

> [I]n Hindu thought creation is possible only against the background of chaos and death; life is born out of darkness; Death himself is the creator. . . . Chaos gives way to order, and to a new birth; but this new life needs still to be nourished by the violent forces from which it was born. (Shulman 1980: 42–43)

The *jagya* is also a remarkable example of the "conceptual overflow," the elasticity of categories and the collapse of boundaries that often characterize the Hindu context, yet what seems to be most arresting is that time, which is trapped within space at the *jagya*'s inauguration (*purbāṅga*), appears to infringe its margins *internally*. Put differently, by creating the potentiality for the perpetuation of life in an ongoing, endless, erotic ritual performance, the *jagya* can be viewed as an attempt to encapsulate *infinity*.

The Thamgharian image of matrimonial encompassment is not only a dimension of an imagined fractal cosmology, an indispensable part of "ritual technology" and general "cosmic efficacy," but it also emerges as the ideal modality of existence and a mental frame for imagining, creating, arranging, and operating in the world, as well as what renders it an auspicious, eternal, "complete," and safe place to dwell in. This will be further demonstrated via the analysis of rice agriculture in the ensuing chapter.

## NOTES

1. According to Turner (1996: 206, 522, 643), *jagya* derives from *yajña* (the Vedic sacrifice) and is the place where the bride and groom sit during the marriage ceremony. *Hom* is the Nepalese word for the Sanskrit *homa* (the central part of the Vedic sacrifice where the offering, be it animals or vegetables, were thrown into the fire), *bedi* is an altar and *jajmān* comes from *jajamāna* (the Vedic sacrificer).
2. On this see also Witzel (1987: 432–34).
3. For example, the animal sacrifices in the Dasaī festival are complemented by the "slaughter" of a pumpkin in the shape of and likened to a buffalo to ensure that the sacrifice is *pūrā* ("complete").
4. Again, this is not without exception. The midnight of *Kalrātri* during the Dasaī festival is the sole occasion where a goat's flesh and blood are offered into the fire at the regional Kāli temple, located on the hill above Thamghar.

5. Thamgharians expressed disgust at my suggestion that, as is often the case with other Hindus (e.g., Das 1976a: 255 and Parry 1994: 151–90), the cremation of the body after death is in fact the final sacrifice (*bali*).

6. Beck (1969: 562); Khare (1976a: 3, 1976b: 3–5); Marriott (1976: 110); Malamoud (1996: 23–53, esp. 24).

7. The term *kāco* (*kācho*) is employed to refer to both uncooked food and a young man who is not sexually mature.

8. Unless cooked with notable pollution antidotes, namely pure milk or (preferably) *ghyu*.

9. Not to mention more general anthropological discussions of sacrifice such as Firth (1963) or Turner (1977).

10. Bennett (1983: 77) provides a detailed drawing of a wedding *jagge* (*jagya*), yet this is accompanied by only scant commentary about its meaning and significance. See also (Gray 1979: 90), Enslin (1990: 134–37), Rutter (1993: 290–358) and Ishii (1999: 42).

11. I agree with Lewis (1980) who advocates that one must obviously exercise caution while attempting to interpret ritual actions, particularly when those who perform them cannot clearly explain and articulate what they are doing and why. I also share with Das (1991) the contention that there are clear limitations to any attempt to employ the comparative method for interpretation of almost anything within the "amorphous mass" that Hinduism is often said to comprise. Yet, as I believe and practice throughout this book, this should not mean renouncing the practice of *etic* interpretation. On the contrary, the exegetic and comparative exercise is not only potentially illuminating and fascinating but is also worthwhile in its own right, and should form an integral part of any anthropological or Indological scholarly work.

12. Daniélou (1964: 132).

13. Kuiper (1975); Daniel (1984: 12); Handelman (1987: 133).

14. See, for example, Daniélou (1964: 352–53) and Snodgrass (1994: 133).

15. My use of "gender" follows Strathern (1988: ix), referring to it as "those categorizations of persons, artifacts, events, sequences, and so on which draw upon sexual imagery."

16. See also Beck (1976: 237–40), Inden (1985b), Fuller (1988: 51–52), Knipe (1988: 118–19), and Eck (1998a: 59–63).

17. This is despite the general scholarly interpretation of the Hindu temple as a gendered living entity—the embodiment of Puruṣa (the primordial male in Vedic cosmogony) possessing a darkened sanctuary, namely, the *Garbhagṛha* (a womb). I will return to this issue and discuss the Hindu temple in more detail in chapter 7.

18. Such as *godān*, *kanyādān*, and *cāmal*.

19. See also Eck (1981, 1998b).

20. A salient example may be the south Indian Tamil *Poṅkal* festival, where the sacred *poṅkal* rice is cooked within a pot thought of as the goddess's womb containing divine semen, as eloquently analyzed by Good (1983: 234–40).

21. This myth was discussed in more detail in chapter 3.

22. For an additional example, see Kapadia (1995: 211–12).

23. Only a newlywed couple may sleep together in a special, tiny windowless room built for them on the house's veranda, and this only lasts until the birth of their first child. Otherwise, parents sleep separately, alongside their children.

24. According to householders' meager testimony, sexual intercourse is usually limited to *brief* encounters at night on the sole initiative of the husband, and should be confined to the privacy of the first floor of the house. Watching such an act is regarded as a grave *pāp*.

25. Namely, Fuller (1980), Shulman (1980), and Good (1989).

26. Mainly involved in and responsible for the creation of houses and the production of the staple—rice.

27. This interpretation gains confirmation from the close association Hindu imagination makes between a tree trunk and fertility, particularly male fertility (Handelman 1987: 144–45), and its often assumed role of a male partner, husband, or lover (Beck 1981: 120–22).

28. Five *liṅgas* are employed for the inner *jagya* and nineteen are used for the external one.

29. See also Bennett (1983: 79) and Rutter (1993: 313).

30. Similar procedures take place during the establishment of a *jagya* in other Brāhmaṇ-Chetri communities (Bennett 1983: 64; Rutter 1993: 301–2), as well as in central India (Babb 1975: 83). Good (1983: 236) mentions a corresponding south Indian notion implied by the Tamil practice of using only the rice fields' pure water for cooking the sanctified *poṅkal*.

31. As will be recalled, the other three deeply intertwined fractal dimensions are: (1) a living entity with an inherent energetic, radiating, burning core of life (fire); (2) temple-deity; and (3) body-*ātmā*.

32. The fractality of the *jagya* is not confined to the dimension of matrimonial encompassment and may also be found, for example, in its complex liturgy. However, these aspects are of secondary relevance at present and will not be elaborated upon here.

33. Namely, the *Puruṣa Sūkta* hymn in Renou (1961: 64–65).

34. Kramrisch (1976: 357–61).

35. As in other *jagyas* the *khāldo*'s dimensions (length, width, and depth) should each be equal to the length of the *kartā*'s hand or forearm.

36. Namely, *kul deutā* (clan deity) *pūjās*.

37. This is similar to a *bartoon* or marriage, where the father is the *kartā* but the main subject (and actor) in the *pūjā* is his son (or daughter).

38. Bāstu *liṅga* is made of a strong elongated unpolished *sāl* tree branch, but I have also seen it embodied by a *sāl* trunk carved by the *kartā* into a long, square pole (see photograph 6.3).

39. The priest takes some soil from the bottom of the *khāldo* and places it at the center of the would-be *rekhi* as its *bedi*, thus establishing a strong bond—in effect an identification—between the two.

40. This is close to the south Indian Tamil perception of union as birth (Trawick-Egnor 1978: 87–90).

41. And it should be performed for them (in the form of a *purān*) at least once after death.

42. One is reminded here of the rite that marked the end of the Vedic *soma* sacrifice where the *jajamāna*, his wife and all ritual paraphernalia were plunged into water (Hubert and Mauss 1964: 47, 139 nn307–8).

43. In striking similarity to the Thamgharian *snān*, this included the affusion of the king, employing a golden strainer typically possessing one hundred holes, with water containing various perfumes, medicinal herbs, fruit and flowers—in short, the entire cosmic powers (Inden 1978: 48–51).

44. In the *jagya*, this includes (among others) offering food to the deities, giving away *dān*, *dos*, and *dakṣiṇā* and the shedding of impurities.

45. As will be recalled, Thamgharian Brāhmaṇs employ many different types of *ṭikā*, and the following discussion is confined to the *ṭikā* (made of *achetā*) given as *prasād* at the end of every *pūjā*, which is of paramount importance and is probably the most significant of all village *ṭikās*.

46. Let alone the employment of other forms of *prasād* such as protective threads (*rakṣā-bandhan*) or ribbons that are transitory in nature and have only a limited effect.

47. See also Cameron (1998: 243–44, n15).

48. *Achetā* (*aksate* or *akshat*) also appears to be the general term for consecrated rice grains used in Indian Hindu rituals, as may be exemplified by Hanchett (1988) and Carter (1995: 133–35), and is also mentioned by Humphrey and Laidlaw (1994: 205) regarding Jains.

49. Bennett (1983: 150–64); Gray (1995: 66–74); Ramirez (2000a: 186–111).

50. For example, Kondos (1982: 278–79, note 27), Bennett (1983: 271), Rutter (1993: 406–16) and Gray (1995: 64–67).

51. See chapter 5.

52. As will be recalled, the most prominent ones are the holy cow (*godān*), *kanyā* (*dān*), *cāmal* grains, coins, and the river.

53. In Thamghar, this would be thought of as a sign of involvement in an inauspicious ritual action.

54. Kramrisch (1976: 43); Burckhardt (1986: 20–21); Snodgrass (1994: 184–85); Malamoud (1996: 84).

55. 6.2.2.7.

56. The two interconnected themes discussed above, appear to be only partially present in the late medieval (1561) and contemporary (1956, 1975) coronations of Nepalese Hindu kings analyzed by Witzel (1987).

57. I would like to draw attention to the shift in the notion of a ritual womb and rebirth evident here: the latter, explicitly celebrated in the Vedic sacrifice (including the *rājasūya*), can no longer be found in the transformative medieval royal rituals except in a rather implicit and abstract form.

58. See also Witzel (1992) who, analyzing a Hindu Newāri ritual from the Katmandu valley, eloquently argues that the ritual *mesocosm* connects with and enables the management of the macrocosm (the deities) by the microcosm (man).

59. Volwahsen (1969); Michell (1977); Kramrisch (1976); Burckhardt (1986); Hudson (1993) and Snodgrass (1994). About the cosmic plan of the city of Bhaktapur in the Katmandu valley see Levy (1992). Similar ideas are widespread throughout south and southeast Asia, and one notable example from the margins of the Hindu

world is found in Geertz's (1980) treatise on the nineteenth-century Balinese theatre state.

60. Beck (1976); Das (1976a: 252); Daniel (1984).

61. This mainly follows Handelman's earlier work on ritual (1990: 3–41), in which he argues that among non-Western peoples, ritual efficacy and ideas of causality often depend on correspondences of analogy, homology, or iconicity, while the ritual is perceived as a microcosmic model of the lived-in world.

62. Hocart (1970: 60–71) emphasizes that the identity of micro- (man) and macro- (world) cosmoses is established "so that by acting on one you can act on another" (ibid. 64).

63. This is hard to establish since ancient architectural manuals mainly explicate *how* to build in a particular manner and are far less concerned with *why*.

64. Later and throughout his book, Gell prefers to employ the term "*volt* sorcery" instead of "image sorcery" (Gell 1998: 32), yet it is the latter which seems to be in more common general use and will thus be employed here.

65. This is part of his more general endeavor to delineate an anthropological theory of art (where, in fact, according to Gell (1998: 7) "*anything* whatsoever could, conceivably, be an art object from the anthropological point of view, including living persons") [my emphasis], which, rather than focusing on the aesthetic properties, values, and meanings represented in art, is centered around action and agency, or in Gell's words, is about "social relations in the vicinity of objects mediating social agency" (ibid.).

66. Here Gell appears to pursue a very similar line to the 'actor network theory' (Latour 1991, 1993; Law 1991), which belongs to the sociology of science and technology. I prefer to follow Gell's theory, which seems to be much closer in spirit, terminology, and subject matter to the present discussion.

67. In the *soma* sacrifice (*Śatapatha Brāhmaṇa* 3.2.1.5.) it is the *jajamāna* who explicitly likens himself to an *artist* (Smith 1989: 76–77).

68. 9.29.

69. See Malamoud (1996: 186) and Miller (1985: 207).

70. My emphasis.

71. See also Marriott (1976).

72. What may seem like the Melanesian "reincarnation" of this idea is analyzed by Strathern (1988), who finds Melanesian persons to be *partibles*, that is, capable of detaching parts of themselves to others.

73. See figure 6.1.

74. In Gell's terms, a prototype is the entity represented in the index ("material entity"—a thing, an object, or image).

# III

# RICE: FROM MUST TO TRUST—AN ONTOGENY OF RICE

This book began with an exploration of how the dynamics of kinship and social relationships are interwoven with Thamgharian Brāhmaṇs' imagining of marriage and the conjugal bond, and the manner in which attaining the state of "completeness" via matrimonial encompassment triggers complex dialectics between a variety of images and perspectives, with dire consequences for Thamghar's fragile social fabric.

In light of the full potential and import of the construct of matrimonial encompassment as unfolded in part II, the striking shortcomings of the real vis-à-vis the ideal clearly do not pertain solely to the realms of kinship and social relationships, but are also equally applicable to the ritual domain. Notwithstanding the considerable effort villagers put into embodying and manipulating a divine matrimonial encompassment and its assumed potential within the village *jagya*, they remain unable to convert their personal and terrestrial existence, punctuated by illness and death, into an everlasting, auspicious, and perpetually revitalizing *hom*. In the present and final part of this work, the prolonged and complex process of the rice-cultivation cycle will emerge as the ultimate path villagers follow for rectifying their emotional landscape and social reality, in what appears as an annual regeneration of the entire community.

Chapter 7 will analyze rice cultivation as an agricultural *jagya*, designed to harness the divine cosmic forces for the mass production of virginal rice, as a primary example of the employment of matrimonial encompassment as a template for creation and production. This will lay the foundations for advancing the argument of chapter 8, namely that in addition to rice, the Thamgharian fields appear to yield trust, a particularly scarce and prized commodity in

Thamghar, thus shedding additional light on the Thamgharian social landscape portrayed in chapter 2. This will be demonstrated through a detailed analysis of the inversion event of *dāī hālnu*, the final task of threshing the rice and bringing it home. In this manner, the final thread in the detailed examination of a number of the cardinal realms of life where the Thamgharian notion of a male encompassing a female is manifested, and has a profound import, will be tied in place. This will offer additional support to one of the overall arguments of the present work, namely, that this fractalic mental construct dominating Thamgharian life plays a leading role in local perceptions of self and the other, action and interaction, society, and the cosmos.

The gender-sensitive analysis of both the village *jagya* (in chapter 6) and the *kunyā*, its three-dimensional agricultural projection (in chapter 7), greatly illuminate the architectural development of the north Indian Śikara Hindu temple, providing an additional, gendered perspective for its comprehension as discussed in the final section of chapter 7. This provides a final substantiation to one of this book's main contentions, that the gendered fractalic image of matrimonial encompassment is not a Thamgharian peculiarity but embodies a previously neglected Hindu fractal dimension, which may considerably enhance our understanding of general Hindu thought and practice.

# 7

# An Agricultural *Jagya*

## INTRODUCTION

This chapter will explore how Thamgharians employ the template of matrimonial encompassment for the production of their staple and measure of wealth and prestige—the organic gold of rice—in an agricultural *jagya* accommodating a fireless *hom*. For simplicity and in accord with the natural progression of the cultivation process, the discussion is divided into a number of parts.

Following brief introductory remarks about the general perception of rice agriculture, the fields, and cultivation, I will focus on the initial stages of the agricultural work, namely *biu rākhnu* (*rākhne*, sowing) and *dhān ropnu* (*ropne*, transplantation).[1] An examination of the *kunyā*, a three-dimensional solid projection of a village *jagya* made of rice, and an analysis of its inherent transformative process will follow. Next, the examination of the gender of the rice grain and the nonmarital encompassment it embodies will demonstrate how the notion of the gendered rice grain and the latter's "biography" may serve to illuminate the prominent role and immense local significance of the auspicious (*prasād*) *ṭikā* made of *achetā*. Finally, I will suggest that the analysis of both the village and agricultural *jagyas*, taken together, may provide an additional perspective for the comprehension of the north Indian Hindu temple.

### A "Culture of Rice"

Rice is considered the most important crop in Nepal, as it is in Asia as a whole (Chandler 1979: 9), and is the staple and prototypical food, not solely for

*Ripened rice.*

Brāhmaṇ-Chetris[2], but for the majority of the population.[3] In contrast with most Tibeto-Burman-speaking people for whom, as exemplified by the Gurungs (Macfarlane 1976: 25–34), rice cultivation is a relatively recent pursuit, high-caste Hindus are thought to have introduced this ancient art to central Nepal upon their arrival from India (Aubriot 1997: 69, 534; Ramirez 2000b: 104).

Arguably, high-caste Nepalese in general and Thamgharian Brāhmaṇs in particular, as well as their Newāri *jyāpu* (farmers) counterparts,[4] share what may be called "a culture of rice." By this I mean that for them rice is not only the staple quantitatively but also qualitatively,[5] that is, in terms of its central meaning and significance for personal, ritual, and social life.[6]

As is often the case in places where it is a staple, for example in north and northeast India,[7] Japan (Ohnuki-Tierney 1993: 40–42),[8] or Bali (Howe 1991: 457), rice is also the defining element of the high-caste Nepalese "meal."[9] Moreover, in Thamghar, as Parry (1985: 613) notes regarding north India (ibid.), the fasting required prior to and during the performance of almost any *pūjā*, does not normally entail refraining from *all* food, but solely implies the abstention from eating cooked rice. Hence, in view of the central significance of rice, it may be surprising that the Brāhmaṇ-Chetri cycle of rice cultivation attracted only modest scholarly attention; more often than not it is mentioned only briefly or in passing, with the notable exceptions of Rutter (1993: 359–78)[10] and Aubriot (1997).[11]

| Crops | Baiśākh (Apr – May) | Jeṭh (May – June) | Asār (June – July) | Sāun (July – Aug) | Bhadau (Aug – Sep) | Asoj (Sep – Oct) | Kārtik (Oct – Nov) | Maṅgsir (Nov – Dec) | Pūs (Dec – Jan) | Māgh (Jan – Feb) | Phāgun (Feb – Mar) | Cait (Mar – Apr) |
|---|---|---|---|---|---|---|---|---|---|---|---|---|
| **1. Rice** | | | | | | | | | | | | |
| ***Sali / Maṅgsire*** | | Sowing – *biu rāknu* | Transplantation – *dhān ropnu* | | Hoeing / Fertilizing | Weeding / Irrigation | | Harvesting – *kunyā+ dāī hālnu* | | | | |
| ***Asāre*** (Spring rice) | Hoeing | Harvesting | | | | | | | | | Sowing – *biu rāknu* | Transplantation – *dhān ropnu* |
| ***Ghaiyā*** (dry rice) | Hoeing | | 2nd Hoeing | | Harvesting | | | | | | | Transplantation – *dhān ropnu* |
| **2. Corn** | Hoeing | 2nd Hoeing | Harvesting | | | | | | | Fertilizing | Plantation | |
| **3. Wheat** | | | | | | | | Plantation | | Fertilizing | Harvesting | |
| **4. Millet** | | | Sowing | Transplantation | Hoeing | | Harvesting | | | | | |
| **5. Potato/ Onion** | | | | | | | Fertilizing | Plantation | Irrigation | Fertilizing | Harvesting | |

Figure 7.1. Annual cycle of main village crops

Rice cultivation is not only the longest agricultural process found in Thamghar, as clearly shown in figure 7.1, but is also the most complex and labor-intensive of all agricultural tasks, upon which almost every village family solely depends for its livelihood. Its various stages are loaded with rich ritualistic activity and numerous technicalities, and its local importance and significance cannot be overestimated.

It is common village knowledge that *sāli* (*maṅgsire*) does not constitute a particular *jāt* (type, here botanical genus) of rice. Rather, it refers to the inherent properties of the grains growing in a flooded field during the auspicious monsoon season, which undergo the vital ritual performances involved in what, as I shall argue below, is an agricultural *jagya*. This is consistent with the often expressed Thamgharian view that rain in particular and flowing water in general are highly pure and energetic symbols of *pragati*, that is, prosperity and continuous progress. It is also in accord with the general Hindu perspective of rain, a river and water in general as an auspicious, sexual fluid (O'Flaherty 1980a).

The elevated, revered status accorded to *sāli*, as Thamgharians steadfastly emphasize, is not extended to other semidry or dry rice crops, the rice grown at other times of the year, or to certain inherently *biṭulo* types of rice (all of which are negligible in terms of yield and importance), which are solely employed for consumption.

According to Gaborieau (1982), the Indo-Nepalese (Brāhmaṇ-Chetris included) year oscillates between an ordered, auspicious period and a highly dangerous, inauspicious time of evil and chaos. The latter, he argues, comprises the four months of the *cāturmāsa*[12] when Viṣṇu withdraws to sleep, supposedly leaving the earth to the evil rule of demons. This period, heralded by danger and disorder, is said to end only with the triumphant return of Viṣṇu and the other gods following the victory of Devī over the demons, celebrated in the festival of Dasaī.[13]

As Gaborieau (1982: 15–16) notes, the *cāturmāsa* overlaps with both the monsoon rains (normally lasting from late June to early October) and the rice cultivation season. Although, in Thamghar at least, the latter stretches well beyond the limited period of the *cāturmāsa*,[14] the aforementioned correspondence seems to suggest that high-caste Nepalese perceive most of or even the entire cultivation season (excluding the harvest) as a particularly inauspicious and dangerous period

However, on the basis of his comparative research among Brāhmaṇ-Chetri, Newār, and Mithila communities, Ishii (1993 passim, particularly 42–49) cogently challenges Gaborieau's thesis, drawing much of the latter's argument into question. In a royal manuscript dating from Dravya (Drabya) Saha's (the founder of the kingdom of Gorkhā) coronation in 1561, the monsoon season is depicted as an auspicious one, "the time of the thriving of the year" (Witzel 1987: 420).

Thamgharian views seem to be in accord with the latter; while familiar with the concept of the *cāturmāsa* itself, both priests and lay villagers alike were totally unaware of any dangerous or inauspicious aspects associated with it. Concomitantly, they could find no significant connection between the *cāturmāsa* and rice agriculture. As Megh-Nath Sharma (35) explained, the *barkhā* (rainy season), which sets the time frame for rice cultivation, is the hardest yet also "the best," most joyous, and happiest part of the year.

## The Rice Fields

Before proceeding to discuss the agricultural *jagya*, a few words about its "ritual arena"—the rice fields—are necessary. The spatial organization of the latter seems to reflect the fact that unlike all other village tasks or agricultural procedures, rice cultivation is never the project of the nuclear or even the extended family, but requires communal effort and collaboration on various levels. Indeed, there are no isolated rice terraces in Thamghar. Cooperation is particularly crucial for irrigation, which is regulated on an egalitarian basis, as well as for the major agricultural procedures, mainly the transplantation and harvest, normally regulated via *parma*.[15]

Situated on the relatively moderate slopes of the Thamgharian ridge, the numerous rice terraces are clustered together, with no gaps or visible boundaries to demarcate plots belonging to individual households. Each cluster constitutes one named "field," about six of which surround the village, forming islands of cultivated land amid the forest engulfing it from three sides.[16] Each field's complex mosaic of plots is well inscribed in a sort of "mental map" that all villagers (with the exception of young children) appear to retain in their minds. During the first stages of cultivation, the sky is often reflected as if "captured" within the numerous flooded terraces, which may be one reason why a rice field/plot is locally called *akāsi* (of the sky, heavenly) or *pāni* (water, irrigated) *kheti* (field).

The rice fields are highly revered by villagers as evinced, for example, by the taboo on emptying one's bowels there, in sharp contrast with the village's (mostly internal) *bāris* (nonirrigated land or fields) that are regarded as suitable venues for this purpose. The particular sanctity attributed to the Thamgharian threshing floor, called *khalo* (*khale*, clean) *garo* (*gara*, terrace) is highly reminiscent of that described in a number of colonial accounts from south India and Ceylon.[17]

Usually one of the largest and centrally located terraces in a given plot, the *khalo garo*, as Sarala Pokharil (43) commented, is often described as being

*A Bhūme tree-temple situated just above a rice field.*

analogous to the field-owner's house. In addition, Thamgharians explicitly refer to it as Bhūme's (the earth goddess, to whom all crops are dedicated) *mandir* (temple), thus bestowing the elevated status of sacred *prasād* to the entire harvest. Similarly to any other temple, cooking, eating, whistling (due to its connotations with evil spirits), or the wearing of shoes inside the *khalo garo* are strictly forbidden. Nonetheless, the village "Untouchables" are not barred from its premises.

Each particular field's manifestation of the tutelary earth goddess is believed to dwell in a tree-temple situated above that field, overlooking its upper terraces.[18] From the vantage point of such Bhūme temples, the cluster of the numerous, dense terraces and the intimately positioned *khalo garos*, each embodying a field-owner's house, could be deemed to embody an inverted reflection of the fragmented village reality. This unified universe on the village periphery seems to enjoy the permanent presence of Bhūme, who had fled the house (and the village) in times immemorial, as narrated in the ensuing story.

## The Origin of Rice Cultivation

In Thamghar, the question "why" is often not merely a query about the reasons behind an action or custom but actually challenges its execution. The answer, in accordance with the general spirit of indirect village discourse, is only rarely straight and to the point and is often couched in terms of *hāmro paramparāgat anusār*—that is, "[this is] according to our ancient tradition."

Alternatively, the answer may take the form of syllogistic reasoning embedded in a narrative, and while priests may go into much detail and indulge in prolonged, complex mythical sagas, lay villagers often provide simpler stories. Such is the following one about the origin of rice cultivation, told by Radhakrishna Dhakal (63) in front of his fellow workers in the small hours following the completion of the nocturnal rice threshing in his field:

> According to our tradition, a long time ago people did not have to cultivate rice in the fields. In those days, rice was planted in the lump of *rāto-māṭo*[19] [Bhūme] situated under the shelf of the *gāgri* [glossed as the sky or the rain], which provided it with water. Imbued with *saha*, the rice grown there sufficed for the whole year and one grain grew in quantity and could fill an entire cooking pot. However, one day a woman living on her own was visited by her *dāju-bhāi* [brothers, relatives belonging to her *māiti*], and instead of placing one grain of rice, she placed one for each of her visitors. This time, at the end of cooking she found only three single cooked rice grains in the pot, as [the angered] Bhūme had fled the house together with the *saha*. Since that day our forefathers began cultivating rice.

This story seems to portray the Thamgharian primordial paradisiacal house in terms that bring to mind the description of the precreation united cosmos in the *Śatapatha Brāhmaṇa*,[20] where "heaven was so near to the earth that it could be touched by the hand" (Kuiper 1970: 104), and the earth goddess, as was the case in ancient India (Malamoud 1996: 54), was clearly synonymous with limitless abundance or *saha*. Moreover, as will be demonstrated below, this story also appears to capture the major objective of the prolonged process of rice agriculture, namely, bringing Bhūme back into the house and replenishing the rice with its lost *saha*. In addition, it also sheds some light on the village custom of fashioning the *jagya*'s *bedi*, embodying Bhūme, out of mud brought from the rice fields.

## On Being a Farmer or Imagining Agriculture

Unlike, for example, south Indians (Dumont 1980: 96; Good: 1983: 239), Vaiṣṇavite Chetris (Rutter 1993: 287), or the Brāhmaṇs in central-west Nepal (Aubriot 1997: 302–3), Thamgharians see no discrepancy between their high ritual status and the practice of agriculture. Moreover, most Thamgharian men over the age of forty who have not been exposed to the influence of higher education in Katmandu, seem unaware of the urbanized elite rhetoric in which farming equals "villageness" and is looked down upon (Pigg 1992). Hence, with the exception of only a few rich village householders who prefer to lease their plots in return for half their yield, the majority of Thamgharians cultivate them in person. Although agriculture as the epitome of a rural existence is, by and large, not deemed as noble a profession as in a number of ancient Hindu texts (Wojtilla 1985; Krishnamurthy 1993: 41), most Thamgharian priests and householders alike exhibited considerable contentment with and some were even proud of being *kisāns* (farmers).

In accordance with Thamgharians' general stance toward animal sacrifice and meat consumption, agriculture does not seem to entail any immoral, polluting "violence." Resembling many other high-caste Nepalese (Müller-Böker 1987: 283) but quite unlike the Brāhmaṇs in the neighboring Dhading district (Ishii 1993: 133–34) or other high-castes in west Nepal (Cameron 1998: 76, 161) who completely abstain from plowing, in Thamghar, Manu's famous plea against the involvement of high-castes (particularly priests) in the "deadly" and violent mode of life that is farming (Manu 4.2–5, 10.83–84),[21] may only be echoed in a number of Thamgharian priests' reluctance to plough. Yet this is not accounted for in terms of *ahiṁsā*,[22] rather, it is in order to avoid inflicting *dukkha* (sorrow, pain) on the bull, the sacred *bāhān* (carrier, throne) of Śiva.[23] For this reason, a number of the village's practicing priests delegate this task to their sons, hire a lower-caste *hali* (ploughman), or obtain his services via *parma*. Moreover, as will be demonstrated below, plowing is not considered polluting, and the *hali* plays a central, priestlike function in the performances accompanying the agricultural work.

Like many other Hindus (and people the world over) who employ agricultural metaphors in reference to human procreative processes and vice versa,[24] in Thamghar, *bājho* (a field that has never been cultivated) *jotnu* (to plough), denotes intercourse with a virgin. In addition, women are barred from plowing as it is said to cause barrenness, and merely *nāghnu*, stepping over a plough, is believed to result in serious childbirth complications and even death. *Nāghnu* (stepping over parts or the whole of a superior's body), which may be regarded as the reversal of *ḍhognu*, is highly offensive, inauspicious, and even dangerous. For example, an "Untouchable" climbing on the roof of a Brāhmaṇ house in order to fix it, commits *nāghnu*, thus giving the entire habitat his own low purity status as if he had entered it. Hence, *nāghnu* can be seen as a kind of inauspicious intimate involvement, which includes penetration albeit not necessarily of a sexual nature.

This suggests that the plough, a saliently masculine entity, is perceived as superior to any woman and further implies that Thamgharian Brāhmaṇs view the act of plowing not so much as a potentially violent act against the earth, but as a form of "sexual labor" as discussed in chapter 6. As may be expected, villagers never explicitly referred to it in these terms, yet this seems to emerge from their reflections and taboos regarding the plough. The superiority of the latter over women is not merely associated with its male gender but, as I would like to suggest, is also related to the distribution of one's agency and the identification of the actor with his action; the householder is identified with his plough (whether he himself does the plowing or not), hence the latter is his bodily extension, in effect it may be imagined as his phallus.

*Amrit Bhandari plows the flooded terraces prior to transplantation.*

In the same vein, there may be little wonder that one of the main and well-observed taboos regarding plowing prohibits it from being carried out on a day when *hom* is performed at one of the *kul*'s households. This is probably since Bhūme cannot be subjected to an erotic engagement in the field, while she is preoccupied in divine coitus within a village *hom*. Hence, it strikes me that what elsewhere may be merely a mutual figurative association between agriculture and sex is perceived in quite literal terms in Thamghar (and probably in many other Hindu communities elsewhere),[25] as was already noted regarding metaphors, symbols, and representations in general. This will be further illustrated below.

## FASHIONING A *JAGYA* OF RICE

### *Biu Rākhnu* (Rākhne)[26]—Sowing

Although Brian Smith's (1989: passim, particularly 82–119) argument for the necessarily sacrificial nature and origin of the Hindu "second birth"—the Upanayana—may not hold,[27] his insight into the Vedic perception of the need to rectify the first blemished cosmogony via ritual creation, appears to capture what has become a general Hindu truism.[28] This, as Smith suggests, seems to be primarily reflected in the Hindu notion of *saṁskāras*. In Thamghar, these rites celebrate, anticipate, reenact or bring about the transformations involved in organic human processes, notably birth, sexual maturation, marriage, and death.

As Shulman (1980: 348) puts it, "impurity breaks into every life at the most crucial moment," and this pollution, inauspiciousness, or even evil must be brought under control. Although never explicitly couched in these terms, the above ideas also seem to govern the major stages of the rice agricultural cycle or the rice's *saṁskāras* as they are sometimes called in Thamghar, as also appears to be the case in south India (Good 1983; Moreno 1987), beginning with the sowing of rice at *biu rākhnu*. This low-key, unceremonious "insemination" of the earth goddess, to which I shall turn below, has its festive ritual reenactment in *dhān ropnu* (transplantation), which takes place approximately one month later.

### From My Field Notes[29]

Just when the hot, hazy, and dusty days of Baiśākh[30] seemed almost unbearable, in the middle of a stormy night of spectacular lightning, quite abruptly as though the skies were suddenly ripped open, the long-awaited monsoon rains finally arrived. Sita Lamichane (42), the mother in one of my host families, awoke well before sunrise. Instead of setting out to fetch water to fill the house's *gāgris* as she normally did (a task that was left on this morning to her eldest daughter), she swiftly revived the fire in the *culo* and hurried, carrying a hoe in her hand, to the family's rice field in the midst of the torrent. Awakened by the storm and unable to sleep, I rushed after her.

Half an hour later, barefoot and soaked, Sita arrived in one of the family's upper terraces and began clearing the corn plants that still grew there. Halfway through hoeing the terrace, she decided that the land was wet enough and that as this was Monday,[31] it was an appropriate day for *biu rākhnu*. She quickly made her way back to the village and called the other family members to come and complete the preparation of the *byāṛ* (*byāṛd*). Literally a male beast, the *byāṛ* is the rice nursery, from which seedlings are later transplanted throughout the field. A short while later, the flooded *byāṛ* was ready to be sown.

At this point, Surya Ram (45), Sita's husband, holding a basket full of *biu* (also called *bij* [semen] or *puruṣ* [male]) in his left hand, began *hānnu* (to hit, throw, thrust and impregnate), that is, to sow using his auspicious right hand. *Hānnu* belongs to a particular vocabulary applied only to rice, while for other crops the simple *charnu* (to sow) is used. Within less than an hour the work was complete and, still under a heavy downpour, the family returned to the village.

*Biu rākhnu* marks the brief conception of rice—the first stage in Bhūme's pregnancy. That in Thamghar the latter is perceived in quite literal terms is revealed by the imperative that in the *byāṛ*, *dui paṭak jotnu hũdaina*—the *byāṛ* should not be ploughed twice. As Surya Ram explained, if the *byāṛ* is ploughed during *biu rākhnu*, villagers view a second plowing (during transplantation) as equal to having intercourse with a heavily pregnant woman (which should be avoided), or worse, cutting open a woman's womb during pregnancy and removing the fetus. Here the "sexual labor" of plowing, when performed during the goddess's pregnancy, emerges as an act of murder.

Therefore, it is perhaps little wonder that I did not observe even a single *byāṛ* that was ploughed even once; instead, villagers preferred to hoe it by hand only. Concomitantly, this implies that *dhān ropnu* (transplantation), which takes place about five weeks after *biu rākhnu*, is thought of as merely altering the seedlings' location within the already pregnant goddess's womb.

## *Dhān Ropnu* (*Ropne*)[32]—Transplantation

Throughout the two months of *dhān ropnu*, as is also the case later during the harvest, the village tea shops are closed as all householders are busy working

Byāṛ—*rice seedling nursery, three weeks after* biu rākhnu.

in the fields. While only men plough and prepare the terraces for transplantation, uprooting the seedlings from the *byāṛ* and transplanting them throughout the field is an exclusively female task. Like human embryos (and newborn babies), the seedlings (unlike the final product—*dhān*) are treated with considerable ambivalence, underlining their embryonic state; as will be seen below, seedlings are placed behind the ears and on the top of the head as a substitute for, auspicious flowers in a *pūjā*, yet a bunch of seedlings thrown over the head of a person (*nāghnu*) is deemed highly inauspicious and, unless it is "undone" by being thrown backward may cause illness or even death.

In contrast with *biu rākhnu*, *dhān ropnu* is a highly festive event that usually lasts for a number of days. Elderly villagers recall that until quite recently, Damāi ("Untouchable" musicians) used to accompany the work all day long,[33] and women sang and danced in the mud. Nowadays, it seems that only very rarely are Brāhmaṇ families able to afford to hire Damāis for the occasion, and I did not witness any women singing and dancing either. Yet villagers claim that women still dance,[34] although the festivities are clearly depicted as more modest than those, for example, in the Katmandu valley Chetri community Rutter (1993: 119, 369–70) eloquently describes. The high point of the transplantation is the celebration of *bhakāri bādhnu* (*bādhne*) that usually takes place on its final day.

*A young* buhāri *during transplantation. The picture was taken on a rainy day, hence her head is covered with a plastic sheet.*

*Flooded rice-fields during transplantation. Closer, three green terraces of* byāṛs *can be seen among others in various stages of preparations prior to transplantation.*

## *Bhakāri Bādhnu*—a Cosmos of Rice Conceived

*Bhakāri* is a large bamboo rice container, a number of which are located on the first floor of every house, and *bādhnu* is to tie, to bind, to protect, or to establish. Hence, implicit within this title is the objective of its performance, namely, to ensure the future filling of the household's *bhakāris* with rice.

## From My Field Notes

It was the third day of transplantation, carried out under continuous heavy rain. In the early afternoon the rain gradually subsided, then stopped completely, and the sun's strong reflection flickered in the flooded terraces, dazzling the groups of *khetālo (khetāla*, workers) working at different corners of the vast field. Muktinath Adhikari, a major landowner who also enjoyed an additional steady income,[35] was sufficiently affluent to avoid participating in *parma* himself, thus while his wife and daughters took an active part in it, he hired all the other male workers needed through *nivek*.

On this afternoon, wearing spotless yellow shorts and a blue shirt, he was walking from terrace to terrace to oversee the progress of the work. In early afternoon Gita (40), Muktinath's wife, provided the workers with a lavish *khājā* (light meal) and an hour later, two-thirds of the *khalo garo* were already

Bhakāri bādhnu—*Muktinath Adhikari with a mud* ṭikā *and seedlings (flowers) behind his ears.*

planted. Thereafter Gita took a large sheaf of seedlings and planted it in the middle of the remaining empty area of muddy water. Around it, she planted seven additional smaller sheaves, which were tilted so that their tips pointed toward the center. She later explained that each enveloping sheaf represents Bāstu Puruṣa (the god of the house), while the central one is where the earth goddess comes to dwell.[36] Located together, she added: "Bāstu and Bhūme make the *bhakāri* (all these sheaves taken together) *purā* (complete)." From this moment onward, the entire plot is believed to be secure and no evil spirit, ghost, pest or the harmful (embodied) intention of a fellow villager, can harm the rice. When the *bhakāri* was in place, Gita took some mud in her right hand and placed it in the center of Bhūme's sheaf, relating to this as *bheṭi garnu* (*garne*).[37] Thereafter, she applied a *ṭikā* to her own forehead, employing the mud from Bhūme's sheaf.

At the same time, Krishna Ram Bastakoti (38),[38] who served as a *hali* during this season was working in an adjacent terrace. Leaving the plough, he entered the *khalo garo* and following Gita, he offered some mud to Bhūme, placed a mud *ṭikā* on his own brow and turned to place a similar *ṭikā* on Muktinath's forehead. He then put a small mark of mud on Muktinath's front teeth and placed a number of green rice seedlings behind the latter's ears and on top of his head, as if they were the auspicious *phul* (flowers) accompanying a *phul-achetā ṭikā*. At this point, the *hali* together with two additional *khetālos*, lifted Muktinath a few inches above the water and thrust him, legs down, back into the mud. This marked the beginning of a short but amusing *hilo khel* (*khelne*, literally, mudplay)—a light-hearted fight in the muddy water between the men and women present.

Reflecting on *bhakāri bādhnu*, Krishna later explained that the mud employed for the *ṭikā* is the goddess's *prasād*. The field-owner is fed mud and later thrust in it in order to create *sambandha* (lit. connection) with the goddess, and the good-humored mud play and general merrymaking during this day are for *khusi banāune*, to make Bhūme happy. Other *khetālos* added that the latter increases the *sāit* (auspiciousness) of the entire event and ensures abundant yields. It is significant to note that more than a simple connection, *sambandha* is an involvement or alliance and is often used in Thamghar to denote marriage.[39]

By sunset, the remaining terraces were all planted, and the workers went to Muktinath's house for the customary *bhāter*. This festive meal, which included goat's meat—a relatively rare delicacy for ordinary villagers—marked the end of the day's celebrations.

As I was repeatedly told, and in sharp contrast with ordinary village behavior, if a householder finds himself short of seedlings at the end of *dhān ropnu*, he may take them from the *bhakāris* of others. This is since the *bhakāri* is the goddess's temple, and the seedlings belong only to her. Within the next few weeks, all the village rice fields became replete with young green seedlings, and a month later the numerous *bhakāris* together with the *khalo garos* were virtually indistinguishable as the low mud boundaries between terraces were no longer visible.

It must be stressed that villagers do not explicitly spell out the import of *bhakāri bādhnu* in the terms employed below. In addition, most villagers I approached on this matter were quite reluctant to talk about the stages of rice cultivation in terms of a *pūjā*, probably due to the frequent and intimate involvement of "Untouchables" in the agricultural process. Nonetheless, I believe that the correspondence between the agricultural performances and a "proper" village *jagya* is couched rather lucidly in visual and dramatic terms. The latter, together with the recurrence of matrimonial encompassment constructed of rice seedlings (or of ripened rice bunches in the *kunyā* as will be elaborated later), provide sufficient grounds to support the suggestion that in *bhakāri bādhnu* villagers establish an agricultural *jagya*—a vibrant mesocosm of rice.

Accordingly, *bheṭi garnu*, which as I wish to recall is an ordinary act of honoring a deity, person, temple, or a sacred book, is another manner for denoting the worship of the goddess. Along the same vein, the mud *ṭikā* given to the householder is for *baran garnu*, nominating him as Bāstu, since only by assuming this cosmic role is he allowed to proceed and come into an intimate *sambandha* with the goddess. Like a *biu* (male seed/seedling) thrown into the mud, the field owner is thrust into Bhūme in a manner reminiscent of Bāstu penetrating the *khaldo* during his village (Bāstu) *pūjā*.

The cheerful mud-fight brings to mind the royal *hilo khelnu* which is believed to affect the prosperity and fertility of the land (Rutter 1993: 119, 363–64), and like the mutual feeding of a bride and groom as part of their wedding, the playful association of men and women in *bhakāri bādhnu* is redolent with sexual overtones. If, as I would like to argue, the relationship of the field owner and the goddess is not simply one of a worshipper and his deity, the mud he is fed must obviously be more than *prasād*. Thus, via the mediation of the *hali*, Bhūme and the householder seem to feed each other, celebrating their marital bond prior to its ludic consummation. Alternatively, Bhūme may be viewed as a lawful wife who feeds her spouse, yet rather than doing so with a rice meal she has cooked, Bhūme seems to feed Bāstu with her own flesh. Far from an act of cannibalism, this may be another actualization of Bhūme becoming matrimonially encompassed by Bāstu.

In sum, the agricultural *jagya* places Bhūme and Bāstu in a prolonged procreative posture that comes very close to the erotic ideal of Bāstu *pūjā*. Like a village *jagya*, the agricultural one is clearly an embodied manifestation of the macrocosm, yet in the field the entity being acted upon, that is, the rice, is not simply identified with but actually *personifies* the "ritual enclosure" of the *bhakāris*. Concomitantly, the gradual fusion of the latter, the *khalo garos* and the plots into one, can be viewed as the earth goddess's pregnancy swelling up until it incorporates the entire green organic universe of the rice fields.

Finally, *bhakāri bādhnu* provides another salient example of Thamgharian Brāhmaṇs' attempt to encompass the future via the merger of a cause with its effect; through the association of the agricultural *jagya* with the hoped-for overflowing *bhakāris* of rice on the house's first floor, the householder (as Bāstu) encompasses his potential ripened future prestige.

The considerable might of the earth goddess while *in situ* in the rice fields, is clearly reflected in her relative size vis-à-vis her Bāstu guardians/consorts. Bhūme's excessive *śakti* is given an even more salient expression in the following stage where, following the harvest, the cut rice is kept for an additional ten days in the *kunyā*.

## THE *KUNYĀ*: A FIRELESS *HOM* IN BHŪME'S PROCREATIVE TEMPLE OF RICE

A number of weeks before the rice becomes *pākca* (*pakeko*, cooked, mature, or ripened), all irrigation is stopped, or *pāni marne*, "killing" the water, as it

Rice fields a few weeks prior to the harvest.

is locally termed. After it is cut, the rice is left to dry for three days before being tied into sheaves, which later form the building blocks of the *kunyā*.[40] This round pile of rice (up to four meters in height with a diameter of approximately three to four meters), established in the *khalo garo* as close as possible to the exact place where the *bhakāri* was planted a few months earlier, is viewed as Bhūme's temple. That the *kunyā* is imagined as another embodied manifestation of the cosmos may be inferred from Ram Pandey's (a middle-aged householder) assertion that he worships the *kunyā*'s lower and upper parts as *pattāl* (hell) and *akās* (heaven) respectively.

Numerous *kunyās* dot the autumnal Nepalese landscape[41] in almost every area where rice is cultivated, with the (main) exception of Newārs' fields, yet this has failed to attract much scholarly attention with one notable exception, namely its Newāri counterpart, termed *phutuchha*. The latter is a round, flat-roofed heap of rice, about one meter in height, which is used for the production of *hakuwā*—black fermented rice. Nowadays, no more than 30 percent of the total crop is turned into the *hakuwā*, which although tasty and easy to digest, is also said to be somewhat malodorous.[42]

In a report dating from the early nineteenth century, Hamilton (1819: 73–75, 222–36) describes how the Katmandu valley Newārs convert most of their rice into *hakuya* (*hakuwā*) in order to correct "the wholesome quality of the grain." The remaining crop is threshed beforehand and employed as the following year's seeds. More recently, Nepali (1988: 47) mentions that the *hakuwā* rice has a particular ritual value and is believed to increase in quantity during the cooking process. Likewise, Toffin (1977: 79–80, 103) notes that the *hakuwā* is taken to be half cooked, is believed to be particularly healthy and provides much *tagat* (vigor). He also mentions the particular sacred atmosphere during its threshing. This bears a remarkable resemblance to present-day Thamgharian notions regarding the *kunyā* as will become clear below. Nonetheless, fermented rice is viewed in Thamghar as *biṭulo*, is treated with much aversion, and is never eaten or used.

## From My Field Notes

It was the last day of the harvest. On this mid-November morning, *Pokhare* (from *pokhari*, a pool or pond), one of Thamghar's southern rice fields, was scattered with groups of *khetālo*, and Saubhagya Aryal (45) had already been hard at work with his fellow workers, wife, and children since daybreak. Gazing anxiously at the clouds that began rolling off the white Himalayan summits to the north, Saubhagya encouraged his men to bring the rice sheaves into the *khalo garo* without delay. Once the *kunyā* is constructed, most of the grain is quite safe and protected from pests or the occasional shower, which is not uncommon at this time of year when the rainy season draws to a close. Nonetheless, a torrential downpour remains a hazard since it can break the *kunyā*, induce untimely germination or fermentation and thus destroy the entire annual crop.

Taking off his hat (as he would otherwise do for a *pūjā*), Saubhagya performed *dhup halnu*,[43] "in order to provide Bhūme with pleasant fragrance" as he later explained. *Dhup halnu* marks not only each and every stage in the *kunyā*'s construction, but also every step of its dismantlement during *dāī hālnu* (the threshing of the rice discussed in chapter 8), and as Thamgharians often note, *dhup halnu* belongs to a large array of actions and gestures performed throughout rice cultivation that are designed to maintain auspiciousness and exhibit signs of and engender plenty and prosperity.[44]

On a small *sal* leaf referred to as a throne (*āsan*) he placed the minor, elongated stone manifestation of Bhūme (Bhūme's *ḍhuṅgo*). Next, Saubhagya performed a modest *pūjā*, offering the goddess a few flowers brought from the village and, assisted by two *khetālo*, he forcefully placed the *buṛi biṭo* over Bhūme. The *buṛi biṭo*[45] is a round, hollow cylinder-like construction (approximately one meter in height) made of seven or nine (must be an odd number) rice sheaves tied together. Sometimes referred to as *pothi*—a female animal and a vulgar term for a woman[46]—it is regarded as Bhūme's *mandir*.

*The* buṛi biṭo.

The tips of the rice sheaves constituting the *buṛi biṭo* were carefully tilted toward its center in order for the grains to point toward and fall over Bhūme. Now the remaining sheaves from the plot were placed around it in a *phero* (*pherā*, a circle),[47] with the grains pointing toward the center, and soon the *buṛi biṭo* was covered.

When the round construction of rice was approximately three meters high, only Prayag Uperty (29) remained on top, shaping the *kunyā*'s upper pointed roof using some straw. At the same time, Radha (Saubhagya's wife) was busy preparing the *juro* destined for the *kunyā*'s pinnacle. The *juro* (about sixty centimeters in height) is literally a straw man made of seven threshed rice sheaves, tied together to fashion the figure of a five-handed male. Each hand, according to Radha, stands for one of the *pañcayan deutā* (*deotā*) or one of the house's *bhakāris* of rice. Glossing on the *juro*, Prayag explained that our body is a *mandir* or *darbar* (palace) with nine *ḍhokā* (gates: the two ears, two nostrils, mouth, and so on). The *sir* (head) is its *gajur*[48] and likewise, the *juro* is the *kunyā*'s *sir*, and must not be placed if the head of the household or joint family (the field owner or his father) passed away during the previous year. A number of the other men present added that the *juro* is a *bhāle*, literally a male beast, and that together with the *buṛi biṭo* it makes a couple. For this reason they should be the last of the *kunyā*'s elements to be simultaneously threshed, in order to render *dāī hālnu* complete (*purā*). When the *juro* was

*During the construction of a* kunyā, *the* buṛi biṭo *is almost totally covered by the surrounding rice-sheaves.*

ready, it was taken around the *kunyā* three times before being placed at its pinnacle and adorned with red ribbons and flowers. Now that the *kunyā* was complete, Saubhagya explained, Bhūme's enormous *śakti* (power) protects the rice against evil until it is threshed in *dāī hālnu*.

Prior to and during the first stages of the *kunyā*'s construction, two men were busy threshing some rice nearby, to prepare the following year's *biu* (seeds), as the grain's germination power is believed to be impaired if it becomes part of the *kunyā*. One *biṭulo* type of rice was also threshed, since its presence in the *kunyā* may lead to the latter's defilement. The three men who assisted in carrying the sacks of threshed grain home at the end of that day's work were invited to an evening meal in Saubhagya's house.

Village priests draw attention to the need to worship the *kunyā* and the *nāg* (serpent deity) together with his consort, *nāgini*, who dwell nearby and assist in guarding it. Yet only a handful of villagers and the local priests seem to possess the knowledge required for the performance of an elaborate *nāg pūjā*, and most simply offer the serpents half a glass of milk. In contrast, all householders with whom I discussed it were unusually forthcoming about the *kunyā*'s *jaro* (*joro,* heat or fever) and the inherent process of cooking, through which its *dhān gumsincha* (*ghumsincha*, is heated) and *pākcha* (is cooked, ripens) during the ten days that pass until *dāī hālnu*. Very often, I was told,

*A field owner's wife prepares the* juro *and adorns it with red leaves and flowers. On the right is the wall of the* kunyā *and a sack full of threshed rice grains that will be used as the following year's seeds.*

*bāph* (vapor, steam) can be seen rising out of the *kunyā*. Not only is this said to result in considerable growth in the quantity of grain and is believed to make threshing easy, but this inherent process of cooking also endows the rice grains with intense *saha* and with the capacity to withstand the husking process, as well as with future good taste and aroma.

As glossed by Mukunda Banjara (38), the general view is that this process occurs because *kunyāmā Annapurnako śakti cha*: the *kunyā* is pervaded by the *śakti* of Annapurna, the goddess of grain who is often equated with Bhūme. Householders also commented that often this power (i.e., the *kunyā*'s inherent heat) could not be felt by touching the structure from the outside, but was particularly intense during *dāī hālnu*, when the *kunyā* was threshed.

As I see it, being analogous to the householder's house, the *khalo garo* with the *kunyā* at its core comes very close to the ideal primordial house where heaven and earth were almost joined together and rice, imbued with *saha*, multiplied in an incredible cooking process in the presence of Bhūme. Although not at all explicit within the mythical habitat, the generation of *saha* within the *kunyā* and its internal cooking process seem to be engendered via a divine matrimonial encompassment, in which the male body of the *kunyā* (the embodiment of the field owner) envelops the earth goddess in a manner

*Threshing an impure type of rice and the seeds for the next season before the* kunyā *is completed.*

that creates a heated process of reproduction and gestation of the entire harvest. The *jaro* appears to be the local avatār (reincarnation) of the highly transformative ancient *tapas*, permeated with salient sexual overtones, hence this is another instance where "cooking" emerges as a euphemism for procreation. Accordingly, the *kunyā*'s ten days of existence appear to be equivalent to the ten months of pregnancy, as Thamgharians (like other Hindus)[49] count them.

Moreover, the identical location of the *bhakāri* and the *kunyā* seems to obviate the time-gap that divides them, evincing their complementary nature as parts of one whole. Hence, if a *jagya* of rice is established in *bhakāri bādhnu*, the *kunyā* should be imagined to embody its fireless *hom*, marking the second birth of rice following its divine gestation. This follows the "natural birth" or emergence of the grain from the stalk, followed by its natural ripening while still rooted in the ground—equally referred to as "cooking."

Thus the *kunyā* seems to employ, actualize, and take the framework of a village *jagya* to its limits; the relatively minor *bedi* becomes the elaborate *buṛi biṭo*, a temple in its own right encapsulating the *murti* of the earth goddess, and the unity of the village Bāstu *liṅgas* or the *bhakāris* is realized by the consolidated bulky *kunyā*, embodying the field owner as Bāstu. Likewise, in the *kunyā* the identification and assimilation of cause and its effect, of actor with

*Sharmananda Subedi prepares* nāg pūjā *near the* kunyā *prior to* dāī hālnu. *Both the* nāg *and* nāgini *(made of dough) are placed on a banana leaf with various offerings in front of them.*

his act and the outcome, and of the subject of the ritual procedure (rice) and the mesocosmic manifestation, is realized to the full. More generally, it may be suggested that at least in a number of high-caste communities, rice cultivation is imagined to be enabled via divine *tapas*, but while Rutter's (1993: 370) Vaiṣṇavit Chetris view this as the ascetic *tapas* of the withdrawing Viṣṇu, in Thamghar rice is gestated and reborn as a result of an erotic *hom* engendered by divine coitus.

These notions not only seem to be in accord with the Newāri views of the *hakuwā* mentioned above, but also bring to mind the Tamil *Poṅkal* harvest festival, the most important of all household rituals in south India.[50] According to Good (1983), the *Poṅkal* encapsulates and reenacts the entire cultivation process within a ritual cooking of rice, thought of in terms of gestation, while the cooking pot embodies the goddess's womb.[51]

Apart from the employment of matrimonial encompassment as a template for the mass production of *saha*-imbued rice, a further unique aspect of the Thamgharian *kunyā* is villagers' apparent ability to touch and feel the divine *tapas* at work. Anthropology, as Gell (1998: 32) argues, has to deal with fiction as well as with real situations, and these are often hard to tell apart; the *kunyā* is clearly a case in point. The unmediated contact with the *kunyā*'s *jaro*

*A* kunyā *in the morning fog during early November.*

or *tapas*, I would like to suggest, is the ultimate element behind villagers' willingness to risk what for most of them is the sole basis for their subsistence. Otherwise, Thamgharian Brāhmaṇs, who tend to avoid any risk as far as possible, could more simply and swiftly thresh their rice either in the field immediately after it is cut,[52] or near their homes,[53] without the establishment of a *kunyā*. For its village architects, feeling the *kunyā*'s divine inner processuality seems akin to what in Gell's (1998: 68–72) terms is the experience of being "captivated." Captivation here is not due to the *kunyā*'s dimensions, as is the case in many Hindu and other temples the world over, but is a corollary of the arresting, tactile contact with its inherent, divine, heated agency.

The ensuing discussion of the gender of rice is not unrelated to the above issues, and is mainly intended to shed more light on villagers' perceptions regarding the nature of the transformation of rice within the *kunyā* and its effects.

## ANOTHER ENCOMPASSMENT

As may be expected from its relative importance and central significance in their lives, and in contrast with all other crops or grains grown in the village, Thamgharians explicitly refer to rice grains as gendered entities. This is not to say that they ever raise the subject or discuss its gender *per se*, but that this matter appears to be common knowledge and largely goes without saying, unless a specific enquiry is made. Moreover, Thamgharians' gender ascriptions for rice are embedded in the ordinary Nepali titles they employ to denote the grain, whether it is husked or not. Thus, as I first heard from Laksmi (a young Thamgharian mother), *dhān*, which is unhusked rice (singular or plural), is *puruṣ* (male), while *cāmal*, the husked but uncooked grain(s), is *nāri* (female).[54] Ostensibly therefore, the unhusked rice grain embodies the ideal complete state of a male encompassing female, forming the last element in the Thamgharian cosmological jigsaw dominated by the fractal of matrimonial encompassment.

However, I would like to suggest that the rice grain represents an additional form or denotation of the cardinal cultural template of a male encompassing a female. Put differently, the encompassment of the feminine *cāmal* by the masculine *dhān* is nonmarital and asexual. A brief reiteration of the gendered biography of rice from *biu rākhnu* to its employment in a *pūjā* or for cooking will render this issue more lucid. Prior to this, I shall note that unlike in Japan (and perhaps also in other places where rice is a staple), where as long as the grain is within the husk it is regarded as being alive (Ohnuki-Tierney 1993: 550), the real life of much of the Thamgharian crop only seems to begin when it is "released" in the violent, even sacrificial act of husking.

Although a large number of Thamgharian households can afford the nominal fee required for employing one of the village's two privately owned diesel machines designed for this purpose, husking, termed *dhān kuṭnu* (hitting the *dhān*), is still mainly performed at home (exclusively by women) using the *ḍhiki*. The latter is a heavy wooden lever held within two upright posts. By stepping on one end, the other, equipped with a metal-covered pestle (*musal*, gendered male) is lifted above a stone-bottomed mortar (*okhli*, *okhal*, perceived as female). When the pestle falls over the *dhān* placed in the mortar, its sheer weight separates the husk from the grain. When questioned, quite a few villagers described husking as the removal of the grain's germination power or even the death of *dhān*. A number of householders drew an explicit analogy between *dhān* and a man, who is believed to die, once drained of his fertile bony substance through ejaculation of semen. Others remarked that being *īśwar* (god), rice never dies but merely modifies its form.[55] However, all men and women with whom I spoke about this seemed to agree that the removal of the husk does not harm (most of) the *cāmal* grains since *cāmalmā śakti cha*, that is, the *cāmal*, is imbued with inner *śakti* (or *saha*).

*Kanikā*, the small amount of *cāmal* that does break during the husking process is said to be tasteless and *biṭulo* (impure). Moreover, villagers seem to view the idea of cooking or employing it in a *pūjā* with considerable aversion and offer it only to the household's beasts. I noticed that this was the case even in families who are considered poor, that is, suffer an annual shortage of rice grains. Although generally speaking no more than 10 percent of the *cāmal* breaks into *kanikā*, it seems quite remarkable that people would rather be short of rice (and risk humiliating loss of face) than simply use their broken grains for cooking. I shall revert to this issue below.

Before it can be used, the grain that emerges intact from the mortar must be carefully washed three times in cold water. Only then, village women explain, when all vestiges of the (masculine) husk are removed, does *cāmal* become truly female. Only now it may be destined for the cooking pot, for use in *dān* transference[56] or may constitute the offerings in a *jagya*. Alternatively, the *cāmal* can be mixed with red *abir* powder, thus becoming sacred *achetā*. As I wish to recall, the latter embodies the goddess in a *pūjā* and is distributed as auspicious *prasād* (*ṭikā*) at its conclusion.

In contrast with all other contexts where the union of the sexes is explicitly referred to as a manifestation of the ideal complete state of marriage, villagers never mentioned this regarding rice. Moreover, confronted with this possibility, a number of villagers vehemently opposed it. Indeed, if the masculine *dhān* must be killed (at least removed) in order to give the feminine *cāmal* a life of its own, being married to the former would render the *cāmal* an inauspicious state of widowhood. What therefore is the type of relationship inherent within a grain of rice? In the absence of a more detailed local exegesis, I believe that the literal meaning and the etymology of the word *achetā* may offer an additional avenue for interpretation.

In Thamghar, *achetā* denotes unbroken, husked rice grains, that is, *cāmal*, mixed with auspicious red *abir* powder, as well as "the indestructible one." This may imply that the wholeness or resistance of this feminine grain may be simply related to its physical ability to withstand the *ḍhiki*'s violent strikes. However, according to Turner (1996: 3), *achetā* is the Sanskrit version of *akṣatā*, "the rice offered in worship, the *ṭikā* put on the forehead at the *dasaī* festival," while *acche* is enduring and indestructible. The nature of *achetā-akṣatā,* particularly if we bear in mind that the Sanskrit prefix *a* often serves to negate the meaning that follows,[57] is implied by the meaning of *kṣata-yoni* or *chata* in Sanskrit, which denotes "a woman who has had connection with a man," or "having a violated womb" (Turner 1996: 110). Likewise, according to Monier-Williams (1995: 3, 325), in Sanskrit *á-kshata* means uninjured, unbroken, whole, unhusked grain, but also a virgin and an unblemished maiden, while *kshatá* means broken, impaired, "produced by a wound" and a violated girl.

It thus seems reasonable to surmise that the endurance and wholeness of the rice grain were, at least in earlier times, perceived in terms of virginity and were related to the grain's supposedly "intact hymen." It may be impossible to conclude whether this notion still lies behind Thamgharians' aversion toward *kanikā* (and they would probably be too shy to admit it using these terms), or whether in rejecting the *kanikā* they simply follow a local custom. Although the latter may well be the case, the ancient meaning of *achetā* seems to bear a striking similarity to its present ritual use as the embodiment of the goddess in *pūjās*. Hence, I would like to suggest that it is the *cāmal*'s well-bounded, enormous power of *virginity* that is evinced in its incredible ability to withstand the considerable virile might of the *ḍhiki*'s *musal* (male pestle). Furthermore, its inherent *śakti* is later harnessed for carrying the ominous load of *dān*, exorcising ghosts (in Thamghar) or combating demons (Ramirez 2000a: 110–11), and it is this power that may also account for the *ṭikā*'s remarkable intrinsic capacities discussed in chapter 6. Unlike the fearsome fierce goddess, *cāmal* and *achetā* seem to be on a par with the all-benevolent young village *kanyā*, the epitome of purity, sacredness, and intense *śakti*. From this perspective, every *dān* transferred via *cāmal* is in effect a *kanyā-dān*.

The interpretation of the *achetā* as *virgo intacta* rice grains vis-à-vis the *kanikā* as violated virgin grains explains why the latter is unsuitable for any human application, and is in line with the general local attitudes toward defloration (external via coitus or internal, i.e., menarche). This may also provide the background for the otherwise unexplained severe danger that eating cooked *kanikā* poses for men, noted by Stone (1977: 61–62), which as implied by Ramanujan (1992: 234–36) may in effect be a more pervasive Hindu notion. Moreover, the virginity of the Thamgharian *cāmal* lends support to the analogy implied by Khare (1976b: 159–60) between the general status of Hindu food and a *kanyā* (a virgin). It now also becomes apparent why Thamgharians insist on threshing the rice destined for seeds prior to it becoming part of the pregnant *kunyā*, as the excessive femininity (the *śakti* of virginity) it might acquire therein, would harm its virility (germination capacity).

However, as the title *achetā* is only attributed to consecrated *cāmal*, that is, when mixed with auspicious red *abir* for ritual purposes, it now remains to consider what is its significance and how this red powder is apparently able to convert simple virginal *cāmal* into a goddess. Keeping in mind village men's deep-seated attitudes toward menstruation and female sexuality, it is clear that *abir*'s red color is not a token of female (woman or goddess) menstrual blood as was suggested, for example, by Bennett (1983: 271) and Rutter (1993: 415). Indeed, Thamgharian men and the few women I discussed this with expressed considerable aversion when I raised this possibility. Concomitantly, on the single occasion when sacrificial blood is mixed with a *ṭikā*, during Dasaī, it is not thought of as the goddess's menstrual blood but as her highly auspicious and potent *prasād*.

In view of the *cāmal*'s gender and the strong general association between married women and the color red, which often serves as an emblem of their married state, I would further like to suggest that the vermilion *abir* mixed with *cāmal* is a sign of marriage. Stated differently, *abir* may be the divine counterpart of *sīdure* (*sindur*), the vermillion color village women put on the parting in their hair to display their married status.

Thus before it can be employed in a *pūjā* and worshipped as a goddess, the *cāmal* should first be bounded through marriage.[58] Once well-contained, its *śakti* is made accessible, auspicious, and benevolent. Only then are the *cāmal* grains considered to personify the goddess. Accordingly, as the embodiment of a *married* virgin goddess, *achetā* appears to reflect the ultimate, ideal state of auspiciousness that was associated with the (now obsolete) stage of a pre-menstrual *married kanyā* in her father's house, as discussed in chapter 3.

Notwithstanding its pervasive use, lay villagers and priests alike are rather parsimonious in their remarks regarding *abir*, viewing it simply as a mark of auspiciousness. Therefore, the question pertaining to the exact identity of the Thamgharian *achetā* and the significance of the *abir* seems, at this stage at least, to remain open. However, it seems legitimate to conclude that the relationship inherent within a single grain of rice is not one of matrimony but is more akin to "parenthood" since the virgin *cāmal* is "born" out of the (male) *dhān* in the husking process. This is highly reminiscent of the Vedic notion of male-pregnancy and indeed, rice husking in Thamghar, with its simultaneous death (of the *dhān*), procreation (the union of the [male] *musal* and the [female] *okhal*), and birth (the emergence of the *cāmal* out of the *dhān*), bears a striking similarity to the Vedic sacrificial process discussed in chapter 4.

I will now turn to a short comparison between the *kunyā* and the north Indian Hindu temple, in an attempt to suggest that the image of matrimonial encompassment is also shared by the latter.

## FROM THE VEDIC FIRE ALTAR TO THE HINDU TEMPLE VIA VEGETAL CONNECTIONS

Reflecting upon the previous discussions of the *jagya* and the *kunyā*, I would like to make one additional and final remark of a more abstract nature, relating to Kramrisch's (1976: 158–59, 207–9, 214–21) discussion of the origins of the north Indian Śikara Hindu temple. This is the curvilinear temple found throughout four-fifths of the Indian subcontinent from the north to its approximate southern boundary at the river Kistna (Kṛṣna) (Kramrisch 1976: 205).[59] The

Śikara temple is also popular in Nepal, though the country is mainly renowned for its unique Newāri pagoda-style temples (Slusser 1982: 141–49). Kramrisch's main contention is that the Hindu temple evolved from the Vedic fire altar (Kramrisch 1976: 146–47 and passim). She further argues that the direct prototype for the north Indian Śikara temple is the "arch of vegetation," that is, the Bihari and Bengali[60] square vegetal tabernacles or ritual structures (Kramrisch 1976: 158–59, 207, 214). From her brief description, it appears that the Thamgharian (and other Nepalese) *jagyas* may well be the counterparts of the north Indian vegetal ritual enclosures.

However, to my mind, even more than a village *jagya*, it is the *kunyā* that displays striking correspondence to the northern brick and stone Śikara temple. Furthermore, I am inclined to suggest that it is the Nepalese *kunyā* that may well embody the intermediate stage that leads to the latter from the *jagya* or the aforementioned vegetal ritual enclosures. This seems particularly remarkable if we bear in mind that the regional Kāli temple near Thamghar was designed (in the 1970s) by a Katmandu architect in a (modest) Newāri pagoda style. Moreover, its single room lacks any *Garbhagṛha* (the inner-most, womblike chamber of a Hindu temple), and as may be expected, Thamgharian Brāhmaṇs are no experts on Hindu architecture and this Kāli temple is probably the only one with which the majority of them are closely familiar. To substantiate the above proposition, I wish to recall that the *kunyā* is explicitly fashioned as a standing male, an embodiment of the householder in his cosmic manifestation as Bāstu. In a typically fractalic manner, the *juro*, shaped as a miniature male, seems to be a small-scale replica of the entire edifice. Moreover, it is Thamgharians themselves who draw the parallel between the *kunyā*'s *sir* (head) or *juro*, and the *gajur* at the pinnacle of a Newāri-style temple. Therefore, the *kunyā* in its entirety corresponds to the superstructure of the brick and stone Hindu temple built in the form of the primeval male—Puruṣa (Kramrisch 1976: 133, 357–61). The resemblance between the *kunyā* and the Hindu temple does not stop at their superstructure, but is also evident in the no-less striking architectural parallelism between the *buṛi biṭo* and the *Garbhagṛha*. The *buṛi biṭo*, the elaborate incarnation of the *bedi* that serves as both the house of the earth goddess and her womb, where the entire crop is gestated and cooked, is highly reminiscent of the feminine, darkened *Garbhagṛha*.

By objectifying the householder's prestige, the *kunyā* comes very close to the role of the Hindu temple as a tangible symbol of the success and reputation of its patron in medieval India (Inden 1985a). Furthermore, by noting that "builders [the donor or patron] and buildings are one; the building is a test of the health and probity of the builder," Kramrisch (1976: 52) implies that this is the case with all builders of Hindu temples.

Ultimately, the notional similarity between the Śikara and the *kunyā* seems to gain confirmation by both temples embodying the image of a *high-caste* male. While this is clear regarding the *kunyā*, which as discussed above embodies a high-caste initiated male Brāhmaṇ, it is less obvious regarding the Śikara temple. According to Mallaya (1949: passim and particularly 9, 267–72) the Hindu temple in general is constructed as a human body, while the term Śikara derives from *śikhā* (also *ṭupi* in Nepali, the tuft of hair that orthodox, initiated high-caste Hindu males have on top of their heads), and also denotes a "head." Moreover, Mallaya (ibid.) further argues that Śikara should relate only to the uppermost part or the temple's roof (also termed *Śiras*), which gave this type of temple its title. He therefore implies that the Śikara not only embodies the cosmic Puruṣa as Kramrisch (1976: 133, 357–61) contends, but that this primordial male is a *high-caste* persona.

My point here is that the interpretation of the *jagya* and the *kunyā* as gendered mesocosms, embodying a divine matrimonial encompassment where either the *kartā* or his rice are gestated and gain their conditional immortality (or *saha*), may offer an additional gender and agency-sensitive perspective for examining the Hindu temple, particularly its north Indian Śikara type.[61] Viewed in light of the Thamgharian *jagya* and *kunyā*, I do not think that the erotic *mithunas* carved on the doorjambs of a Hindu temple's *Garbhagṛha* or on its external walls should be taken simply as (Tantric) symbols of *mokṣa* as Kramrisch (1976: 346) suggests, or merely auspicious aesthetic markers as claimed by Michell (1977: 76). Rather, I contend that these are scaled-down images of the temple itself and an integral part of its general salient fractalic structure. Put differently, it is the Hindu temple itself that embodies a *mithuna* in the image of matrimonial encompassment, hence the encompassment of the feminine *Garbhagṛha* by the masculine superstructure, that is, the body of a "high-caste" Puruṣa, also denotes their engagement in eternal, architectural, divine, and procreative coitus.[62]

Concomitantly, the Hindu temple is clearly not intended to embody an androgyne as was suggested, for example, by Knipe (1988: 119). This is made clear if we consider that "the androgyne is that form of union in which conventional sexual activity is impossible" (Shulman 1980: 351). Likewise, the Tantric androgyne represents a state where sensuality and sexuality are entirely internalized (within each sex) and completely satiated, and thus "lust is truly conquered" (O'Flaherty 1980a: 319–20). If so, the fundamental transformation undergone by a worshipper entering the Hindu temple, as Kramrisch (1976: 156–65) herself implies, may be imagined as gestation and rebirth in a fireless *hom*. The latter is engendered via the inherent potency of the divine coitus embodied by the temple itself, by the *śakti* of the (male or female) deity residing in the *Garbhagṛha*, or else it is the result of the threefold circuitous causal pathway[63] discussed in chapter 6.

I would further like to argue that the vestiges of the erotic and procreative *tapas* that pervaded the Vedic fire sacrifice and possibly also the earlier days of the Hindu temple, may still be found in the divine, transformative heat that is believed to pervade the core of many Hindu rituals to this day (Beck 1969; Handelman and Shulman 1997: 188–89).

## ENDNOTE

The present chapter has attempted to demonstrate the decisive role of both the matrimonial and nonmatrimonial versions of the notion of male encompassing female within the realm of rice agriculture. This has provided a final and crucial building block to the fractalic image I have attempted to construct in the reader's mind throughout this book. This notion was found to be a decisive element in the manner in which Thamgharian Brāhmaṇs construe the self, the other, and the cosmos. It also appears to have profound meaning and significant implications about each of its cardinal scales of manifestation, from the single minuscule rice grain, through the family unit of a husband and wife, to the house and the village and agricultural *jagyas*, reflecting the manner in which Thamgharians imagine their entire cosmos. Furthermore, it was also suggested that this notion of a male encompassing a female is not merely a unique local concept. Rather, it may also be found elsewhere, is implicit in ancient Hindu thought, cosmology, ritual and temple architecture, and may well prove to constitute an additional key perspective for their interpretation.

It would thus seem appropriate to draw the entire book to a close at this point. However, the agricultural *jagya* seems to end only with the dismantling of the *kunyā*, the threshing of rice, and bringing it home,[64] which although not directly concerned with the notion of matrimonial encompassment, have significant, pertinent, and far-reaching social implications that involve the entire Brāhmaṇ community in Thamghar.

Chapter 8 will therefore briefly address these issues via the analysis of the major inversion event of *dāī hālnu*, demonstrating the manner in which village men, unable and probably unwilling to "exorcise" their womenfolk, are nonetheless capable of obviating the looming image of a demon dwelling in a fellow householder's inner self from their social landscape. In this final chapter, I shall endeavor to tie together the social threads interwoven throughout this work, in an attempt to do justice to and shed more light on additional aspects of the Thamgharian social universe.

## NOTES

1. Although normally Nepali root verbs end with the suffix *nu*, in local pronunciation many root verbs end with the suffix *ne*.

2. Stone (1978: 490); Rutter (1993: 168 and passim); Aubriot (1997, esp. 68–69).

3. Mallick (1982); Löwdin (1986: 17); Upadhyaya (1996: 193); Karan and Ishii (1997: 77–78); and Krauskopff (2003: 78).

4. Toffin (1977); Löwdin (1986); Nepali (1988); and Levy (1992: 64 and passim).

5. Following the conceptual distinction drawn by Ohnuki-Tierney (1993: 42–43).

6. This is implied in or emerges, with minor exceptions, from all the major works discussing high-caste Nepalese life. See in particular Stone (1978), Bennett (1983), Rutter (1993), and Gray (1995: 132–4).

7. Parry (1985: 613) and Lindenbaum (1986: 255).

8. Notwithstanding the fact that in Japan rice is no longer the staple quantitatively, only qualitatively (Ohnuki-Tierney 1993: 39–43).

9. Stone (1978: 49) and Rutter (1993: 168).

10. In the main, this part of Rutter's work discusses Viṣṇu's prominent ascetic role in the rice cultivation cycle in a Chetri village.

11. Aubriot presents a comprehensive account of the unique and highly complex rice irrigation system employed in a Brāhmaṇ village in west-central Nepal. In particular, she develops a comparative discussion of the technical aspects of rice cultivation and attends to the social dimensions of this particular water-management system.

12. From the eleventh day of the clear fortnight of Āṣadha (Asār in Nepali, June–July) until the eleventh day of the clear fortnight of Kārtika (Kārtik, October–November).

13. Gaborieu's division of the year seems to be consistent with north Indian (Raheja 1988: 57; Fuller 1992: 109–10), and Vaiṣṇavite Chetri (Rutter 1993: 362–78) notions.

14. See figure 7.1.

15. Most village "Untouchables" do not own much land suitable for rice cultivation, and they do not participate in the Brāhmaṇ *parma*. They do, however, take part in the *nivek* system of paid (in cash and kind) labor.

16. The fourth is a large, barren landslide.

17. For example, Lewis (1884) and Le Mesurier (1885).

18. The *murti* of the goddess (made of a small rock or an elongated pebble) often lies within a small rectangular or square niche cut in the ground level of the tree trunk.

19. Literally, red mud. Mixed with fresh cow dung, it creates a paste employed by the householder's wife to purify the kitchen after every rice meal.

20. I.4.1.22.

21. Doniger and Smith (1991: 74, 245).

22. The doctrine advocating nonviolence and nonkilling.

23. Cf. Lecomte-Tilouine (1996: 32) about the Brāhmaṇs of Gulmi who avoid plowing since it is regarded as wounding their "mother."

24. Trawick-Egnor (1978: 32–47); Dube (1986); Moreno (1987); Manu 9.33–51 in Doniger and Smith (1991: 201–2); Cameron (1998: 161–62).

25. Namely, Moreno (1987) and Kapadia (1995: 211–12).

26. Lit. sowing rice; *biu* is a seed as well as seedling and *rākhnu* is to put. This expression is also employed to denote lighting a fire and as slang for human impregnation.

27. See chapter 4.

28. See also Parry (1994: 152).

29. Adapted.

30. Mid-June (1996).

31. Sowing can take place only on Monday, Thursday, or Friday, but villagers could not explain why.

32. *Dhān* is unhusked rice and *ropnu* is to plant.

33. The village Damai are part of the ordinary auspicious background in marriage, be it human or divine (such as the marriage of Kāli prior to the Dasaī sacrifices).

34. It may simply be that women were too shy to perform in my presence.

35. An Upadhya Brāhmaṇ, forty-five-year-old Muktinath was a senior teacher in the local Sanskrit high-school. Endowed with a strong political sense, he had just left the Congress party to join the RPP, *Rashtriya Prajatantra* (National Democratic) Party.

36. A number of villagers said that the central sheaf is the sacred cow while the encompassing ones are her *goṭhālo* (herdsmen) guarding it. One householder viewed the latter as the house's *pañchāyan deota*.

37. See chapter 5 and below.

38. A rather poor Jaisi Brāhmaṇ.

39. See also Turner (1996: 589).

40. Also *kunyū̃* or *kunyo*. Villagers were unable to elaborate on the origin or exact meaning of the word, and Turner (1996: 98) does not prove helpful either.

41. See, for example, the picture of a huge Tarāi *kunyā* built by Tharus in Guneratne (1999: 137).

42. When no reference is noted, the data provided herein is based on three short-term studies of Hindu Newār customs of rice cultivation I conducted in Bhaktapur in July 1995, autumn 1995–1996, and 1999.

43. *Dhup halne*, literally, to offer incense. This is a minor *pūjā*, where a few drops of *ghyu* are sprinkled over a burning coal situated on a leaf or stone.

44. Likewise, much attention is paid throughout to avoid any implicit association with death or evil spirits.

45. *Buṛi* literally means an old lady and is a common manner of referring to one's wife. *Biṭo* is simply a sheaf of rice.

46. See also Turner (1996: 392).

47. Turner (1996: 410) notes that to make *phero* is also to encircle, to embrace.

48. This is the pinnacle of a Newāri pagoda-style Hindu temple (Slusser 1982: 148) of which the Kāli *mandir* above Thamghar is a minor replica.

49. For example, Vajracharya (1997: 6–7).

50. Beck (1969: 557); Dumont (2000: 412–19); Hamilton (2003: 156); Krishna (2003: 406).

51. In the terms employed here, this is clearly a manifestation of a ritual fractal.

52. As I witnessed during a short stay in a Brāhmaṇ hamlet on the southern outskirts of the Katmandu valley in the autumn of 1995.

53. As is done with the *biṭulo* spring crop that a number of families are able to grow, and as also reported by Aubriot (1997: 58, 547).

54. In parenthesis I would like to note that the notion of rice as a gendered entity is by no means unique to Thamghar. According to Dr. Inga Britt Krause (p.c.) these are also found among high-caste Hindus in western Nepal, while Levy (1992: 321–22, 642) reports on very similar notions regarding consecrated rice among Bhaktapurian Newārs. The latter view the husk as Śiva while the kernel is likened to Śakti—his consort. Regarding other high-caste communities, Kondos (1982: 278–79, n27) notes that the husked rice grains are gendered male and associated with Brahma, and likewise Gray (1979: 93) talks about "a male meaning" to husked rice. Although the Thamgharian notion that husking removes the grain's reproductive potential (as will be discussed below) is shared by one Chetri community (Rutter 1993: 412, 421), the latter view the husked grain as a symbol of ascetic celibacy, in clear contrast with its vital femaleness in Thamghar.

55. Here Thamgharians come very close to the *Upaniṣhadic* metaphorical employment of rice as a token of the *karmic* process of rebirth (Potter 1980: 245–48; O'Flaherty 1980c: xvi–vii).

56. Thamgharians explicitly note that *cāmal* (not *dhān*) is the optimal grain for conveying *dān*.

57. Coulson (1992: 106–7).

58. Like Kāli, whose marriage is performed in her regional temple on the hill overlooking Thamghar prior to the climactic Dasaī sacrifices.

59. The pyramidal Dravidian type is mainly found in the remaining southern fifth of the subcontinent (Kramrisch 1976: 205).

60. Campbell (1976: 85) describes similar ritual enclosures in Kangara, north India.

61. Cf. Volwahsen (1969), Brown (1971), Kramrisch (1976), Michell (1977), Burckhardt (1986), Meister (1991), Snodgrass (1994), and Hardy (1995, 1998).

62. A very similar imagery is evident in a Tamil love poem dated from A.D. 100–300, which tells about the Pāṇḍya queen awaiting the return of her husband and king Neṭuñceḷiyan from battle. The king's palace is described as if it was a temple having both an external and internal houses. The inner house is where Aṇaṅku, Śrī (the goddess), dwells and is called the "house of the embryo" (*Garbhagṛha*), which in effect was a bedroom. There, the queen sat on her large round marriage bed, replete with symbols of marital and fertile power, coitus and pregnancy (Hudson 1993: 132–34).

63. This clearly only pertains to the temple's patron.

64. In effect, the agricultural *jagya* may be seen to conclude only when rice is cooked at home. Although the grain's gender, as implied above, may have significant implications on the manner in which cooking is perceived, this subject falls beyond the scope of the present work.

# 8

## The "World Upside-Down" or a Himalayan Inversion of Hierarchy and Trust

### INTRODUCTION

Following Gambetta (1988b), who intimates that every human society is located somewhere on the continuum between cooperation and competition, trust and mistrust, while simultaneously maintaining a balance between these qualities, the present chapter demonstrates how the Brāhmaṇ community of Thamghar is embroiled in a process that engenders an oscillatory annual shift of its position on the above continuum. Furthermore, I would like to argue that it is probably this recursive movement that may account for the resilience evinced by what would otherwise appear to be the village's rather fragile social fabric.

Notwithstanding the dominance of the antisocial tendencies characterizing the Thamgharian social milieu, and despite the involvement of strong negative feelings behind the amiable façade exhibited in everyday social interactions, Thamgharian villagers are well aware of their mutual dependency. This is encapsulated in the saying that *tai pani milāera baschan*,[1] meaning that despite everything, people know that they must live together and make appropriate mutual concessions, although this is hardly noticeable and rarely admitted. The high point of this social ideology, so to speak, as Thamgharian Brāhmaṇs steadfastly reaffirm, and a notable exception to what is otherwise the ordinary state of village social affairs (discussed in chapter 2), occurs in *dāī hālnu* (*hālne*), the climactic finale of the rice cultivation season, when the rice is threshed and brought home. This liminal time of the year usually sees the end of the rainy season, and *dāī hālnu* also concludes a series of festivals, notably the most exciting and joyous Dasaī and Tihār.[2] In fact, *dāī hālnu* also marks the year's end since in Thamghar, as in other rural Hindu communities,[3] the year is dominated by the cycle of rice cultivation, and very little attention is paid to the new year according to the official calendar that follows the Indian Bikram Sambat era.[4]

These autumn days are characterized by abundance; the village *dhārās* (springs) are overflowing with water, a variety of vegetables grow in the kitchen gardens, and the lush forest seems ready to invade the village, thus making the daily search for forage an easy task. With the ripening of rice, householders express greater self-assurance and mutual tolerance, and any minor controversies that occur in relation to irrigation die out when it is stopped a number of weeks prior to the harvest. During this joyful period, both men and women show unparalleled mutual generosity and solidarity, and vegetables, various commodities, or assistance may be handed to the needy "for the sake of *dharma*,"[5] as villagers put it, without expecting anything in return. The village changes its appearance as all houses are repainted and decorated before the celebrations of Dasaī and Tihār, and many village people, especially children, walk around wearing new clothes.

Dasaī involves a great deal of reconciliation within the *kul* and unless a conflict is totally beyond rectification, all *kul* members will come to receive their *ṭikā* and blessing from its elderly senior member. Throughout the last days of Tihār, householders spend much time together, playing and gambling cheerfully at the few village teashops. These festive weeks, relying no doubt on the growing cooperation throughout the rice cultivation season, gradually seem to alleviate the mutual suspicion, tension, and anxiety present in ordinary village social interaction. As I shall argue below, these sociable expressions of geniality culminate in *dāī hālnu* to create a total inversion of the social atmosphere and behavior found at all other times of the year.

## *DĀĪ HĀLNU*

*Dāī hālnu*, the final and one of the most laborious stages of rice cultivation, is usually executed through *parma*. In a medium-sized field, a group of fifteen to eighteen men would suffice to carry out the task. Householders account for the nonparticipation of women in *dāī hālnu* by saying that "they are not suitable for the hard work needed" and "the night is not for them to wander out of the house."[6] Likewise, the work itself is performed at night since "this is our *calan* [custom]," and the night's moisture eases the work by reducing the amount of dust.[7]

As with any other work, Thamgharians ensure that *dāī hālnu* commences at an auspicious moment; it is symbolically inaugurated a few days before the actual work begins, by the householder tilting the *juro*. *Dāī hālnu*'s various stages are laden with numerous ritualistic, verbal, and bodily gestures performed in order to avert the evil eye, deter ghosts and demons, ensure the safe arrival of the rice into the house and bestow fertility. It is important to note that each household performs *dāī hālnu* in a slightly different manner, but variations are mainly stylistic and of little relevance here. The description herein includes *dāī hālnu*'s basic and most common features.

## From My Field Notes

Two to three hours after normal bedtime (around 8:30 at this time of the year), the field owner walks from house to house, waking and calling his coworkers to join him. The threshing begins late at night and is marked by high spirits and an unusually festive atmosphere; the men stand in a large circle, beating the sheaves against the ground while the field owner's children spur on the oxen that walk over the rice for a final threshing. The oxen are tied in a row to a special bamboo pillar called *miyo*, with a sheaf of rice tied toward its top. As in a *jagya*'s *liṅga*, this sheaf of rice is said to be the *miyo*'s head and is a token for it being alive.

The execution of any work in Thamghar is normally strongly associated with *dukkha* (trouble, effort, or literally: sorrow, pain, suffering), yet in the darkness of the night men tend to be at their best, expressing much joy and

Dāī hālnu—*the men stand in a circle and thresh the rice by beating the sheaves on the ground. Behind them, on the left (with a man standing on top) is the half-threshed* kunyā.

*Children spur the oxen walking around the* miyo *(right).*

amicability, and jokes, rarely heard in public at other times, are told all night long. People feel free to laugh and gossip about any subject and tell many stories of love, women, and even sex. In the background, the children scream and play unrestrained and the general atmosphere is of a cheerful, light-hearted festival. Ordinary boundaries and rules concerning caste and purity status are ignored and Brāhmaṇs overlook what in the village would otherwise be considered "polluting" contact with the "Untouchables" working alongside.

Within a few hours, a mound of rice appears in the middle of the human circle while heaps of straw pile up around the *khalo garo*. The symbolic joining of the *buṛi biṭo* and *juro*, which are the last to be threshed, marks the completion of the work. A fire is now lit and most *khetālo* (workers) come to smoke and keep up the festivity around its warmth, while two Brāhmaṇ householders begin to cook *jhyauko* (a particularly pure rice dish cooked in pure *ghyu*) nearby. Prior to the meal, a portion of *jhyauko* is placed on a double *sāl* leaf at the top of the rice heap as an offering to Bhūme, along with a special sickle symbolizing Śiva.

The meal is eaten communally and in high spirits, without the usual clear separation between high and low castes. Afterward, people remain by the fire for a while, Brāhmaṇs tell jokes about "Untouchables" and the latter make fun of Brāhmaṇs, sharing humorous stories about women and sex, and nar-

*Retiring to sleep in the straw around the pile of rice.*

rating past tales about evil spirits' attempts to attack the precious rice mound. The festivities draw to a close toward the small hours of the night when *āgo marcha* (the fire dies) and the workers enter the heaps of straw surrounding the *khalo garo* for a few hours of sleep before daybreak. The straw provides a warm and convenient bed for all, and in order to feel more secure, the men

*The morning after* dāī hālnu*: carrying the straw to the village in the morning mist.*

and children are careful to maintain physical contact, sleeping head to toe or side by side. The field owner sets fire to a single bundle of straw with which he creates a circle of ashes around the pile of rice in order to protect it from evil spirits and *birs*, and he is the last to retire to sleep.

During the first set of events in which I participated, the mornings were quite chilly and misty, and visibility was limited to only a few meters; the image of the men emerging from the piles of straw touched by the first dim rays of sun in the morning haze was indeed surreal. During the morning, two men stay behind to guard the rice while the others carry the straw to the village. There, the field owner's wife offers them a special festive meal, which in contrast with ordinary meals, is quite a leisurely affair. Following this and prior to retiring to sleep, the workers sit in front of the house enjoying the household's hospitality.

Around noon, all return to the field and the next few hours are devoted to a careful public quantification of the "organic gold"—the threshed rice—in the *khalo garo*. In stark contrast with the distribution of grain on the Indian threshing floor, which was often thought of as a token of the moneyless *jajmāni* system (Dumont 1980: 97–108; Good 1982; Fuller 1989: 32), Thamgharian householders maintain that no grain should ever be given away in the field so that it does not lose any *saha* before it is brought home. Once the rice has been measured, the *khetālo* carry the heavy sacks on their backs to the village

*A pile of rice (with the* miyo *behind it) at the center of the* khalo garo.

where, following a lavish *khājā*, they bring the rice inside the *talo* (first floor) and pour it into the empty *bhakāris*. At the bottom of the first one, the householder places the embodied manifestation of the earth goddess in the form of a small stone that was employed inside the *kunyā*. The first floor of the house, opened to fellow householders for this brief visit, is closed once again until

*Pouring the rice into a* bhakāri *on the first floor.*

next year, and *dāī hālnu*'s festivities conclude with a final festive meal at the field owner's house. Over the next six weeks, *dāī hālnu* is performed at every household until all village *bhakāris* are once again overflowing with grain.

Following *dāī hālnu*, men take leave of all agricultural work, saying that they need to rest "like a woman who has given birth." During these first days of the dry season the nights are crisp and cold but during the day, when the sun is up in the crystal-clear sky, temperatures outside are moderate and pleasant. Throughout this period, Thamgharian men continue their mutual social visiting and those who can afford to, spend most of the day together in the sun, gambling and playing cards.[8] However, this practice gradually declines as householders become increasingly preoccupied with the preparations of their fields for winter crops. Old disputes soon reemerge into the forefront of social life and interpersonal tensions, mistrust and hostility gradually replace the relatively short-lived sociability and cohesion.

## ANALYSIS

### Inside-Out: Inversion and Liminality in the Field

For clarity, the ensuing analysis will be divided into two parts in accordance with the key stages of *dāī hālnu*. The first considers the events taking place at night in the field while the second deals with the events back home in the village.

*Dāī hālnu*'s first stage is characterized by inversions of time, space, and speech. In contrast with all other occasions during the year, on this night people willingly face the nocturnal evil dangers together, working, eating, and sleeping outside. Moreover, the jovial meal at the end of the night's work often takes place around midnight, a highly ominous hour, particularly afflicted by evil spirits and *birs*. Similarly, contrary to the usual taboo regarding any reference to sexual matters and the general shyness people feel about such personal affairs, not to mention the scarcity of benign humor in village discourse, during the night of *dāī hālnu* men exercise remarkable freedom of speech, which includes joking and exchanging tales about women and sex. However, the aforementioned inversions have only marginal significance in *dāī hālnu* and are best viewed as accompanying markers of the more significant inversion that takes place during this night, to which I now turn. Viewing the night's activities in liminal terms assists in making sense of their meaning, as will be demonstrated below.

My general argument is that the labor and sleep in the field should be viewed as a transformation through liminality and inversion, which takes place in several stages. First comes separation—the men leave their houses and the village and descend to the field. Next follows a transformative stage, equivalent to marginality or limen in Van Gennep's terms, which in itself may be divided into three parts: the collective work, the communal meal following completion of the threshing, and sleeping side by side in the heaps of straw surrounding the rice mound. The latter two parts merit a brief elaboration.

At this point it is necessary to refer back to the discussion regarding the significance of food in chapter 2, where it was noted that sharing a meal (cooked rice) is a trust-building action and a prologue for social intimacy. This is exemplified by the instance of a relative coming to sleep in one's house: no matter how late he or she may arrive, the guest must first eat a full rice meal before retiring to bed. By the same token, in *dāī hālnu* where the *khalo garo* is viewed as analogous to the field-owner's abode, it is imperative for the workers to have a meal of cooked rice before retiring to sleep, otherwise at this late hour of the night, a *khājā* could obviously suffice. In order to guard the rice, the men sleep side by side (as family members usually do) in the heaps of straw surrounding the *khalo garo*. I often heard that *lās kurna bhandā rās kurna gāro huncha* that is, it is harder to watch (guard from evil spirits) the pile of *dhān* than to guard a dead body (considered extremely vulnerable to evil and pollution). This, I argue, is a further stage in the growing intimacy between the *khetālo* on this night, enhancing their transformation from hostile and suspicious neighbors to close friends, even members of one big family. Indeed, this part is clearly characterized by feelings and an atmosphere of "communitas" in Turner's terms. Moreover, the nocturnal transformation can be seen as a mini *jagya* in itself, in which the group of men, encompassing Bhūme and the pile of rice in the hub of the *khalo garo*, undergo a "cooking" process within the *parāl* (*pharal*, straw) and are awakened or reborn the next morning as members of one large family.

The final liminal stage of aggregation is enacted the following morning when the straw is carried up to the field owner's house, where the workers are treated to a festive meal. This acknowledges their altered status, as on almost no other occasion of *parma* are people invited home for a proper rice meal, let alone a festive one.

Let me summarize some of the points argued hitherto. During the night of *dāī hālnu*, householders undergo a certain "rite of passage" or transformation. Together they engage in an inversion of their daily behavior, relationships, and norms, and collectively they share the perils of work and sleep outside, thus facilitating their metamorphosis from dangerous strangers to trustworthy friends or kin.

In fact, *dāī hālnu* (re)creates a moral community that (as far as possible) will be reunited, year after year, to thresh each other's rice.[9] Only after this transformation has taken place do the men become qualified for the two main inversions of *dāī hālnu*: the public display of the scale of wealth and prestige (the amount of rice produced), concealed at all other times of the year, and the revelation of the householder's heart and presentation of his inner self to the inquisitive eyes of fellow householders.

## Divine Hierarchy and the Revelation of Scale

There are various reasons that induce villagers to conceal their scale of wealth and prestige (rice), and preclude them from granting fellow villagers access into the first floor of their house, as discussed in detail in chapter 2. Thus it may be expected that every family would thresh its rice during the (safe hours of the) day, possibly by themselves (as is done by the few families that manage to cultivate the spring crop or as is the normal practice in the Katmandu valley) in the field or near the house, without turning its measurement into a semipublic event. What therefore is the significance of the display and public manipulation of scale during *dāī hālnu*, and what are its implications?

I would like to argue that the significance of this public display of wealth stems from the special place the people of Thamghar accord rice; in addition to being their staple diet it also embodies the scale of a householder's wealth and is a major component of his prestige. Moreover, as will be recalled, unlike all other Thamgharian hierarchies that are mainly perceived as unalterable and divine,[10] a change in one's position on the scale of wealth and prestige is mostly perceived as demonic, representing evil and danger, blurred categories and confused positions, giving rise to grave feelings of uncertainty and dread.

People thus feel the need to control change, as there is no real way of preventing it. I would therefore like to suggest that in *dāī hālnu*, the continuous process of economic change is incorporated and co-opted through the mechanism of inversion and public exposure of scale. To my mind, this is not an occasion where hierarchy itself is inverted, opposed, or challenged. Instead it is the presentation of a new, altered hierarchy, perceived as "natural" or divinely ordained, beyond any question or protest. Furthermore, it is little wonder that the scale, concealed at all other times of the year, is exposed during *dāī hālnu* when all seem to possess sufficient wealth, and such a revelation entails pride not shame. It may also temporarily satisfy the immense mutual curiosity that has accumulated during the year. What further seems to facilitate the general acceptance of the hierarchy thus established is the psychological effect involved in the communal work and mutual help throughout *dāī hālnu*, where householders actively share and are part of each other's success.

The hierarchy of wealth seems to follow a cyclical annual process as summarized in figure 8.1 below. By this process, the values attributed to the

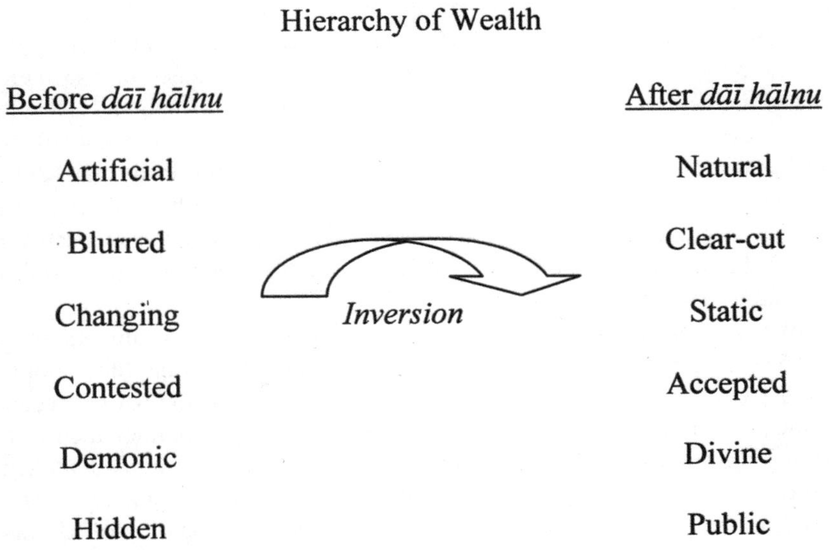


**Figure 8.1. Hierarchy of wealth**

above hierarchy during *dāī hālnu* are gradually altered over the course of the year, as suspicion, doubt, and mistrust grow, until they give way to their opposites some time prior to *dāī hālnu*, when they are again negated, inverted, and transformed while hierarchy is reestablished.

The discussion hitherto suffices, I believe, to suggest that both the hierarchy of wealth and prestige, and that of trust and sociability undergo annual processes that parallel one another, yet do not completely overlap. The Thamgharian social universe may be seen to make a recursive oscillatory movement along the continuum stretching between cooperation and competition, and trust and mistrust, without ever reaching either pole.

## Outside-In: Operations on Perceptions of Self and Community in Inversion

In the first stage of *dāī hālnu*, "inside" goes "out" and aspects of life normally exclusively confined to the inner recesses of the house—meals, sleep, and sex—are taken outside. In the second stage, the house opens up its heart (the first floor) to accommodate, if only briefly, it's new temporal extended family, and thus outside comes in. This second major inversion, in fact the climax of *dāī hālnu*, occurs when fellow workers bring the rice inside the *talo*, thus breaking the taboo which normally precludes such access. Its significance becomes apparent if we keep in mind that the first floor of a Thamgharian house is analogous to the householder's heart, and contains the rice, the scale of his wealth and prestige and the actual means for sustaining his family, as well as his potentially dark emotions embodied in the form of a *bir*. In entering the *talo*, not only do village men check for the possible existence of *birs* there, but they are also offered a glimpse into the householder's inner self. Despite the low level of mutual confidence in Thamghar, ten to fifteen householders who personally witness and inspect the first floor are not easily gainsaid.[11] The enactment of spatial inversion by entering into the forbidden *talo* symbolizes a mental equivalent of reaching and revealing one's heart; in effect it encodes the moral and emotional transformation that took place in the fields, in spatial terms.

This inversion addresses the deep concern the people of Thamghar share regarding the perceived duality in the world; the perception that, for example, there are two distinct aspects to both houses and humans: the benign façade and the potentially evil interior. This is, indeed, one of the main elements underlying the village social milieu, characterized by strong fragmentary tendencies, mutual anxiety, and mistrust. The discrepancy between the inner self and the social person is resolved during *dāī hālnu* via a negation of its existence. By inviting fellow householders to enter and inspect the *talo*, villagers demonstrate that the everyday equivalence between inner self as authentic but negative or evil, and the positive but inauthentic social person, is invalid. In such a manner, *birs* are collectively "exorcised" and the very possibility of the existence of evil in the *talo*, as well as the householder's inner self, is obviated.

Instead of the daunting perception of a multifaceted personality, *dāī hālnu* brings to the fore a new equilibrium in which exterior = interior = authentic = positive. Thamgharians appear to believe that throughout the year they all wear masks and, playing on Goffman's (1959) back and front stages, conceal their authentic but dangerous "hearts." In *dāī hālnu*, not only are these masks removed, but the existence of masks as such is denied.

### Leaving Out the Women

One question, which has thus far remained unanswered, is why, in sharp contrast to all other stages of the agricultural *jagya*, are women totally excluded from participating in *dāī hālnu*? Although village men bring forward other reasons as mentioned above, I believe that the main, twofold reason is as follows. First, householders obviously do not wish their wives to be associated with, witness, or take part in the multiple inversions of everyday norms and behavior that take place in *dāī hālnu*. This may well be tied to the perceived difference in the permeability of the male and female bodies to pollution and evil.

Primarily however, I believe that village men, who are obviously unfamiliar with the classical theory of inversion as a "steam valve" for preservation, renewal, and consolidation of the social order,[12] feel that they cannot risk even a brief inversion of their unique position derived from the imagery of matrimonial encompassment. This may indeed seem dangerously undermining, as revealed by the tendency of many village men to evade the inversion entailed by Tij. There, as I wish to recall, wives assume the status of the goddess, but husbands are often reluctant or simply do not wish to acknowledge the situation in which women are "on top," to borrow from Zemon-Davis (1978), by refusing to cook and cater for their wives before and after their fast.

## PLAYING WITH IMAGES: ILLUSION OF DEMONS AND TRUST

Comparison between some of the main features of the inversion in Newfoundland (Handelman 1984, 1990: 139–59) and those of Thamghar draws attention to further peculiarities of the latter. Taken together, the two events provide a striking example of communities that share very similar social conditions, which have however evolved from radically different backgrounds.

Notwithstanding the major differences in ecological conditions, religion, and other social characteristics,[13] marked similarities can be found. These include the constant use of masks in social interactions and the general atmosphere of mutual fear and suspicion. In both communities, inversion is a medium for operation on the dual perception of personhood, which forms a significant concern for individuals and society. Through inversion, members of both communities reconstitute social relationships on a basis of mutual trust as opposed to "must," which otherwise lies at the root of everyday (minimal) instrumental relationships.

The mechanism by which trust is created in each community is, however, quite different. In Newfoundland, through inversion and reversion, masks are worn to conceal the social person in order to allow controlled revelation of the "demon" (inner self), only to deny its existence afterward. In Thamghar, through the mechanism of inversion and analogy, masks are removed (the doors of the first floor are opened) to reveal the benign interior and to deny the existence of any demon underneath/inside. While in Newfoundland this is done in a ludic way, during the liminal, festive period of the year's end (Christmas), in Thamghar it is the end of the agricultural year (the rice harvest) that provides the setting for inversion. Concomitantly, the benign humor present in the latter has only marginal significance and is unrelated to the major inversions of the event, as subversive humor might jeopardize the precarious element of trust achieved.

Following Handelman (1984: 270) who notes that inversions are fictive events, I would like to argue that the above processes involve a great deal of social imagining and that illusion plays a crucial role in the course of *dāī hālnu*. Although men enter the first floor in person to witness firsthand that no *bir* lurks therein, this does not give them direct evidence about the householder's inner self. Furthermore, despite entering the *talo*, they are not allowed to wander around the house or go up into the second floor that many houses possess.

In addition, although every villager visits no more than twenty houses, people seem to take it for granted that every house in the village has been inspected and found to possess a benign interior. Thamgharian Brāhmaṇs exude the impression that they overlook these issues and prefer to indulge in a communal *māyā* (illusion). Likewise, I would like to suggest that illusion is not only necessary for sustaining the fragile fabric of trust thus created, but perhaps also forms the backbone of trust elsewhere. During *dāī hālnu*, individuals and collectivity seem to view themselves through an invertive mirror that reflects a mirage of what they deem to be an ideal state of social affairs.

## CONCLUSION

The inversion of *dāī hālnu* forms a cultural arena where imaginings of self, the other, and collectivity are operated upon, manipulated, altered, and re-established. On a broad level, the analysis of *dāī hālnu* suggests that the year's end, a period that people may feel to be liminal, is prone to generating events that include inversions, and these may be part of or are disguised within different circumstances, such as the harvest of rice. In addition, it advocates that inversion events should also be examined in liminal and processual terms, that is, as possible parts of a process or as a process in themselves. Viewed in this way, inversion may also be found to be an expression of or a way for generating change in different realms of life and to different degrees, thus calling into question the view that inversion is necessarily conservative in nature,[14] and subscribing to the socio-historical analysis of Zemon-Davis (1978), Gell (1997) and Muir (1997: 103).

That the main agricultural cycle and the annual regeneration of food mark the rejuvenation of society and cosmos is obviously not unique to Thamghar and may be found in places where agrarian cosmology or ideology prevail, such as in Japan (Ohnuki-Tierney 1993: 44–62, 81–98). Yet in Thamghar, this is also capable of generating an annual fluctuating social process, which to a certain extent may be responsible for holding its community together. *Dāī hālnu* may be viewed as marking the dramatic and triumphant return of Bhūme with her potent *saha* (imbued within the rice) into the house, thus drawing the rice fields back into the orbit of the village. In my view this constitutes a serious attempt to alleviate the unintentional, negative social implications of matrimonial encompassment (discussed in part I) and turn the entire village into an embodied representation of the ideal primordial house. Thus the house, householder, the village, and its community are remade from the outside. Indeed, in view of the fragmented Thamgharian community and the lack of a permanent, enlivening presence of the earth goddess, it may come as no surprise that the village's regeneration cannot come from within. For this reason householders must go outside into the creative, fertile, auspicious, and feminine grounds of their rice fields, which emerge as the village's "background potentiality" in Hirsch's (1995) terms.[15] Only there, so it seems, can the kernels of amicability and harmonious relationships be sown. The latter's gradual growth and ripening, marked by the progression of the rice cultivation cycle as well as the gradual execution of *dāī hālnu* throughout the village, results in a short-lived blooming of the Thamgharian social universe.

Ostensibly, the agricultural *jagya* requires going *out* of the village and *down* the hill, yet I would suggest that it actually entails an *upward* movement in scale and an *inward* one cosmically, toward the primordial, encompassed, energetic feminine grounds of the village background potentiality. The periodic regeneration, operation upon, and maintenance of the invigorating, gendered, fractalic image of matrimonial encompassment is no less crucial for the well-being of the Thamgharian social body than it is for human bodies, village houses, and mesocosmic ritual arenas.

The vibrant colors of the somewhat elusive ideal social and cosmic scene painted during *dāī hālnu* is bound to gradually fade away until they are totally erased with the arrival of the monsoon rains, only to be repainted once more, albeit for a short term, during the following year's rice harvest.

## NOTES

1. *Milāera* comes from *milāunu*, to adjust, to bring together, or to mix, and *baschan* derives from *basnu*, to sit, dwell, and settle.
2. These are normally celebrated during the months of September to November.
3. Good (1983: 236); Moore (1983: 298); and Ishii (1993).
4. Which began in 57 B.C.
5. This complex concept literally means religion, but here is used more in the sense of religious duty or ethical behavior that enhances one's accumulation of merit.
6. Notwithstanding the fact that, for example, it is mainly women who take part in the modest communal "thanksgiving" evening *pūjā* held for Bhūme a number of days prior to *dāī hālnu*.
7. Nonetheless, the few families that manage to cultivate the spring crop (and appear to be able to thresh their rice without help) always tend to do so during the day.
8. Thamgharian women, whose main opportunity for social gathering is limited to proper religious contexts (such as Tij, *purān*, or a wedding), hardly seem to take part in this short spurt of sociability.
9. I know of one family who repeatedly invited a man for the festive meals of *dāī hālnu*, many years after he was crippled and could no longer take part in the work.
10. Namely, those of ritual purity, caste, age, and *kul* seniority.
11. There are, however, a number of villagers who report the existence of *birs* in *talos* they have entered, but such claims are not normally taken seriously.
12. Notably Gluckman's rituals of rebellion (1954) and Turner's early work on rituals of status reversal (1969).
13. The Egalitarian Christian society on the Newfoundland coast versus the Hindu patriarchal and hierarchical society of the Himalayan community in Thamghar.
14. Gluckman (1954); Marriott (1966); Turner (1969: 172–203); Handelman (1984: 247, 1990: 158–59). In contrast with his earlier writing, Turner (1978: 295) seems to acknowledge the possibility that inversion, as a liminal situation, may generate change.
15. In his conceptualization, Hirsch (1995) suggests viewing landscape as a cultural process, that is, the relationship between two poles of experience, namely the "foreground actuality" and "background potentiality."

# 9

## Conclusion

Vale de Almeida's (1996: 165) recent observation regarding the term "conclusion" is highly pertinent at this point, drawing our attention to the false call for a final, authoritative endnote that seems to be embedded within this title. This, he maintains, is because the main objective of the study thus concluded ought to be the identification of new paths for others to improve and challenge one's work, build upon and employ parts of it elsewhere. With a similar aim in mind, I shall reflect upon the journey made throughout this book and attempt to raise, albeit in a more succinct and abstract manner, some of the major points discussed hitherto.

Understanding the cardinal, shifting contexts in which the fractalic and varied image of a male encompassing a female operates and is of significance, clearly opens up a number of innovative avenues for the exploration, discussion and comprehension of the organization of life and experience in Thamghar. This in turn has crucial implications for our comprehension of Hindu culture and society, and further advances modern anthropological theories of ritual, action, and agency.

I have attempted to demonstrate the capacity of this image, although apparently rather an abstract and distant construct, to become densely imbued with meaning, and serve as a highly powerful mental frame for reflection, interpretation, and manipulation of reality. In fact, as though possessing a life of its own, this image was found to be capable of molding people's emotions, perceptions of self and the other, kinship, and social relationships in a rather unpredictable manner. Succinctly put, the fractalic image of male encompassing female in Thamghar, particularly via marriage, embodies a way of life and a path for survival, a means of exercising control, as well as a route for creating a family, producing offspring, and accumulating wealth and prestige. At times, however, it is imagined as a gateway to possession and potentially being sacrificed from within. It is a salient element that generates doubt and interpersonal suspicion, thus reproducing the slope of mistrust, which once again carries the village community to the brink of dissolution. Yet it also embodies the main path to rebirth and the attainment of (conditional) immortality, and ultimately, though rather indirectly, it provides a primary route for building social relationships and cultivating mutual trust.

In probing the significance and implications of the Thamgharian gendered fractal and the image of matrimonial encompassment, rather than focusing on a relatively narrow domain of study with only minor diversions, I have opted for a broader approach that places equal emphasis on an examination from a number of perspectives and well-trodden paths of scholarly inquiry, which mutually inform, elucidate, and reinforce each other. I believe that only such a synthesis of perspectives, which pays heed to the ways in which various domains of human existence are potentially interconnected or may influence one another, can fully illuminate the power that indeed, as Wagner (1986: 56) notes, lies within an image (particularly a fractalic one), namely, its ability to fuse together and condense entire realms of possible ideas and interpretations.

If through the employment of the above strategy the previous pages managed to convey or create in the mind of the reader a somewhat coherent picture of life in Thamghar, it is mainly an artifact of the Euro-American tendency to look for consistency and Western modes of understanding. A well-known adage states that what is probably the only major consistency found within the Hindu universe in general, and the Thamgharian world is obviously no exception, is the omnipresence of complexity (often referred to in hydrodynamic terms) and inherent paradoxes. Much of the latter is captured and elucidated via the general concept of fractality, yet this obviously neither obviates the inherent complexity nor does away with the paradoxes that govern Thamgharian reality or Hindu life more generally. In addition, as is applicable to all similar research and written accounts, the picture presented herein is, inevitably, an incomplete one. Not only have I presented only part of the field material I have amassed, but I believe that any research cannot exhaust the myriad ramifications of a certain local notion or idea. This obviously calls for further future study and writing. For example, I have merely implied but did not focus on or analyze in detail the pertinence of the notion of matrimonial encompassment to the local perception of landscape, how it affects women's self-perception or its implications regarding cooking in general and cooking rice in particular, vis-à-vis the general Hindu management of food. Likewise, there appears to be much scope for additional comparative research, which would take as its starting point the various manners in which Hindus imagine the bond created between a husband and his wife via marriage.

This journey began with a detailed exploration of the intricacies and dynamics of the Thamgharian social world, overshadowed by the perception of a dual personhood; a lurking ominous inner self encapsulated within the inauthentic mask of amicability donned by the social person. It thus became apparent that neither the misappropriation of the framework for action and interaction with the nonhuman cosmic forces (in a hungry and emotive yet rather mechanical cosmos) onto the human social realm, nor the more general explanations previously proposed by others, are fully able to account for the intensity of doubt and suspicion that typifies the Thamgharian social world as it is locally perceived.

Unraveling the complexity of women's roles and positions as well as the dynamics of their alteration throughout their life trajectory, I argued that rather than being unitary in nature, these should be thought of as constituting parts of a mental hybrid of ideas about kinship and women. Concomitantly, it became apparent why the "ideology of the patriline" or "public male discourse" in relation to women should not be taken at face value. Instead, my point was that there are multiple, contextualized and varied perspectives at work, and that these must be examined simultaneously if one is to discern their import and overall impact.

This formed the prologue for evaluating the construct of human matrimonial encompassment and the analysis of the dialectics of images, perspectives, and cardinal village notions it elicits, in order to decipher the latter's rather unintended and undesirable social implications. In the main, I suggested that the above notional confluence, led by the image of matrimonial encompassment, plays a central role in fashioning the local perception of fellow villagers' daunting "moral geology," thus stimulating the village's self-perpetuating vicious cycle of mistrust and hostility. Exploration and extrapolation via the properties of the kinship-born image of matrimonial encompassment thus proved to be highly illuminating within the social realm.

A comparison with the workings of a similar south Indian notion demonstrated how, by drawing upon and making selective use of a similar pool of cultural images, albeit within a different kinship milieu, the image of matrimonial encompassment may become imbued with quite contrasting imports, which are highly significant for the local comprehension of experience and perception of relationships in both contexts.

Through the examination of the multifarious and fluctuating trajectories that sexual imagery and divine marriages had undergone within the historical body of Hindu myth and ritual, it was possible to map and locate the notion of matrimonial encompassment within its correct metaphysical context. This offered a new angle from which to view the decline of (mainly divine) reproduction that characterizes post-Vedic Hinduism, highlighting the contemporary idiosyncrasy of the Thamgharian notion of matrimonial encompassment, which seems largely oblivious to the major change in the Hindu mind-set toward procreation evident from *Upaniṣadic* times onwards.

Although it may be impossible to determine the origins of the aforementioned Thamgharian imagery, the historical and comparative discussions of ancient Hindu myth and ritual may suggest that its roots are ancient indeed and may go back to Vedic times. This may be the result of the fact that Thamgharians (like many other Nepalese Brāhmaṇs) hail from Hindus who had fled India in the middle-ages, thus experiencing little if any Muslim influence. In a similar vein, the worship of Vedic gods (e.g., Indra) and other Vedic practices (such as the worship of frogs)[1] are still vivid among Nepalese Newārs. Hence Nepal's avoidance of any long-term Muslim occupation may render it a "cultural museum" where Vedic and other ancient Hindu practices and concepts still exist. In addition, Thamgharian notions, although not Tantric per se, may also be the outcome of living in the midst of a highly influential Himalayan Tantric milieu. Here it is important to recall that Thamgharian Brāhmaṇs trace their physical origins and believe that the founder of their village was the mysterious and wonder-making *siddha* Damodar who was probably a Tantric priest and practitioner.

Proceeding to the ritual realm, the performance of the *jagya* and *hom* served as a magnifying glass for a detailed examination of the way Thamgharians present and operate upon their fractalic, gendered, and animated cosmos, governed throughout its various scales by the notion of male encompassing female, more often than not via matrimony. This illustrated how the employment of matrimonial encompassment within the mesocosmic *jagya* is a striking attempt to dissolve the Vedic-born solid nexus of life (procreation and rebirth) and death (sacrifice). In effect, it was shown to be a local endeavor to reinvent the ancient Hindu transformative mechanism of sacrificial rebirth via a divine matrimonial encompassment within a *pūjā*-like context.

A comparison with other, similar Hindu ritual and architectural undertakings demonstrated how the *jagya*'s efficacy depends on the particular manner in which the display and reification of the otherwise rather implicit fractality embedded in the cosmos, sets into motion a tripartite circuitous causal pathway for affecting and transforming the primary participants. It thus becomes apparent how cosmic fractality forms the backbone of the Hindu world's causality, why from a Hindu point of view the cosmos *must* be organized as a fractal, and what lies behind the fundamental importance Hindus ascribe to maintaining alignment with their fractal world. This further suggests why many other people in traditional societies find it appealing to imagine a fractal universe and opens a new perspective for understanding ritual efficacy elsewhere.

The Thamgharian tendency to fuse what Westerners would often separate was highlighted by the notion that matrimonial encompassment embodies the ideal modality of existence, divinely harmonious and overflowing with life and potency for man, the gods, and the cosmos alike. Redolent with auspicious erotic dynamism, matrimonial encompassment's primary virtue emerged as constituting an ideal passage to divine immortality, a means to conquer and envelope eternity.

The analysis of the village and agricultural *jagyas*, which mutually reflect and substantiate each other, demonstrated that in the Thamgharian world, both man and rice are gestated within a divine matrimonial encompassment and born from a similar ritual womb, which ultimately is their own. Thus, matrimonial encompassment appears to embody a crucial *modus operandi*, a template people use for creation, action, and production in the world.

Just as the implications of the image of matrimonial encompassment within the village's social milieu seemed rather unpredictable, it is particularly the divine matrimonial encompassment of the agricultural *jagya* that emerged as the ultimate lever of sociability in Thamghar. The rice fields, embodying the village's auspicious, energetic feminine "background potentiality," come into the fore as the arena where Thamgharian Brāhmaṇs are not only able to regenerate their annual food supply, but also where householders refurbish their prestige and reestablish their social standing. Moreover, the rice fields emerge as the loci in which the never-ending process of economic change is incorporated through the display of a new hierarchy of wealth, one that is based on an uncontested divine foundation, where perceptions of self and community are fundamentally operated upon in an apparent effort to rectify the village's social landscape.

Taken together, the analysis of the village and agricultural *jagyas* sheds new light on the evolution of the Hindu temple and suggests an innovative manner for its comprehension. This further consolidated one of the prime theses underlying the present book, namely, that the gendered fractal dominating Thamgharian imagery, perception, and praxis is not a unique local notion, as may ostensibly be the case, but is also implicit in general Hindu modes of life and imagination. Thus the fractal of matrimonial encompassment offers itself as a key notion for the comprehension and interpretation of Hindu culture and society elsewhere.

Much of the present study may be thought of as an attempt to respond to Gell's[2] plea for an anthropology that makes sense of (often irrational) behavior and utterances in the context of social relations, including those that exist in the vicinity of objects or artifacts mediating social agency, from a local high-caste Hindu Nepalese perspective. Indeed, if any single, major analytical vantage point has been employed throughout, it was strongly inspired by

Gell's action-centered approach, with its particular concern for the manner in which nonhuman entities are manipulated for eliciting specific effects, as well as their agency, inherent capacities, and personlike qualities. This was refined by the concept of embodied representation, by paying particular attention to the gender of nonhuman entities in the Thamgharian world and, at times, the sexual import of action. Thus I hope I have been able to demonstrate how in Thamghar, the gender and agency of objectlike entities are flip sides of the same coin.

To me, this approach has proved invaluable not only for the analysis of the Thamgharian material, but also for the examination of an array of other myths and rituals, allowing fresh light to illuminate certain more general Hindu practices and fundamental convictions. All of these are concerned in one way or another with a number of deep-seated notions that Thamgharian Brāhmaṇs appear to share with many other Hindus throughout the ages, namely, that we live in a fractal world, where agency and personhood may be distributed and dispersed, and these may result in or be channeled into a circuitous causal nexus. Concomitantly, this relates to the perception that entities may mutually affect one another and that humans maintain close, at times even intimate relationships with the nonhuman beings that exist in their world, as well as with the landscape they fashion around themselves. As was primarily demonstrated regarding the pervasive and crucial employment of rice, being attentive to local sensitivities regarding the gender and agency of nonhuman entities affords making sense of their significance, inherent faculties, and potencies, and elucidates their usage. The key to their social and ritual significance or particular roles in Thamghar lies not in seeking to decode such beings' "symbolic" value, but in understanding their personlike properties, identity as "social agents," biography, and gender. This approach, applied to the examination of the import and significance of the fractal of matrimonial encompassment in and beyond Thamghar, further highlights how Thamgharian perception, like much of Indian culture more generally, takes its major cues and is largely predicated upon the understanding of the nature of human bodies and persons, the biological processes of life, and the fundamental difference and relationships between the sexes.

It is my hope that this book may offer an additional gender and agency-sensitive perspective to those already occupying a central place in the analysis of Hindu contexts, and that the prominent images encapsulated within the various Thamgharian fractals, particularly that of matrimonial encompassment, will allow new insights into and prove helpful for the examination of additional issues within the corpus of Hindu thought and experience.

One general theme that has been recursively implied at different junctures throughout this work and was particularly evoked by its comparative discussions, concerns the relationship between content and form, as highlighted by the fractalic image of a male encompassing a female, the *jagya*, *hom* and the concept of sacrifice, the Nepalese *ṭikā*, the Hindu temple, and inversion in social life. Together, these all seem to suggest that the meaning and content of an image, construct, act, or mode of social behavior are not inherently embedded within their structure or form. Instead, these are determined by their specific context and are often ephemeral, shifting, negotiable, and open-ended in nature.

Finally, I would like to recall Geertz's[3] personal testimony, inspired by Kalidasa's *Sakuntala*, which narrates the tale of a sage who at first, while squatting before a real elephant, denied its existence and only later, when looking down at the beast's footprints after the animal had disappeared, doubted himself and finally declared with certainty that "an elephant *was* here." Geertz states that for him, anthropology is "trying to reconstruct elusive, rather ethereal and by now wholly departed elephants from the footprints they have left on my mind"; the Nepalese elephant that lumbered in my imagination and that I have attempted to reconstruct and research in the present book took the shape of a gendered fractal, governed by the image of a male encompassing a female.

I have employed this fractal to instill some order into the otherwise seemingly unrelated spheres of life and cultural phenomena within the complex and, what at times may appear chaotic, context of a Himalayan Brāhmaṇ village, which offers a new perspective to examine and comprehend Hindus and Hinduism elsewhere. Primarily, however, it appears that this Thamgharian fractal, like others dominating the Hindu imagination, is a fascinating expression of man's efforts to address the overwhelming complexity and unpredictability inherent in life and the cosmos—a primary way out of chaos. Notwithstanding the apparent abyss that lies between Western and Hindu horizons, and unlike so many Western concepts that are at odds with Hindu perceptions of reality, the Thamgharian fractal seems to come very close to the general place fractals occupy in our natural sciences, and thus supports Latour's[4] thesis regarding the general symmetry between the supposedly modern and premodern worlds.

## NOTES

1. Vajracharya (1997).
2. Gell (1998: passim, particularly 1–11).
3. Geertz (1996: 167).
4. Latour (1993).

# Appendix A

## A Personal Note

Upon reading this book, two questions that relate to the practicalities of the field research may arise in the minds of both the uninitiated and professional readers alike, particularly if one is familiar with Nepal and its people. The first question, pertaining to any anthropological fieldwork, is how the people and location of the study were determined. The second one, more specific to the present book, is how it was possible to go through what appear to be the closed social doors and hearts of the host community, and penetrate the walls of suspicion and purity barriers of a high-caste Hindu community. In addition, some readers may be interested to read a brief personal account recounting a few of the author's field experiences. The following lines attempt to address these issues.

### FINDING A HOME IN THE HIMALAYAS

As mentioned in the introduction, my initial plan was to study Brāhmaṇs living in the Katmandu valley. Following my arrival in Nepal, I gradually found this to be both less personally appealing and also very difficult to pursue in the hub of the Valley's rather busy urban center, where Brāhmaṇs did not comprise any localized community. Having completed an intensive language course and following a two-week field trip to Gulmi district in west-central Nepal, I made up my mind to seek a rural community to study. My first attempt to live in a tiny Brāhmaṇ hamlet situated on a hill on the edge of the Katmandu valley was very interesting, yet I found the place unsuitable for extended research and left after a fortnight. Now, I was looking for a medium-sized rural Brāhmaṇ community away from the Katmandu valley, where I would be able to live with a priestly family.

During this time, I was staying as a paying guest in the large house of an affluent Newāri family in the old part of Katmandu. As in other neighboring Newāri households, they often hired the services of a young Brāhmaṇ village boy who would work for them morning and evening. In return, the family provided him with lodging, food, and minimal educational expenses, and he was given the opportunity to study during the day. Shortly after I began to live with this family they took on a new servant. Ramesh, a seventeen-year-old Brāhmaṇ, arrived at this house extremely tired and rather overwhelmed by the trip he had just made for the first time in his life. He quickly fell ill and spent his entire first two weeks in the capital on a mattress in a side room. Long before I could converse with him properly in Nepali, I had the opportunity to give this clever boy a hand in a number of minor matters and we became friends. Later, when Ramesh heard that I was seeking a Brāhmaṇ priestly family to stay with, he immediately suggested I should come and live in his father's house in Thamghar.

During the following weekend, Ramesh took leave from his work and accompanied me on a brief visit to his parents' house. After a day-long bus journey we embarked on the twenty-six-kilometer, rather strenuous walk up and down hills and ridges to reach Thamghar. Apart from the fact that it took us almost a whole day of walking to reach the place (instead of the three to four hours of easy walking I had been promised), Thamghar was very much as Ramesh had described it; the high majestic peaks of the Himalaya dominated the landscape to the north, his parents lived in a community mainly populated by Brāhmaṇs and his father, Sharmananda Subedi (71), was a respectable practicing priest.

At the end of this successful initial visit, it was agreed that I would return within a week for a longer stay, and in spite of the protests of Ramesh's parents, I insisted on paying them in order to cover any extra expenses that might be incurred during my stay.

I can remember clearly that even on this first visit, unlike my previous brief experience in another Brāhmaṇ hamlet near the Katmandu valley, I was granted free access to the family's pure water container and other corners of the house, including the pure recesses of the kitchen where rice is cooked. Relatives of Thamgharians living in nearby settlements told me that they don't let the trekkers, who sometimes go past their hamlet and are invited for a meal, enter their houses. However, my purity status did not seem to be an issue in Thamghar. As far as Ramesh's parents were concerned, this was perhaps due to the fact that they knew I had shared the same house in Katmandu with their son. For the rest of the village my general acceptance, purity-wise, was probably facilitated by the fact that acceptance by a priestly family entails similar tolerance by all their *jajmāns* (clients), who are taken to occupy a slightly inferior ritual and purity status.

No doubt, it was Sharmananda who set the tone regarding my status, replying to all inquiries that my *jāt* (caste—Jewish—of which no one had ever heard before) is *suddha* (*pure*) and *ṭhulo* (high, big).[1] Thus, in theory, I could enter all Brāhmaṇs' houses, and was even invited by the village *pūjāris* (special priests) serving at the regional Kali temple to pay homage to the goddess there and perform a *pūjā* (worship). However, in reality, I only entered people's houses very rarely, mainly when accompanied by Sharmananda during his house calls. When I was by myself, I was mostly confined to the public domain and was entertained, like village householders, outside on the veranda.

Sharmananda and his wife were extremely patient with the invading stranger in their midst and went out of their way to assist me. However, my first days in Thamghar were not easy, equipped, as I was at the time, with only limited communication and other necessary skills for village life; although I arrived in Thamghar after gaining relative fluency in the Nepali spoken in Katmandu, it took some time before I could understand the local spoken dialect.

I do not wish to dwell here on the subject of physical hardships (e.g., sleeping, like the rest of the family, outside the house's walls under the thatched roof of the open veranda; the numerous insects I made acquaintance with, or the mice that roamed the roof beams at night and would occasionally drop off onto one's bed)—being accustomed to camping from an early age and having gone through Israeli army training made for reasonably painless adjustment.

Though I gradually became accustomed to the local way of speaking and began to understand the answers to my questions, the main difficulty was that the latter always took the form of counter-questions, often with complete disregard to the issues I wished to discuss. Alternatively, villagers simply wanted to know why I needed to know what I had asked, and saw little reason to cooperate.

In fact, whenever I appeared in public, I became the center of attention and the target for endless interrogation. The latter, so it seemed, left no area of my previous life and past experiences, as far as Thamgharian imagination goes, unexamined. Moreover, village men assiduously compared my answers with those given to my host family and later also to my research assistant. It seemed that all the topics I attempted to raise, the questions I asked, my gestures, and general behavior in one household were often already known when I came to another, and were bound to become public knowledge by the end of the day when I arrived at one of the village tea shops. While people were proud and quite amused to display their familiarity with my personal history and whereabouts during that day, they were reluctant to provide any information about themselves or much in the way of direct answers to my questions. Living "in the public eye" became my real hardship, far harder than simply sleeping outside the house or taking a shower in public with other men or women in the village *dhāro* (*dhāra*, spring).

Beginning my work with Eknath Lamichane, who became my research assistant toward the end of my fourth month in Thamghar, and whose wonderful family provided me with a second village home, considerably expanded my circle of acquaintances. Yet it did not enable me to push my research much further, since it seemed as if all village people, apart from the families with whom I lived and a handful of others, had joined forces in refusing to reveal much detail about their lives; if I exhibited a particular interest in any subject, this alone seemed a good reason to conceal it.

My breakthrough was gradual. I believe that this took place mainly since householders, many of whom continuously attempted to elicit information about my two host families from me without success, realized that I did not divulge information to anyone. With the passage of time, local people learned that I was quite harmless, many of them benefited from my free distribution of medicines and first-aid care, and slowly I gained the reputation of a trustworthy person who never revealed any detail he had learned. Although this did not prevent villagers (men, women and children alike) from occasionally attempting to obtain information about others, invitations to dine and stay at people's houses began to pour in. Often, I was indirectly reminded that I was invited because people trusted me, or was "put to the test" at an initial stage of my visit by being asked for details about my two host households. As I later realized, such hospitality is never part of the ordinary Thamgharian social scene and is reserved for respected *outsiders* only, such as the highly esteemed priest who comes to Thamghar once or twice a year for prolonged visits during which he performs exorcism sessions against *birs* (demons).

The frequent leisurely visits to various village households expanded my circle of acquaintances, deepening what would otherwise have probably remained a rather superficial set of relationships, and provided me with ample opportunity for private conversations on almost any topic. Understandably, householders became reluctant to comment publicly on any of the subjects discussed with me in their houses, and my presence no longer attracted much public attention.

Throughout my fieldwork, I remained relatively healthy, and when I did become ill, Sharmananda's diagnosis always pointed toward natural causes such as eating the wrong food or combinations of it but, unlike all his other clients, I was assured that no *boksi* (sorcerer) sought to harm me since I had no local enemies. However, just before leaving Thamghar I fell ill once more and this time, Sharmananda's diagnosis became quite ambiguous. Finally, he decided to *phuknu* (exorcise the evil spirit) out of my body, which was perhaps a sign that it was indeed time for me to leave.

## NOTE

1. Brāhmaṇs see themselves as belonging to a *ṭhulo jāt*, compared with the *sānu* (*sāno*, small) "Untouchable" *jāts*.

# Appendix B

## Glossary

When local pronunciation differs from Turner's or from Sanskrit it appears in parenthesis.

*abhiṣekah* (*abhiṣek*): Enlivening (Vedic) bath in potent water, often said to effect rebirth. Anointment.

*abir (avir)*: Highly auspicious vermilion powder.

*achetā*: Unbroken rice mixed with red vermilion (*abir*) powder.

*āghan* (*āgan*): The yard in front of the house.

Agni: The god of fire.

*agnicāyana*: Literally the piling up of fire. This is a year-long sacrificial process, explicitly conceptualized as gestation, in which the Vedic fire altar embodying Prajāpati is built.

*akās*: Heaven, the seven upper worlds.

*akāsi/pani kheti*: One of the titles given to the rice fields in Thamghar. Lit. meaning heavenly, fields "of the sky," fields irrigated with rain water.

Ananta: Lit. eternal, endless. The boundless cosmic King serpent.

*aṅga*: Limb, organ.

Annapurna: Lit. full of grain, also the earth goddess and a massive Himalayan mountain range in Nepal.

*ardhanārīśvara*: Śiva who is half female (Pārvatī/Śakti).

*ardhāngini*: Conceptualization of marriage as a state where the wife becomes the left half of her husband's body.

*ātmā*, *ātman*: Individual soul, mind, which is also regarded to be an *aṅsa* (*aṅgsa*, shared, one of many other equal parts or limbs) of god.

*auṣadh* (*auṣadhi*): Medicine.

avatār: Manifestation or incarnation of a god, deity, particularly of Viṣṇu.

*bāhān*: The throne of a deity, often the animal carrying it.

*baikuṇṭha*: Viṣṇu's heaven.
*bali*: Animal sacrifice, the "sacrifice" of a pumpkin in lieu of an animal, the offering of vegetables in lieu of meat, mainly to demons.
*balini (baleṇḍāri)*: The invisible sphere fashioned on the ground by the raindrops falling off the roof around the house, which protects it against evil spirits.
*ban* Devī: A particularly fierce territorial forest goddess.
*bandhu*: Vedic equations, correspondences, and connections.
*baran garnu*: To nominate, appoint, choose. Brāhmaṇ *baran* – nominating a priest for a ritual performance.
*bāri*: A nonirrigated field.
*bartoon*: Male initiation, high-castes' "second birth." Also termed *vratabandha* and *bartaman*.
Bāstu *pūjā*: Thamghar's most elaborate *jagya*, celebrating the major *saṁskāra* of the house—its marriage.
Bāstu Puruṣa: The (male) house god, embodied in a branch or curved tree trunk (*liṅga*).
*bāsudhāra*: The main rite of the *purbāṅga*, in which the family priest pours *ghyu* on the seven *Mātṛkās* situated above the door's lintel so that it drops over the plate overflowing with *naibedde,* which the *kart*ā's wife holds on her head.
*bedi*: The square mud "altar" at the *jagya*'s center, which represents Bhume and on which *hom* is performed. Its dimensions must be proportional to those of the *kartā*'s arm or palm.
*Bhagavad Gītā*: A major philosophical treatise and the heart of one of the two major Hindu epics, the *Mahābhārata*.
Bhai Ṭikā: Lit. the younger brother's *ṭikā*. The final, climactic day of the Tihār festival (celebrated during the dark half of the month of Kārtik, October/November), named after its main focus, the attribution of a colorful *ṭikā*—an auspicious forehead mark on each brother's forehead.
*bhaīsi*: Female buffalo.
*bhakāri*: Rice container made of bamboo.
*bhakāri bādhnu* (*bādhne*): Lit. to tie the *bhakāris*. The establishment of Bhūme's temple of rice seedlings during transplantation.
*bhakti*: Religious devotion, a general name for Hindu devotional movements.
*bhānsā*: The kitchen situated on the ground floor, the cooking and eating area of a house.
*bhāt*: Cooked rice.
*bhāter*: Festive communal meal.
*bheṭi garnu* (*garne*): An act of honoring a deity, person, temple, or a sacred book, as well as a gift to a higher entity (usually a coin).

Bhūme (Bhūmī): The earth goddess, also called *Annapurna*
*bhūt*: Evil spirit, often the ghost of a person who died a violent death and did not turn into an honorable ancestor (*pitṛ*).
*bir*: A horrific devouring demon.
*biṭulo*: Mildly impure.
*biu rākhnu* (*rākhne*): Lit. *biu* is seed or seedling and *rāknu* is to put. Sowing rice.
*boksi*: Sorcerer/witch.
Brahmā: The creator, in Thamghar he is worshipped as a tall *liṅga* as part of an elaborate *jagya* only.
Brahman: The undefined cosmic soul.
Brāhmaṇ: A member of the highest caste (Bāhun in Nepali).
*buhāri*: Daughter-in-law.
*buṛi biṭo*: *Buṛi* literally means an old lady and is a common manner of referring to one's wife. *Biṭo* is simply a sheaf of rice. This is a round, hollow cylinder-like construction (approximately one meter in height) made of seven or nine (must be an odd number of) rice sheaves tied together. Generally referred to as *pothi* and regarded as Bhume's *mandir*.
*byāṛ* (*byāṛd*): Lit. male beast. The nursery of rice seedlings.
*cāmal*: Husked, uncooked rice.
*canduwā*: A colorful cloth canopy that is tied above the *jagya* between its *liṅgas*. Various *naibedde* are usually placed on the *canduwā* and distributed as *prasād* at the end of the *jagya*.
*cāturmāsa*: The four months during which Viṣṇu withdraws to sleep, from the eleventh day of the clear fortnight of Āşadha (Asār in Nepali, June–July) until the eleventh day of the clear fortnight of Kārtika (Kārtik, October–November).
*caurāsī godān*: An elaborate *jagya* in which an elderly person (a man or widowed woman celebrating or approaching the age of eighty-four) donates eighty-four *godāns* to priests and lay Upādhyāya Brāhmaṇs, for avoiding hell after death.
*cautāro* (*chautāra*): This famous feature of the Nepali landscape is found along mountainous footpaths throughout the hilly areas of the country. A *cautāro* consists of a rectangular stone-coated ramp about 1.5 meters high, with steps surrounding it at a level convenient for porters to support their load and for people to sit. At its center, two sacred trees are planted and ceremoniously married, making it an auspicious, safe resting place for travelers.
*ched*: A general term for evil spirits.
Chetri: The caste below Brāhmaṇ – Kṣatriya in India.
*cokho*: Pure.

*culo*: The domestic hearth where rice is cooked.

*dahi* (*dai*): Curd.

*dāī hālnu* (*hālne*): The event and action of threshing the harvest of rice.

*Dāijo*: dowry.

*dakṣiṇā* (*dacchinā*): Priestly fee, normally paid in cash. Usually given together with *dān* thus forming a complex called *dān-dakṣiṇā*.

Damāi: The "Untouchable" tailor-musician caste.

*dān*: Lit. a (unreciprocated) gift. Negative moral substance, that is, sin, *dos*, inauspiciousness, and other such influences one must dispose of to maintain a balance of merit.

*darśana (darśan)*: Vision, the exchange of sight with an image of god, anticipating the reward of divine grace. Experience of the divine in a *pūjā*.

Dasaī: A major Nepalese festival celebrating the goddess's victory over the demons. Also called Durgā *pūjā*, this is the climax of the high-castes' annual festivals, held in honor of the goddess Durgā (Kāli) during the light fortnight of Asoj (September–October).

*deutā* (*deotā*), *īśwar*: A god, deity, divinity.

Devī: The goddess. During a *pūjā* it is manifested by *achetā* placed in a small leaf plate.

*dhāgo*: A thread, cord.

*dhān*: Unhusked (uncooked) rice.

*dhān ropnu* (*ropne*): Rice transplantation.

*dharma*: A multifarious concept that mainly denotes religious/moral duty, ethical behavior, and righteousness, as well as religious work, devotion, fasts, the performance of *pūjās*, or religion in general.

*Dharmaśāstra*: A set of Brāhmaṇ-authored texts written in Sanskrit in north India, dating from around 200 B.C. to A.D. 500.

*dhāro* (*dhārā*): A village spring.

*ḍhiki*: A mechanical wooden device for husking rice.

*ḍhognu*: An expression of deference when one bows down, touching a superior's legs with one's forehead.

*ḍhuṅgo*: A stone manifestation of a deity.

*dhup hālnu* (*halne*): Lit. to offer incense. This is a minor *pūjā*, in which a few drops of *ghyu* are sprinkled over a burning coal situated on a leaf or stone.

*Dikagadge, diggaj* (*diggadj*): A cosmic elephant, carrying the universe on its back.

*Dik-pāl (Dikapālas)*: The Vedic guardians of space.

*dīkṣā*: The preliminary ("initiatory") rite of a number of Vedic sacrifices.

*dos*: Blame, fault.

*dubo*: Sacred grass (*Cynodon Dactylon*) employed in village *pūjās*.

*dukkha*: Sorrow, pain.
*dunu* (*duna*): A small leaf plate used for placing the offering in a *pūjā*.
*gāgri*: The large copper water vessel of the house.
*gajur*: This is the (ideally) golden, uppermost metal part of a Nepalese (Newāri style pagoda) temple.
Gaṇeś: Son of Śiva and Pārvati, the elephant-headed god and renowned remover of obstacles. Manifested in a *pūjā* by a lump of fresh cow dung with *dubo* grass at its center.
Gaṅgā, Ganges: The holiest river in India, personified as a goddess. All rivulets around Thamghar are considered as its tributary streams.
Gaṅgā *jal*: Water consecrated by a few drops taken from the sacred Ganges.
*garbha*: Womb.
*garbhabati*: Being pregnant.
*Garbhagṛha*: Lit. "the house which is the womb," the innermost part of a Hindu temple and the place where its principal deity resides.
*gāũ*: Village.
*gāyatrī* (*gāyantrī*): The sacred mantra whispered into a high-caste initiate's ear at the time of the Upanayana (*bartoon*).
*ghar*: A house, a family, and a wife's natal home.
*ghyu*: Clarified butter.
*godān*: The gift of a cow, *dān* embodied in a cow.
*gotra*: An exogamous agnatic unit whose members may belong to different *kuls* and *jāts* and are believed to be descendant from one of the seven mythical *ṛṣis*.
*guru*: Teacher, priest.
*hakuwā*: Black fermented (Newāri) rice.
*hali*: Ploughman.
*hāṛ-nātā*: Lit. "bone relatives," members of one *kul* who are all thought to share a bony substance (blood relatives).
*hiṁsā/ahiṁsā*: Violent, animal sacrifice/harmlessness, nonviolence, peaceful.
*hom*: Fire worship, offering of *ghyu* and mainly rice grains into the fire.
*ijjat*: Prestige, honor.
*īśān*: The auspicious northeast direction.
*jagya*: A sacred ritual enclosure and the complex ritual performance enacted therein.
Jaisi Brāhmaṇs: The offspring of a union between an Upādhyāya man and an Upādhyāya widow or divorcee; second to Upādhyāya in terms of ritual purity, they cannot serve as priests. Upādhyāya Brāhmaṇs will not eat rice cooked by a Jaisi.
*jajamān* (*jajmān*): A priest's employer, a householder, and his family.
*jajamāna*: The Vedic sacrificer.

*janaï*: High-caste's sacred thread, given in the *bartoon*.
*jānne mānche*: Lit. "the one who knows," a village healer.
*jāt, jāti*: Genus, type, a general classificatory term, caste.
*jaro* (*joro*): The inner intense heat of the *kunyā*. A fever.
*juro*: A straw man made of seven threshed rice sheaves, tied together to fashion the figure of a five-handed male and placed on top of the *kunyā*.
*juṭho*: Extremely polluted, leftovers from a rice meal.
*kalas*: A copper ritual flask.
Kāli: A fierce, blood-craving female manifestation of the goddess Durgā.
Kāli *yug*: The present era of darkness and amorality.
*kanikā*: The rice that brakes during husking, regarded *biṭulo*.
*kanyā (konyā)*: A virgin girl under ten years of age, a virgin goddess.
*kanyādān*: A bride and the central stage of marriage when the bride is transferred from her father to her husband.
*kartā*: Lit. an actor, the main performer or patron of a *pūjā*.
*khājā*: A light afternoon meal that does not include cooked rice.
*khāldo*: A square pit dug in the ground at the *jagya*'s center for the performance of *hom*, regarded as Bhume's vulva.
*khalo garo (khale gara)*: The threshing floor regarded as analogous to the field owner's house.
*khānā*: Lit. a meal, cooked rice.
*khet*: Irrigated rice field.
*khetālo (khetāla)*: Agricultural workers.
*khuṭṭa pāni*: The water used by the wife to wash her husband's feet before every rice meal.
*kul*: Clan, an agnatic descent group whose members are thought to share a bony substance forming the *kul*'s "social body."
*kul deutā*: The clan god.
*kul deutā pūjā*: The communal worship (sacrifice) for the *kul deutā*.
*kuṇḍa*: Pit, pond.
*kunyā, kunyũ*: A round edifice made of rice (up to four meters in height with a diameter of approximately three to four meters), established in the *khalo garo*. Regarded as Bhume's temple.
Lakṣmī: The goddess of wealth and fortune. Viṣṇu's prominent wife.
*liṅga, liṅgam*: Lit. a phallus, the prevalent representation of Śiva. A pole or tree trunk.
*lok*: A world.
*Mahābhārata*: Sanskrit epic recounting the story of the war between the Kauravas and the Pāṇdavas, which includes the *Bhagavad Gītā*.
*māiti*: A married woman's natal home.

*mālik*: Ruler, householder, husband, god.
*man*: Heart, mind.
*mānche*: A man, person.
*maṇḍala*: A mystic meditative diagram made of geometrical shapes focusing on a central point, which is believed to embody a two-dimensional representation of the cosmos.
*mandir*: Temple.
mantra: A highly effective verbal formula used in meditation and religious practice.
*Mātṛkās* (*Mātrikas*): Unmarried (usually seven or ten) mother goddesses. Fierce, powerful, territorial guardian deities.
*mithuna, maithuna*: The state of being a couple. Man and woman in a sexual union.
*miyo*: A special bamboo pillar with a sheaf of rice tied toward its top, around which the oxen walk for a final threshing during *dāī hālne*.
*mokṣa*: Release (*nirvāṇa* in Buddhism) from *saṁsāra*. A state of ultimate divine bliss.
*mul*: Main, chief. Also the name of a particularly inauspicious *nakṣatra*.
*mul kalas*: The main *kalas* in a *pūjā*, usually a *gāgri*.
*murti*: An image, idol, an embodied representation of a god.
*nakṣatra* (*nachetra*): One of twenty-seven "stars," or the lunar mansions dominating Hindu astrology, of which *mul* is considered to be particularly unlucky and inauspicious.
*nāg, nāgini*: The semidivine snake deity and his consort who dwell in the underworld and are believed to appear in all village springs.
*nāghnu*: The offensive action of stepping over parts or the whole of a superior's body/divine entity.
*naibedde, naibedya*: Various offerings, mainly foodstuffs made of rice flour and fried in *ghyu*.
*narka, narak*: Hell.
Newār: A Tibeto-Burman-speaking people, mainly located in and believed to be the indigenous population of the Katmandu valley.
*nivek*: A system where workers are hired on a daily basis.
*nokar*: Servant.
*nuwāran*: Name-giving ceremony, the first *saṁskāra*, celebrated for a newborn baby on the eleventh day after the birth, when he/she is given his/her name and the mother is purified from the extreme birth pollution that affects her until this day.
Nyāya philosophy: One of India's six classical philosophical systems, whose distinguishing feature is the focus on reason and logic.

*paīco* (*paincho*): A well-calculated and symmetrical manner of reciprocity in which goods are meticulously exchanged and returned, as soon as possible, in identical terms of quantity. See also *parma* below.

*pākeko* (*pākcha*): Ripened, cooked (being cooked).

*pāknu*: To cook, mature, ripen.

*pañchāyan deutā*: The five house gods, usually: Viṣṇu, Śiva, Gaṇeś, Sūrya, and Devī.

*pāni*: Water.

*pāp*: Sin.

*parma*: A system of calculated mutual aid in which work is strictly exchanged and returned, as soon as possible, in identical terms of time and workforce or its equivalence in cash or mostly in kind. See also *paīco*.

*parsanu* (*parsine*): A welcoming gesture, mainly toward a groom, which includes the attribution of *ṭikā* on his forehead.

Pārvatī: Śiva's consort.

*pattāl*: The seven underworlds.

*peṭ*: Belly, stomach.

*phal (phol)*: Fruits. The expected benefit from a ritual performance.

*phuknu*: To blow, the common method for exorcising evil spirits.

*phul-achetā pūjā*: The offering of *achetā* and *phul* (flowers)—the most prevalent basic act of worship known in Thamghar.

*pitṛ (pitṛi)*: An ancestor's spirit.

*pothi*: A female animal and a vulgar term for a woman.

*pradakṣiṇā* (*pardacchinā*): Circumambulation while keeping the auspicious right side toward the object/entity being worshipped.

Prajāpati: The Vedic lord of creatures, the creator.

*prakṛtī, prakriti*: Nature, female.

*prasād*: Food leavings of a deity that are full of blessing and grace by virtue of coming into contact with it. It is distributed hierarchically at the end of every *pūjā*.

*pratimā*: An embodied representation of a person, a god, or an entity.

*praveś*: Lit. entrance. The entrance *pūjā* performed for the purification of a newly constructed house.

*prāyaścitta* (*payachista*) *godān*: The rite of giving *godān* for enhancing one's ritual state and ensuring the effects, success, and general fortunes of a ritual procedure.

*prêt*: Ghost, the spirit of a deceased person before he becomes a *pitṛ*. An evil spirit.

Pṛthvi, Prithivī: The earth goddess.

*pūjā*: Worship, homage, mainly the offering of food to the deities.

*pūjāri*: Special appointed priest.

*puṇya*: Merit.
*purā*: Complete.
*purān*: A prolonged *jagya* centered around the public reading of Hindu *Purāṇas*.
*Purāṇas*: The treatises of Hindu mythology. Dating back from the beginning of the first millennium until approximately the tenth century.
*purbāṅga*: The first (preliminary) *aṅga* (limb) of a *jagya* or other elaborated *pūjā*.
*Purohit* (*puret*): Priest.
*puruṣ*: Male.
Puruṣa: The Vedic primordial masculine being, later usurped by Prajāpati.
*rakṣā-bandhan*: A protective thread worn on the right and left wrist for men and married women, respectively.
*rekhi*: A geometrical contrivance reflecting the cosmos, said to "bind" the deities toward their place during the performance of an elaborate *pūjā*. A *yantra*.
*Ṛg Veda*: The most ancient and important Vedic text, this compilation of hymns constitutes the first Veda.
*ṛṣi* (*rishi*): Sage, seer. Thamgharian Brāhmaṇs claim that each of their *gotras* originated from one of seven mythical *ṛṣis*. The Vedas are believed to have been transmitted via ancient *ṛṣis* who first "saw" the eternal truth that they convey.
*sae auṣadh (auṣadhi)*: Lit. one hundred medicines. The prototypical remedy for every possible illness, represented by a plethora of herbs, grains, fruit, and substances. One of the major elements placed within the *mul kalas*.
*saha*: The everlasting power of things (mainly food). Inner potency and power.
*sāit*: A highly desirable yet rather elusive state of affairs, characterized by high auspiciousness and good fortune, and mainly associated with cosmic compatibility in time and action.
Śaiva: Belonging to the Hindu sect that regards Śiva as the supreme deity.
*śakti*: (Mainly female) energy, power.
Śakti: One of Śiva's consorts.
*sāl*: A sacred tree.
*sāli* (*maṅgsire*): The auspicious type of rice, grown during the monsoon season, pure enough to be employed for ritual purposes. Ripens during the month of Muṅgsir (Maṅgsir, November–December).
*sambandha*: Lit. connection, involvement, alliance, or marriage.
*samidhā*: The wooden twigs used for *hom*.
*saṁnyāsin* (*sannyāsi*): An ascetic, *yogi*.
*saṁsāra*: The eternal cosmic cycle of births and deaths, from which Hindus aspire to be emancipated/released and attain *mokṣa*.

*saṁskāra* (*sanskār*): A Hindu "rites of passage" (life cycle ritual), notably *nuwāran*, *bartoon*, marriage, and death.

*śaṅkā* (*śaṅkālu*): Doubt, mistrust, uncertainty (suspicion).

*saṅkalpa*: The *kartā*'s declaration of intention prior to the performance of a *pūjā*.

*saṅkha*: A manifestation of Viṣṇu in the form of a large shell, which is blown during *pūjās*.

Sāṅkhya philosophy: Alongside Yoga, this forms one of the most influential Hindu systems of thought. One of its cardinal themes is the view of existence as the combination of *Puruṣa* (spirit) and *Prakṛti* (nature).

*sansār* (*sangsār*): The world, universe.

*siddha*: Locally glossed as "the one whose words materialize," "successful in all he does." A "perfect *yogī*" (lit. practitioner of yoga, extreme *Tantric*, ascetic) belonging to a late medieval movement that formed a complex amalgamation of various religious ideas, alchemical traditions, and magic.

Śikhara: The north Indian curvilinear Hindu temple.

Śiva: One of the major (male, carnivorous) Hindu gods. Represented in a *pūjā* by a *diyo*, a small oil lamp.

Śivā-*liṅgam*: The pan-Hindu manifestation of Śiva as a *liṅgam* (phallus) emerging out of a *yoni*.

*snān*: The shower in the energetic, invigorating water of the *mul kalas*, at the end of a *jagya*.

*sṛṣti (shristi)*: (Pro)creation, divine emanation.

*suddha*: Highly pure.

Śūdra: A general title for the lower castes of "Untouchables."

*supāri*: Areca nut, often employed in village *pūjās* as a *pratimā* for a living *kanyā*, a virgin goddess.

Sūrya: The (male) sun god.

*swarga*: Heaven.

*Swasthāni Vrata Kathā*: A religious text mainly narrating the story of the marriage of Śiva and Pārvati, which is found in every Brāhmaṇ household. It is worshipped and read, a chapter every night, during Māgh (January–February).

*talo (tala)*: The first floor of the house.

*Tantra*, *Tantric*: The doctrine and cult of the Goddess or Śakti whose texts date from the sixth to the seventh century, a follower of that Hindu sect.

*tapas*: Internal heat, associated with both asceticism and eroticism, as well as ritual transformation, also an entity or substance.

*thar*: A group of people sharing the same family name, yet not necessarily belonging to the same *gotra*.

*ṭhulo*: High, big.

Tihār: One of Nepal's paramount festivals. Celebrated for five days during the dark half of the month of Kārtik, (October–November). Bhai Ṭikā is held on its last day.

Tij: An important women's festival, celebrated throughout Hindu Nepal, which falls on the third day of the bright half of Bhadau (August–September).

*ṭikā*: An auspicious forehead mark, usually made of *achetā* and distributed at the end of every *pūjā*.

*tilāni pāni*: An auspicious mixture of *Gaṅgā jal*, *til* (sesame seeds), sacred *kuś* (*kuśa*) grass, and *jāo* (*jau*, barley).

*tīrtha*: Lit. all the rivers surrounding Thamghar. A Hindu pilgrimage site conceptualized as a ford, connecting heaven to earth.

*titro (titrā)*: A common partridge.

*toran*: An elongated rope with green leaves interwoven with fresh flowers.

*tulasi*: Basil (*Ocimum Basilicum*), a manifestation of Viṣṇu planted in front of every house in a decorated mud pedestal.

*ṭupi* (*śikhā*): The auspicious top strand often tied into a knot at the crown of the skull of a high-caste male.

Upādhyāya Brāhmaṇs: The upper echelon of Brāhmaṇs, priests.

Upanayana: High-castes' male initiation, *bartoon* in Thamghar.

*Upaniṣads*: Speculative texts that were compiled after the Vedas, dating back approximately to the sixth century B.C.

Vaiṣṇava: Belonging to the Hindu sect that views Viṣṇu as the supreme deity.

*Vāstupuruṣamaṇḍala*: This is the square metaphysical plan (grid) of all Hindu architectural forms.

Vedas: The four most ancient and sacred Hindu scriptures, probably complied from the fifteenth century B.C., including the *Ṛg Veda*, Yajurveda, Samaveda, and Atharvaveda.

*vedi*: The Vedic altar.

Viṣṇu: One of the major (male, vegetarian) Hindu gods. Usually represented in a *pūjā* by a *kalas*.

*yajña*: The Vedic sacrifice.

Yama-rāj: The god of death.

*yantra*: Lit. an "object serving to hold," a *rekhi* in Thamghar.

*yogī*: Ascetic.

*yoni*: The female sexual organ.

*yug*: A Hindu era, epoch.

# Bibliography


Appadurai, A. 1981. Gastro-Politics in Hindu South Asia. *American Ethnologist* 8, no. 3, 494–511.

———. 1985. Gratitude as a Social Mode in South India. *Ethos* 13, no. 3, 236–45.

———. 1988. Putting Hierarchy in its Place. *Cultural Anthropology* 3, no. 1, 36–49.

———. 1990. Topographies of Self: Praise and Emotion in Hindu India. In *Language and the Politics of Emotion* (eds.) C. A. Lutz and L. Abu-Lughod. Cambridge: Cambridge University Press.

Appadurai, A., and C. A. Breckenridge. 1976. The South Indian Temple: Authority, Honour, and Redistribution. *Contributions to Indian Sociology* (N.S.) 10, no. 2, 187–211.

Aubriot, O. 1997. *Eau: Miroir des Tensions, Ethno-Histoire d'un Systeme d'Irrigation dans les Moyennes du Nepal Central*. Thesis. University of Provence.

Babb, L. A. 1970. Marriage and Malevolence: The Uses of Sexual Opposition in a Hindu Pantheon. *Ethnology* 9, 137–48.

———. 1975. *The Divine Hierarchy: Popular Hinduism in Central India*. New York: Columbia University Press.

———. 1991 [1986]. *Redemptive Encounters: Three Modern Styles in the Hindu Tradition*. Berkeley: University of California Press.

Bak, P. 1997. *How Nature Works: The Science of Self-Organized Criticality*. Oxford: Oxford University Press.

Basham, A. L. 1985 [1954]. *The Wonder That Was India*. London: Sidgwick & Jackson.

Bateson, G. 1987 [1972]. *Steps to an Ecology of Mind*. Northvale, NJ, and London: Jason Aronson Inc.

Bayly, C. A. 1986. The Origins of Swadeshi (Home Industry): Cloth and Indian Society, 1700–1930. In *The Social Life of Things: Commodities in Cultural Perspective* (ed.) A. Appadurai. Cambridge: Cambridge University Press.

Bayly, S. 1989. *Saints, Goddesses, and Kings: Muslims and Christians in South Indian Society 1700–1900*. Cambridge: Cambridge University Press.

——. 1999. *Caste, Society, and Politics in India from the Eighteenth Century to the Modern Age*. The New Cambridge History of India. Cambridge: Cambridge University Press.

Beck, B. E. F. 1969. Colour and Heat in South Indian Ritual. *Man* (N.S.) 4, no. 4, 553–72.

——. 1974. The Kin Nucleus in Tamil Folklore. In *Kinship and History in South Asia* (ed.) T. R. Trautman (Michigan Papers on South and Southeast Asia, no. 7). Ann Arbor: The University of Michigan.

——. 1976. The Symbolic Merger of the Body: Space and Cosmos in Hindu Tamil Nadu. *Contributions to Indian Sociology* (N.S.) 10, no. 2, 213–43.

——. 1981. The Goddess and the Demon: A Local South Indian Festival and Its Wider Context. *Purusartha* 5, 83–136.

Ben Dov, Y. 2002. *The Journey to India, Second Series: The Indian Fractal* (Hebrew) <http://www.bendov.info./heb/ind/indifrac.htm> (14 Oct. 2005)

Bennett, L. 1976. The Wives of the Rishis: An Analysis of the Tij-Rishi Panchami Women's Festival. *Kailash* 4, no. 2, 185–207.

——. 1977. *Mother's Milk and Mother's Blood: The Social and Symbolic Roles of Women Among the Brahmans and Chetris of Nepal*. Thesis. Columbia University.

——. 1978. Sitting in a Cave: An Analysis of Ritual Seclusion at Menarche Among Brahmans and Chetris in Nepal. *Contributions to Nepalese Studies* 4, no. 1, 31–45.

——. 1983. *Dangerous Wives and Sacred Sisters: Social and Symbolic Roles of High-Caste Women in Nepal*. New York: Columbia University Press.

Berreman, G. D. 1993 [1963]. *Hindus of the Himalaya: Ethnography and Change*. Delhi: Oxford University Press.

Biardeau, M. 1976. *Le Sacrifice dans l'Inde Ancienne*. Bibliotheque de l'Ecole des Hautes, Sciences Religeuses, vol. 79. Paris: Presses Universitaires de France.

Bista, K. B. 1969. Tîj ou la Fête des Femmes. *Objets et Monde* 5, 7–18.

——. 1972. *Le Culte du Kuldevata au Nepal en Particulier chez Certains Ksatri de la Vallée de Kathmandu*. Paris: CNRS.

Bloch, M. 1992. What Goes Without Saying: The Conceptualisation of Zafimaniry Society. In *Conceptualising Society* (ed.) A. Kuper. London: Routledge.

Blustain, H. S. 1977. *Power and Ideology in a Nepalese Village*. Thesis. Yale University.

Bouillier, V. 1982. Si les Femmes Faisaient la Fête . . . A propos des Fêtes Féminines dans les Hautes Castes Indo-Népalaises. *L'Homme* 22, no. 3, 91–118.

——. 1991 [1981]. From the Fountain to the Fireplace: The Daily Itinerary in Domestic Space Among High Indo-Nepalese Castes. In *Man and His House in the Himalayas: Ecology of Nepal* (ed.) G. Toffin. New-Delhi: Sterling Publishers.

Bourdieu, P. 1977. *Outline of a Theory of Practice*. Cambridge: Cambridge University Press.

——. 1992 [1990]. *The Logic of Practice*. Cambridge, MA: Polity Press.

Brown, P. 1971 [1965]. *Indian Architecture*. Bombay: Taraporevala.

Buddhi, S. P. 1994 (2051 B.S.). *Shri Swasthani Brata-Katha*. Kathmandu: Ratna Pustak Bhandar. (Nepali).

Burckhardt, T. 1986 [1958]. *Sacred Art in East and West: Its Principles and Methods*. Middlesex: Perennial Books, Ltd.

Burghart, R. 1987. Gifts to the Gods: Power, Property and Ceremonial in Nepal. In *Rituals of Royalty: Power and Ceremonial in Traditional Societies* (eds.) D. Cannadine and S. Price. Cambridge: Cambridge University Press.

Busby, C. 1997. Permeable and Partible Persons: A Comparative Analysis of Gender and Body in South India and Melanesia. *The Journal of the Royal Anthropological Institute* 3, no. 2, 261–78.

———. 2000. *The Performance of Gender: An Anthropology of Everyday Life in a South Indian Fishing Village* (London School of Economics Monographs on Social Anthropology 71). London: The Athlone Press.

Cameron, M. M. 1998. *On the Edge of the Auspicious: Gender and Caste in Nepal*. Urbana and Chicago: University of Illinois Press.

Campbell, J. G. 1976. *Saints and Householders: A Study of Hindu Ritual and Myth Among the Kangara Rajputs*. Kathmandu: Ratna Pustak Bhandar.

Cantlie, A. 1981. The Moral Significance of Food Among Assamese Hindus. In *Culture and Morality: Essays in Honour of Christoph von Fürer-Haimendorf* (ed.) A. C. Mayer. Delhi: Oxford University Press.

Caplan, L. 1985. The Popular Culture of Evil in Urban South India. In *The Anthropology of Evil* (ed.) D. Parkin. Oxford: Basil Blackwell.

Carsten, J., and S. Hugh-Jones 1995. Introduction. *About the House: Levi-Strauss and Beyond*. Cambridge: Cambridge University Press.

Carter, A. T. 1995 [1982]. Hierarchy and the Concept of the Person in Western India. In *Concepts of the Person: Kinship, Caste, and Marriage in India* (eds.) A. Östör, L. Fruzzetti, and S. Barnett. Delhi: Oxford University Press.

Chandler, R. F. 1979. *Rice in the Tropics: A Guide to the Development of National Programs*. Boulder, CO: Westview Press.

Coomaraswamy, A. K. 1995. An Indian Temple: the Kandarya Mahadeo. In *Ananda K. Coomaraswamy: Essays in Architectural Theory* (ed.) M. W. Meister. Delhi: Oxford University Press.

Copi, I. M. 1972 [1953]. *Introduction to Logic*. New York: The Macmillan Company.

Coulson, M. 1992 [1976]. *Sanskrit: An Introduction to the Classical Language*. Kent: Hodder & Stoughton.

Courtright, P. B. 1995. *Sati*, Sacrifice, and Marriage: The Modernity of Tradition. In *From the Margins of Hindu Marriage: Essays on Gender, Religion and Culture*. (eds.) L. Harlan and P. B. Courtright. Oxford: Oxford University Press.

Czarnecka, J. 1986. Status of Affines Among High Caste Hindu in the Nepalese Hill Area. In *Recent Research in Nepal: Proceedings of a Conference Held at the Universitat Konstanz, 27–30 March 1984* (ed.) K. Seeland. Munchen: Weltforum Verlag.

Daniel, E. V. 1984. *Fluid Signs: Being a Person the Tamil Way*. Berkeley, Los Angeles, London: University of California Press.

Daniel, S. B. 1980. Marriage in Tamil Culture: The Problem of Conflicting "Models." In *The Power of Tamil Women* (ed.) S. S. Wadley. Syracuse, NY: Syracuse University Press.

Daniélou, A. 1964. *Hindu Polytheism*. New York: Pantheon Books.

Daryn, G. 1998. Moroccan Hassidism: The Chavrei Habakuk Community and its Veneration of Saints. *Ethnology* 37, no. 4, 351–72.

Das, R. V. 1991. On the Subtle Art of Interpreting. *Journal of the American Oriental Society* 111, no. 4, 737–67.

Das, V. 1976a. The Uses of Liminality: Society and Cosmos in Hinduism. *Contributions to Indian Sociology* (N.S.) 10, no. 2, 245–63.

——. 1976b. Masks and Faces: An Essay on Punjabi Kinship. *Contributions to Indian Sociology* (N.S.) 10, no. 1, 1–29.

——. 1982 [1977]. *Structure and Cognition, Aspects of Hindu Caste and Ritual.* Delhi: Oxford University Press.

——. 1985. Paradigms of Body Symbolism: An Analysis of Selected Themes in Hindu Culture. In *Indian Religion* (eds.) R. Burghart and A. Cantlie. London: Curzon Press.

Derrett, J. D. M. 1959. *Bhū-Bbaraṇa, Bhū-Pālana, Bhū-Bhojana*: an Indian Conundrum. *Bulletin of the School of Oriental and African Studies* 22, 108–23.

Doherty, V. S. 1974. The Organizing Principles of Brahman-Chetri Kinship. *Contributions to Nepalese Studies* 1, no. 2, 25–41.

Doniger, W. 1999. *Splitting the Difference: Gender and Myth in Ancient Greece and India*. Chicago: The University of Chicago Press.

Doniger, W., and B. K. Smith 1991. *The Laws of Manu*. London: Penguin Books.

Dube, L. 1986. Seed and Earth: The Symbolism of Biological Reproduction and Sexual Relations of Production. In *Visibility and Power: Essays on Women in Society and Development* (eds.) L. Dube, E. Leacock, and S. Ardener. Delhi: Oxford University Press.

Dumont, L. 1960. World Renunciation in Indian Religions. *Contributions to Indian Sociology* (N.S.) IV, 3–62.

——. 1980 [1966]. *Homo Hierarchicus: The Caste System and its Implications.* Chicago and London: The University of Chicago Press.

——. 2000 [1957]. *A South Indian Subcaste: Social Organization and Religion of the Prmalai Kallar*. New Delhi: Oxford University Press.

Eck, D. 1981. India's *Tirthas*: "Crossings" in Sacred Geography. *History of Religions* 20, no. 4, 323–44.

——. 1998a. *Darśan: Seeing the Divine Image in India*. New York: Columbia University Press.

——. 1998b. The Imagined Landscape: Patterns in the Construction of Hindu Sacred Geography. *Contributions to Indian Sociology* (N.S.) 32, no. 2, 165–88.

Eglash, R. 1999. *African Fractals, Modern Computing, and Indigenous Design*, New Brunswick, NJ, and London: Rutgers University Press.

Eliade, M. 1958. *Rites and Symbols of Initiation: The Mysteries of Birth and Rebirth.* New York: Harper Torchbooks

——. 1973 [1958] *Yoga, Immortality, and Freedom*. Princeton: University Press.

——. 1978. *A History of Religious Ideas, Vol. 1: From the Stone Age to the Eleusinian Mysteries*. Chicago: University of Chicago Press.

Enslin, E. M. W. 1990. *The Dynamics of Gender, Class and Caste in a Women's Movement in Rural Nepal*. Thesis, Stanford University.

——. 1998. Imagined Sisters: The Ambiguities of Women's Poetics and Collective Actions. In *Selves in Time and Place: Identities, Experience and History in Nepal* (eds.) D. Skinner, A. Pach III, and D. Holland. Lanham, MD: Rowman & Littlefield Publishers, Inc.

Epstein, T. S. 1959. A Sociological Analysis of Witch Beliefs in a Mysore Village. *The Eastern Anthropologist* 12, no. 4, 234–51.

Firth, R. 1963. Offering and Sacrifice: Problems of Organization. *The Journal of the Royal Anthropological Institute of Great Britain and Ireland* 93, nos. 1&2, 12–24.

Fortier, J. 1995. *Beyond Jajmani: The Complexity of Indigenous Labor Relations in Western Nepal*. Thesis. University of Wisconsin.

Foster, G. M. 1965. Peasant Society and the Image of the Limited Good. *American Anthropologist* 67, no. 2, 293–315.

Fruzzetti, L. M. 1982. *The Gift of a Virgin: Women, Marriage, and Ritual in a Bengali Society*. New Brunswick, NJ: Rutgers University Press.

Fruzzetti, L., Á. Östör, and S. Barnett 1995 [1982]. Bad Blood in Bengal: Category and Affect in the Study of Kinship, Caste, and Marriage. In *Concepts of the Person: Kinship, Caste, and Marriage in India*. (eds) A. Östör, L. Fruzzetti, and S. Barnett. Delhi: Oxford University Press.

Fuller, C. J. 1980. The Divine Couple's Relationship in a South Indian Temple: Mīnākṣī and Sundaresvara at Madurai. *History of Religions* 19, no. 4, 321–348.

——. 1984. *Servants of the Goddess: The Priests of a South Indian Temple*. Cambridge: Cambridge University Press.

——. 1987. Sacrifice (*Bali*) in the South Indian Temple. In *Religion and Society in South India: A Volume in Honour of Prof. N. Subba Redd* (eds.) V. Sudarsen, G. P. Reddy, and M. Suryanarayana. Delhi: B. R. Publishing Corporation.

——. 1988. The Hindu Temple and Indian Society. In *Temple in Society* (ed.) M. F. Fox. Winona Lake, IN: Eisenbrauns.

——. 1989. Misconceiving the Grain Heap: A Critique of the Concept of the Indian Jajmāni System. In *Money and the Morality of Exchange* (eds.) J. Parry and M. Bloch. Cambridge: Cambridge University Press.

——. 1992. *The Camphor Flame: Popular Hinduism and Society in India*. Princeton: Princeton University Press.

——. 1995. The "Holy Family" of Śiva in a South Indian Temple. *Social Anthropology* 3, no. 3, 205–217.

von Fürer-Haimendorf, C. 1960. Caste in the Multi-Ethnic Society of Nepal. *Contributions to Indian Sociology* 4, 12–32.

——. 1966a. Unity and Diversity in the Chetri Caste in Nepal. In *Caste and Kin in Nepal, India and Ceylon: Anthropological Studies in Hindu-Buddhist Contact Zones* (ed.) C. von Fürer-Haimendorf. London and Hague: East-West Publications.

——. 1966b. *Morals and Merit: A Study of Values and Social Controls in South Asian Societies*. London: Weidenfeld & Nicholson.

Gaborieau, M. 1982. Les Fetes, le Temps et l'Espace: Structure du Calendrier Hindou dans sa Version Indo-Nepalais. *L'Homme* 32, no. 3, 11–29.

——. 1991 [1981]. The Indo-Nepalese House in Central Nepal: Building Patterns, Social and Religious Symbolism. In *Man and His House in the Himalayas: Ecology of Nepal* (ed.) G. Toffin. New-Delhi: Sterling Publishers.

——. 1992. Des Dieux dans Toutes les Directions: Conception Indienne de l'Espace et Classification des Dieux. In *Puruṣārtha* 15, 23–42.

Gaddis, J. L. 2002. *The Landscape of History: How Historians Map the Past.* Oxford: Oxford University Press.

Gambetta, D. 1988a. Mafia: The Price of Distrust. In *Trust: Making and Breaking Co-operative Relations* (ed.) D. Gambetta. Oxford: Basil Blackwell.

——. 1988b. Can We Trust Trust? In *Trust: Making and Breaking Cooperative Relations* (ed.) D. Gambetta. Oxford: Basil Blackwell.

——. 1993. *The Sicilian Mafia: The Business of Private Protection.* Cambridge, MA: Harvard University Press.

Geertz, C. 1980. *Negara: The Theatre State in Nineteenth-Century Bali.* Princeton: Princeton University Press.

——. 1996 [1995]. *After the Fact: Two Centuries, Four Decades, One Anthropologist.* Cambridge, MA: Harvard University Press.

Gell, A. 1975. *The Metamorphosis of the Cassowaries.* London: Athlone Press.

——. 1992. The Technology of Enchantment and the Enchantment of Technology. In *Anthropology Art and Aesthetics* (eds.) J. Coote and A. Shelton. Oxford: Clarendon Press.

——. 1997. Exalting the King and Obstructing the State: A Political Interpretation of Royal Ritual in Bastar District, Central India. *The Journal of the Royal Anthropological Institute* (N.S.) 3, no. 3, 433–50.

——. 1998. *Art and Agency: An Anthropological Theory.* Oxford: Clarendon Press.

——. 1999. Strathernograms, or, the Semiotics of Mixed Metaphors. In *The Art of Anthropology: Essays and Diagrams* (ed.) E. Hirsch (London School of Economics Monographs on Social Anthropology, vol. 67). London: The Athlone Press.

Gellner, D. N. 1994. Priests, Healers, Mediums, and Witches: The Context of Possession in the Kathmandu Valley, Nepal. *Man* (N.S.) 29, no. 1, 27–48.

——. 1996 [1992]. *Monk, Householder, and Tantric Priest: Newār Buddhism and Its Hierarchy of Ritual.* Cambridge: Cambridge University Press.

Gellner, D. N., and D. Quigley (eds.) 1995. *Contested Hierarchies: A Collaborative Ethnography of Caste Among the Newars of the Kathmandu Valley, Nepal.* Oxford: Oxford University Press.

Gellner, D. N., and U. S. Shrestha 1993. Portrait of a Tantric Healer: A Preliminary Report on Research into Ritual Curing in the Kathmandu Valley. In *Nepal, Past and Present* (ed.) G. Toffin. Paris: CNRS.

Gellner, E. 1988. Trust, Cohesion, and the Social Order. In *Trust: Making and Breaking Cooperative Relations* (ed.) D. Gambetta. Oxford: Basil Blackwell.

Gleick, J. 1987. *Chaos: Making a New Science.* New York: Penguin Books.

Gluckman, M. 1954. *Rituals of Rebellion in South-East Africa.* Manchester: Manchester University Press.

Godlier, M. 1999. *The Enigma of the Gift.* Cambridge, MA: Polity Press.

Goffman, E. 1959. *The Presentation of Self in Everyday Life.* New York: Anchor Books.

——. 1974. *Frame Analysis: An Essay on the Organization of Experience.* Cambridge, MA: Harvard University Press.

Gold, A. G. 1994. Sexuality, Fertility, and Erotic Imagination in Rajasthani Women's Songs. In *Listen to the Heron's Words: Reimagining Gender and Kinship in North India* (eds.) G. G. Raheja and A. G. Gold. Berkeley: University of California Press.

Gonda, J. 1957. *Some Observations on the Relations Between "Gods" and "Powers" in the Veda, a propos of the Phrase Sūnuh Sahasah*. The Hague: Mouton & Co.

———. 1965. *Change and Continuity in Indian Religion*. The Hague: Mouton & Co.

———. 1969. *Ancient Indian Kingship from the Religious Point of View*. Leiden: Brill.

———. 1970. *Viṣṇuism and Śivaism: A Comparison*. London: The Athlone Press.

———. 1980. *Vedic Ritual: The Non-Solemn Rites*. Leiden: Brill.

Good, A. 1982. The Actor and the Act: Categories of Prestation in South India. *Man* (N.S.) 17, 23–41.

———. 1983. A Symbolic Type and its Transformations: The Case of South Indian Poṇkal. *Contributions to Indian Sociology* (N.S.) 17, no. 2, 224–44.

———. 1987. Divine Coronation in a South Indian Temple. In *Religion and Society in South India* (eds.) V. Sudarsen, R. D. Prakash, and M. Suryanarayana. Delhi: B. R. Publishing Corporation.

———. 1989. Divine Marriage in a South Indian Temple. *Mankind* 19, no. 3, 181–97.

———. 2000. Congealing Divinity: Time, Worship, and Kinship in South Indian Hinduism. *The Journal of the Royal Anthropological Institute* 6, no. 2, 273–92.

Good, D. 1988. Individuals, Interpersonal Relations, and Trust. In *Trust: Making and Breaking Cooperative Relations* (ed.) D. Gambetta. Oxford: Basil Blackwell.

Gough, E. K. 1956. Brahman Kinship in a Tamil Village. *American Anthropologist* 58, 826–53.

Gray, J. 1979. Keep the Hom Fires Burning: Sacrifice in Nepal. *Social Analysis* 1, 81–107.

———. 1980. Hypergamy, Kinship and Caste Among the Chetris of Nepal. *Contributions to Indian Sociology* (N.S.) 14, no. 1, 1–34.

———. 1982. Chetri Women in Domestic Groups and Rituals. In *Women in India and Nepal* (eds.) M. Allen and S. N. Mukherjee. Canberra: Australian National University.

———. 1995. *The Householder's World—Purity, Power and Dominance in a Nepali Village*. Delhi: Oxford University Press.

Gregory, C. 1996. Hierarchical Opposition Revisited. *Oceania* 67, no. 2, 152–55.

Guneratne, K. B. 1999. *In the Circle of the Dance: Notes of an Outsider in Nepal*. Ithaca, NY, and London: Cornell University Press.

Gupta, S. 1988. The *Mandala* as an Image of Man. In *Indian Ritual and Its Exegesis* (ed.) R. F. Gombrich (Oxford University Papers on India, Vol. 2, part 1). Delhi: Oxford University Press.

Gupta, S., and R. Gombrich 1986. Kings, Power, and the Goddess. *South Asia Research* 6, no. 2, 123–38.

Hamilton, F. 1819. *An Account of the Kingdom of Nepal and the Territories Annexed to This Dominion by the House of Gorkha*. Edinburgh: Archibald & Constable Company.

Hamilton, R. W. 2003. Rice Festivals: Community and Celebration. In *The Art of Rice: Spirit and Sustenance in Asia* (ed.) R. W. Hamilton. Los Angeles: UCLA Fowler Museum of Cultural History.

Hanchett, S. 1988. *Coloured Rice: Symbolic Structure in Hindu Family Festivals.* Delhi: Hindustan Publishing Corporation.

Handelman, D. 1984. Inside-Out, Outside-In: Concealment and Revelation in Newfoundland Christmas Mumming. In *Text, Play, and Story: The Construction and Reconstruction of Self and Society* (ed.) E. Bruner. Long Grove, IL: Waveland Press.

——. 1987. Myths of Murugan: Asymmetry and Hierarchy in a South Indian Puranic Cosmology. *History of Religion* 27, no. 2, 133–70.

——. 1990. *Models and Mirrors: Towards an Anthropology of Public Events*. Cambridge: Cambridge University Press.

Handelman, D., and D. Shulman 1997. *God Inside Out, Siva's Game of Dice*. New York and Oxford: Oxford University Press.

Hardy, A. 1995. *Indian Temple Architecture: Form and Transformation: The Karnata Dravida Tradition 7th to 13th Centuries*. New Delhi: Abhinav.

——. 1998. Form, Transformation and Meaning in Indian Temple Architecture. In *Paradigms of Indian Architecture: Space and Time in Representation and Design* (ed.) G. H. R. Tillitson. London: Curzon.

Harlan, L., and P. B. Courtright 1995. Introduction. In *From the Margins of Hindu Marriage: Essays on Gender, Religion and Culture* (eds) L. Harlan and P. B. Courtright, Oxford: Oxford University Press.

Hart, K. 1988. Kinship, Contract, and Trust: The Economic Organization of Migrants in an African City Slum. In *Trust: Making and Breaking Cooperative Relations* (ed.) D. Gambetta. Oxford: Basil Blackwell.

Heesterman, J. C. 1957. *The Ancient Indian Royal Consecration: The Rājasūya Described According to the Yajus Texts and Annoted*. The Hague: Mouton & Co.

——. 1959. Reflections on the Significance of the Daksinā. *Indo-Iranian Journal* 3, 241–58.

——. 1985. *The Inner Conflict of Tradition: Essay in Indian Ritual, Kingship, and Society*. Chicago: The University of Chicago Press.

——. 1992. An Inauspicious World. A Review Article. *Social Analysis* 32, 87–94.

——. 1993. *The Broken World of Sacrifice: An Essay in Ancient Indian Ritual*. Chicago: University of Chicago Press.

Hershman, P. 1977. Virgin and Mother. In *Symbols and Sentiments: Cross-Cultural Studies in Symbolism* (ed.) I. Lewis. London: Academic Press.

Herzfeld, M. 1997. *Cultural Intimacy: Social Poetics in the Nation-State*. London: Routledge.

Hess, L. 1993. Staring at Frames Till They Turn into Loops: An Excursion Through Some Worlds of Tulsidas. In *Living Banaras: Hindu Religion in Cultural Context* (eds.) B. R. Hertel and C. A. Humes. Albany: State University of New York Press.

Hiriyanna, M. 1993 [1932]. *Outlines of Indian Philosophy*. Delhi: Motilal Banarsidass.

——. 1996 [1949]. *Essentials of Indian Philosophy*. London: Diamond Books.

Hirsch, E. 1995. Introduction, Landscape: Between Place and Space. In *The Anthropology of Landscape: Perspectives on Place and Space* (eds.) E. Hirsch and M. O'Hanlon. Oxford: Oxford University Press.

Hocart, A. M. 1970. *Kings and Councillors: An Essay in the Comparative Anatomy of Human Society*. Chicago: University of Chicago Press.

Hoek van den, A. W. 1992. Fire Sacrifice in Nepal. In *Ritual, State, and History in South Asia: Essays in Honour of J. C. Heesterman* (eds.) A. W. van den Hoek, D. H. A. Kolff, and M. S. Oort. Leiden: E. J. Brill.

Höfer, A. 1979. *The Caste Hierarchy and the State in Nepal: A Study of the Muluki Ain of 1854*. Insbruk: Universitätsverlag Wagner.

Höfer, A., and B. P. Shrestha. 1973. Ghost Exorcism Among the Brahmans of Central Nepal. *Central Asiatic Journal, International Periodical for the Languages, Literature, History and Archaeology of Central Asia* 17, no. 1, 51–77.

Holland, D., and D. Skinner 1995. Contested Ritual, Contested Femininities: (re)Forming Self and Society in a Nepali Women's Festival. *American Ethnologist* 22, no. 2, 279–305.

Howe, L. E. A. 1983. An Introduction to the Cultural Study of Traditional Balinese Architecture. *Arhcipel* 25, 137–158.

——. 1991. Rice, Ideology, and the Legitimation of Hierarchy in Bali. *Man* (N.S.) 26, no. 3, 445–67.

Hubert, H., and M. Mauss 1964. *Sacrifice: Its Nature and Function*. Chicago: University of Chicago Press.

Hudson, D. D. 1993. Madurai: The City as Goddess. In *Urban Form and Meaning in South Asia: The Shaping of Cities from Prehistoric to Pre-colonial Times* (eds.) H. Spodek and D. M. Srinivasan. Hanover: National Gallery of Art.

Hugh-Jones S. 1995. Inside-Out and Back-to-Front: The Androgynous House in Northern Amazonia. In *About the House: Levi-Strauss and Beyond* (eds.) J. Carsten and S. Hugh-Jones. Cambridge: Cambridge University Press.

Humphery, C., and J. Laidlaw 1994. *The Archetypal Actions of Ritual: A Theory of Ritual Illustrated by the Jain Rite of Worship*. Oxford: Clarendon Press.

Inden, R. B. 1978. Ritual, Authority, and Cyclic Time in Hindu Kingship. In *Kingship and Authority in South India* (ed.) J. F. Richards. Madison: South Asian Studies, University of Wisconsin.

——. 1985a. Hindu Evil as Unconquered Lower-Self. In *The Anthropology of Evil* (ed.) D. Parkin. Oxford: Basil Blackwell.

——. 1985b. The Temple and the Hindu Chain of Being. In *L'Espace du Temple: Espaces, Itineraries, Mediations* (ed.) J. C. Galey. Paris: Editions de l'Ecole des Hautes en Sciences Sociales.

——. 1992 [1990]. *Imagining India*. Oxford: Blackwell.

Inden, R. B., and R. W. Nicholas 1977. *Kinship in Bengali Culture*. Chicago: University of Chicago Press.

Ingold, T. 1993. The Temporality of the Landscape. *World Archaeology* 25, no. 2, 152–74.

Ishii, H. 1993. Seasons, Rituals, and Society: The Culture and Society of Mithila, the Parbate Hindus, and the Newārs as Seen Through a Comparison of Their Annual Rites. In *From Vedic Altar to Village Shrine* (Senri Ethnological Studies 36). Osaka: National Museum of Ethnology.

——. 1999. A Comparative Study of Life-Cycle Rituals of the Newārs, Parbate Hindus, and the Maithils. In *Anthropology and Sociology of Nepal: Culture, Societies, Ecology and Development* (eds.) R. B. Chhetri and O. P. Gurung. Kathmandu: SASON.

Jackson, W. J. 2004. *Heaven's Fractal Net: Retrieving Lost Visions in the Humanities*. Bloomington: Indiana University Press.

Jamison, S. W. 1996. *Sacrificed Wife/Sacrificer's Wife: Women, Ritual, and Hospitality in Ancient India*. Oxford: Oxford University Press.

Kaelber, O. W. 1976. Tapas, Birth, and Spiritual Rebirth in the Veda. *History of Religions* 15, no. 4, 343–86.

Kapadia, K. 1995. *Śiva and Her Sisters: Gender, Caste, and Class in Rural South India*. Boulder, CO: Westview Press.

Kapferer, B. 1997. *The Feast of the Sorcerer, Practices of Consciousness and Power*. Chicago and London: The University of Chicago Press.

Kaplan, D., and B. Saler 1966. Foster's "Image of Limited Good": An Example of Anthropological Explanation. *American Anthropologist* 68, 202–5.

Karan, P. P., and H. Ishii 1997 [1996]. *Nepal: A Himalayan Kingdom in Transition*. Delhi: Bookwell.

Khare, R. S. 1976a. *The Hindu Hearth and Home*. Delhi: Vikas Publication House.

——. 1976b. *Culture and Reality: Essays on the Hindu System of Managing Foods*. Simla: Indian Institute of Advanced Study.

——. 1992. Introduction. In *The Eternal Food: Gastronomic Ideas and Experiences of Hindus and Buddhists* (ed.) R. S. Khare. Albany: State University of New York Press.

Kinsley, D. 1986. *Hindu Goddesses: Visions of the Divine Feminine in the Hindu Religious Tradition*. Berkeley: University of California Press.

Knipe, D. M. 1975. *In the Image of Fire: Vedic Experience of Heat*. Delhi: Motilal Banarsidass.

——. 1977. *Sapiṇḍīkaraṇa*: The Hindu Rite of Entry into Heaven. In *Religious Encounters with Death: Insights from the History and Anthropology of Religions* (eds.) F. E. Reynolds and E. H. Waugh. University Park: The Pennsylvania State University Press.

——. 1988. The Temple in Image and Reality. In *Temple in Society* (ed.) M. V. Fox. Winona Lake, IN: Eisenbrauns.

Kolenda, P. 1984. Woman as Tribute, Woman as Flower: Images of "Woman" in Weddings in North and South India. *American Ethnologist* 11, no. 1, 98–117.

——. 1990. Untouchable Chuhras Through their Humor "Equalizing" Marital Kin Through Teasing, Pretence, and Farce. In *Divine Passions: The Social Construction of Emotion in India* (ed.) O. M. Lynch. Berkeley: University of California Press.

Kondos, V. 1982. The Triple Goddess and the Processual Approach to the World: The Parbatya Case. In *Women in India and Nepal* (ed.) M. Allen and S. N. Mukherjee (Monographs on South Asia, no. 8). Canberra: Australian National University.

——. 1991. Subjection and Ethics of Anguish: The Nepalese Parbatya Parent-Daughter Relationship. *Contributions to Indian Sociology* (N.S.) 25, no. 1, 113–34.

Kramrisch, S. 1976 [1946] *The Hindu Temple*. Vol. 1 and 2. Delhi: Motilal Banarsidass.

——. 1984 [1981]. *The Presence of Śiva*. Princeton: Princeton University Press.

——. 1991. Space in Indian Cosmogony and in Architecture. In *Concepts of Space Ancient and Modern* (ed.) K. Vatyayan. Indira Gandhi National Centre for the Arts. New Delhi: Abhinav Publications.

Krause, I. B. 1980. Kinship, Hierarchy and Equality in North Western Nepal. *Contributions to Indian Sociology* (N.S.) 14, no. 2, 169–94.

Krauskopff, G. 2003. Body Art and Cyclic Time: Rice Dancing Among the Tharu of Nepal. In *The Art of Rice: Spirit and Sustenance in Asia* (ed.) R. W. Hamilton. Los Angeles: UCLA Fowler Museum of Cultural History.

Krishna, N. 2003. Rice in the Human Life Cycle: Traditions from Tamil Nadu, India. In *The Art of Rice: Spirit and Sustenance in Asia* (ed.) R. W. Hamilton. Los Angeles: UCLA Fowler Museum of Cultural History.

Krishnamurthy, R. 1993. Agriculture in Ancient India and its Social Aspects. In *Agriculture in Ancient India* (ed.) V. Bedekar. Thane: Institute for Oriental Study.

Kuiper, F. B. J. 1970. Cosmogony and Conception: A Query. *History of Religions* 10, no. 2, 91–138.

——. 1975. The Basic Concept of Vedic Religion. *History of Religion* 15, no. 2, 197–220.

Laidlaw, J. 2000. A Free Gift Makes No Friends. *The Journal of the Royal Anthropological Institute* 6, 617–34.

Lakoff, G. and M. Johnson 1980. *Metaphors We Live By*. Chicago: The University of Chicago Press.

Lamb, S. 2000. *White Sarees and Sweet Mangoes: Aging, Gender, and Body in North India*. Berkeley: University of California Press.

Larson, G. J. 1998 [1969]. *Classical Sāmkhya: An Interpretation of Its History and Meaning*. New Delhi: Motilal Banarsidass.

Larson, G. J., and R. S. Bhattacharya 1981. *Sāmkhya: A Dualistic Tradition in Indian Philosophy*. Encyclopedia of Indian Philosophy 4. Princeton: Princeton University Press.

Latour, B. 1991. Materials of Power: Technology Is Society Made Durable. In *A Sociology of Monsters: Essays on Power, Technology, and Domination* (ed.) J. Law. London: Routledge.

——. 1993 [1991]. *We Have Never Been Modern*. New York: Harvester and Wheatsheaf.

Law, J. 1991. Introduction. In *A Sociology of Monsters: Essays on Power, Technology, and Domination* (ed.) J. Law. London: Routledge.

Le Mesurier, C. J. R. 1885. Customs and Superstitions Connected with the Cultivation of Rice in the Southern Province of Ceylon. *The Journal of the Royal Asiatic Society of Great Britain and Ireland* 17, 366–72.

Lecomte-Tilouine, M. 1993a. The Proof of the Bone: Lineage and Devali in Central Nepal. *Contributions to Indian Sociology* (N.S.) 27, no. 1, 1–23.

——. 1996. The Cult of the Earth Goddess Among the Magars of Nepal. In *Diogenes* 44/2, no. 174, 27–44.

Leslie, J. (ed.) 1992 [1991]. *Roles and Rituals for Hindu Women*. Delhi: Motilal Banarsidass.

Levy, R. I. 1992 [1990]. *Mesocosm, Hinduism, and the Organization of a Newar City in Nepal*. Delhi: Motilal Banarsidass.

Lewis, G. 1980. *Day of Shining Red: An Essay on Understanding Ritual*. Cambridge: Cambridge University Press.

Lewis, J. P. 1884. The Language of the Threshing-Floor. *Journal of the Ceylon Branch of the Royal Asiatic Society* 8, no. 2, 237–69.

Lindenbaum, S. 1986. Rice and Wheat: The Meaning of Food in Bangladesh. In *Aspects in South Asian Food Systems: Food, Society, and Culture* (eds.) R. S. Khare and M. S. A. Rao. Durham, NC: Carolina Academic Press.

Löwdin, P. 1986. *Food Ritual and Society: A Study of Social Structure and Food Symbolism Among the Newārs*. Thesis. Uppsala University, Sweden

Macdonald, A. W. 1975. *Essays on the Ethnology of Nepal and South Asia*. Kathmandu: Ratna Pustak Bhandar.

———. 1994 [1976]. Sorcery in the Nepalese Code of 1853. In *Spirit Possession in the Nepal Himalaya* (eds.) J. T. Hitchcock and R. L. Jones. New Delhi: Vikas Publishing House.

Macfarlane, A. 1976. *Resources and Population: A Study of the Gurungs of Nepal*. Cambridge: Cambridge University Press.

———. 1985. The Root of all Evil. In *The Anthropology of Evil* (ed.) D. Parkin. Oxford: Basil Blackwell.

———. 1992. Louis Dumont and the Origins of Individualism. *Cambridge Anthropology* 16, no. 1, 1–28.

Madan, T. N. 1965. *Family and Kinship: A Study of the Pandits of Rural Kashmir*. London: Asia Publishing House.

———. 1995 [1982]. The Ideology of the Householder Among the Kashmiri Pandits. In *Concepts of the Person: Kinship, Caste, and Marriage in India* (eds.) A. Östör, L. Fruzzetti, and S. Barnett. Delhi: Oxford University Press.

Malamoud, C. 1989. Indian Speculations about the Sex of the Sacrifice. In *Fragments for a History of the Human Body* (eds.) M. Feher, R. Naddaff, and N. Tazi. New York: Zone.

———. 1996. *Cooking the World: Ritual and Thought in Ancient India*. Delhi: Delhi University Press.

Mallaya, N. V. 1949. *Studies in Sanskrit Texts on Temple Architecture, with Special Reference to Tantrasauccaya* (Sanskrit series 42). Annamalai: Annamalai University Press.

Mallick, R. N. 1982. *Rice in Nepal*. Kathmandu: Kala Prakashan.

Malville, J. M. 1991. Astrophysics, Cosmology, and the Interior Space of Indian Myths and Temples. In *Concepts of Space Ancient and Modern* (ed.) K. Vatsyayan. Indira Gandhi National Centre for the Arts. New Delhi: Abhinav Publications.

Mandelbaum, D. G. 1968. Family, *Jāti*, Village. In *Structure and Change in Indian Society* (eds.) M. Singer and B. S. Cohn. Chicago: Aldine Publishing Company.

Mandelbrot, B. 1982. *The Fractal Geometry of Nature*. San Francisco: W. H. Freeman and Company.

——. 2005 [1997]. *Fractals and Scaling in Finance: Discontinuity, Concentration, Risk*. New York: Springer.
Mandelbrot, B., and R. L. Hudson 2004. *The Mis(behavior) of Markets: A Fractal View of Risk, Ruin, and Reward*. New York: Basic Books.
Marglin, F. A. 1981. Kings and Wives: The Separation of Status and Royal Power. *Contributions to Indian Sociology* (N.S.) 15, nos. 1&2, 155–81.
Marriott, M. 1966. The Feast of Love. In *Krishna: Myths, Rites, and Attitudes* (ed.) M. Singer. Honolulu: East-West Center Press.
——. 1968. Caste Ranking and Food Transactions: A Matrix Analysis. In *Structure and Change in Indian Society* (eds.) M. Singer and B. S. Cohn. Chicago: Aldine Publishing Company.
——. 1969. Review of Homo Hierarchicus. *American Anthropologist* 71, 1166–75.
——. 1976. Hindu Transactions: Diversity Without Dualism. In *Transactions and Meanings: Directions in the Anthropology of Exchange and Symbolic Behaviour* (ed.) B. Kapferer. Philadelphia: Institute for the Study of Human Issues.
——. 1989. Constructing an Indian Ethnosociology. *Contributions to Indian Sociology* (N.S.) 23, no. 1, 1–39.
——. 1998. The Female Family Core Explored Ethnosociologically. *Contributions to Indian Sociology* (N.S.) 32, no. 2, 279–304.
Marriott, M., and R. B. Inden 1977. Toward an Ethnosociology of South Asian Caste Systems. In *The New Wind: Changing Identities in South Asia* (ed.) K. David. The Hague: Mouton Publishers.
Mauss, M. 1967. *The Gift: Forms and Functions of Exchange in Archaic Societies*. New York: W. W. Norton & Company.
Meister, W. M. 1991. The Hindu Temple: Axis of Access. In *Concepts of Space Ancient and Modern* (ed.) K. Vatsyayan. Indira Gandhi National Centre for the Arts. New Delhi: Abhinav Publications.
——. 2003. Vāstupuruṣamaṇḍalas: Planning in the Image of Man. In *Maṇḍalas and Yantras in the Hindu Traditions* (ed.) G. Buhnemann. Leiden: E. J. Brill.
Menski, W. F. 1992. Marital Expectations as Dramatized in Hindu Marriage Rituals. In *Roles and Rituals for Hindu Women* (ed.) J. Leslie. Delhi: Motilal Banarsidass.
Meyer, J. J. 1995 [1930]. *Sexual Life in Ancient India*. New York: Dorset Press.
Michell, G. 1977. *The Hindu Temple: An Introduction of Its Meanings and Forms*. London: Paul Elek.
Michener, H. A., J. D. DeLamater, and S. H Schwartz. 1990. *Social Psychology*. San Diego: Harcourt Brace Jovanovich Publishers.
Miller, J. 1985. *The Vision of Cosmic Order in the Vedas*. London: Routledge & Kegan Paul.
Misra, B. 1966. Rice Rituals of Orissa. *Asian Folklore Studies* 25, 235–46.
Moffatt, M. 1975. Untouchables and the Caste System: A Tamil Case Study. *Contributions to Indian Sociology* (N.S.) 9, no. 1, 111–22.
Monier-Williams, M. 1995 [1899]. *Sanskrit-English Dictionary*. Oxford: Clarendon Press.
Moore, M. A. 1983. *Taṛavād: House, Land, and Relationship in a Matrilineal Hindu Society*. Thesis. The University of Chicago.

——. 1989. The Kerala House as a Hindu Cosmos. *Contributions to Indian Sociology*, (N.S.) 23, no. 1, 168–202.

Moreno, M. 1987. Agriculture as a Sacrament: A New Approach to the Cycle of Rice in South India. *Lambda Alpha Journal of Man* 18, 53–62.

Moreno, M., and M. Marriott 1989. Humoral Transactions in Two Tamil Cults: Murukan and Mariyamman. *Contributions to Indian Sociology* (N.S.) 23, no. 1, 149–67.

Muir, E. 1997. *Ritual in Early Modern Europe*. Cambridge: Cambridge University Press.

Müller-Böker, U. 1987. Man, Religion, and Agriculture in the Kathmandu Valley. In *Heritage of the Kathmandu Valley: Proceedings of an International Conference in Lübeck, June 1985* (eds.) N. Gutschow and A. Michaels (Nepalica 4/20). Sankt Augustin: VGH Wiessenschaftsverlag.

Nabokov, I. 1997. Expel the Lover, Recover the Wife: Symbolic Analysis of a South Indian Exorcism. *The Journal of the Royal Anthropological Institute* 3, no. 2, 297–316.

Narayanan, V. 2003. Gender in a Devotional Universe. In *The Blackwell Companion to Hinduism* (ed.) G. Flood. Oxford: Blackwell Publishing.

Nepali, G. S. 1988 [1965]. *The Newārs: An Ethno-sociological Study of a Himalayan Community*. Kathmandu: Himalayan Booksellers.

Nicholas, R. W. 1995. The Effectiveness of the Hindu Sacrament (*Saṁskāra*): Marriage, Divorce, and Caste in Bengali Culture. In *From the Margins of Hindu Marriage: Essays on Gender, Religion and Culture* (eds.) L. Harlan and P. B. Courtright. Oxford: Oxford University Press.

O'Flaherty, W. D. 1971. The Origin of Heresy in Hindu Mythology. *History of Religions* 10, no. 4, 271–333.

——. 1980a. *Women, Androgynes, and other Mythical Beasts*. Chicago: The University of Chicago Press.

——. 1980b [1976]. *The Origins of Evil in Hindu Mythology*. Berkeley: University of California Press.

——. 1980c. Introduction. In *Karma and Rebirth in Classical Indian Traditions* (ed.) W. D. O'Flaherty. Berkeley: University of California Press.

——. 1981 [1973]. *Śiva the Erotic Ascetic*. Oxford: Oxford University Press.

——. 1985. *Tales of Sex and Violence: Folklore, Sacrifice, and Danger in the Jaiminīya Brāhamaṇa*. Chicago: The University of Chicago Press.

Ohnuki-Tierney, E. 1993. *Rice as Self: Japanese Identities Through Time*. Princeton: Princeton University Press.

Ortner, S. B. 1981. Gender and Sexuality in Hierarchical Societies: The Case of Polynesia and Some Comparative Implications. In *Sexual Meanings: The Cultural Construction of Gender and Sexuality* (eds.) S. B. Ortner and H. Whitehead. Cambridge: Cambridge University Press.

Osella, F., and C. Osella 2000. Migration, Money, and Masculinity in Kerala. *The Journal of the Royal Anthropological Institute* 6, no. 1, 117–33.

Östör, Á., and L. Fruzzetti 1995 [1982] Concepts of the Person: Fifteen Years Later. In *Concepts of the Person: Kinship, Caste, and Marriage in India*. (eds.) A. Östör, L. Fruzzetti, and S. Barnett. Delhi: Oxford University Press.

Pagden, A. 1988. The Destruction of Trust and its Economic Consequences. In *Trust: Making and Breaking Cooperative Relations* (ed.) D. Gambetta. Oxford: Basil Blackwell.

Pandey, R. B. 1993 [1969]. *Hindu Samskāras: Socio-Religious Study of the Hindu Sacraments*. Delhi: Motilal Banarsidass.

Parry, J. P. 1979. *Caste and Kinship in Kangara*. London: Routledge & Kegan Paul.

———. 1980. Ghosts, Greed, and Sin: The Occupational Identity of the Benares Funeral Priests. *Man* (N.S.) 15, no. 1, 88–111.

———. 1985. Death and Digestion: The Symbolism of Food and Eating in North Indian Mortuary Rites. *Man* (N.S.) 20, 611–30.

———. 1986. The Gift, the Indian Gift and the "Indian Gift." *Man* (N.S.) 21, 453–73.

———. 1989. On the Moral Perils of Exchange. In *Money and the Morality of Exchange* (eds.) J. Parry and M. Bloch. Cambridge: Cambridge University Press.

———. 1994. *Death in Banaras*. Cambridge: Cambridge University Press.

Pigg, S. L. 1992. Inventing Social Categories Through Place: Social Representations and Development in Nepal. *Comparative Studies in Society and History* 34, no. 3, 491–513.

Piker, S. 1966. "The Image of Limited Good": Comments on an Exercise in Description and Interpretation. *American Anthropologist* 68, no. 5, 1202–25.

Pintchman, T. 1994. *The Rise of the Goddess in the Hindu Tradition*. Albany: State University of New York Press.

Pocock, D. F. 1973. *Mind, Body and Wealth: A Study of Belief and Practice in an Indian Village*. Oxford: Basil Blackwell.

Potter, K. H. 1980. The Karma Theory and Its Interpretation in Some Indian Philosophical Systems. In *Karma and Rebirth in Classical Indian Traditions* (ed.) W. D. O'Flaherty. Berkeley: University of California Press.

Quigley, D. 1993. *The Interpretation of Caste*. Oxford: Clarendon Press.

Raheja, G. G. 1988. *The Poison in the Gift*. Chicago and London: The University of Chicago Press.

———. 1994a. Introduction: Gender Representation and the Problem of Language and Resistance in India. In *Listen to the Heron's Words, Reimagining Gender and Kinship in North India* (eds.) G. G. Raheja and A. G. Gold. Berkeley: University of California Press.

———. 1994b. On the Uses of Irony and Ambiguity: Shifting Perspectives on Patriliny and Women's Ties to Natal Kin. In *Listen to the Heron's Words: Reimagining Gender and Kinship in North India*. (eds.) G. G. Raheja and A. G. Gold. Berkeley: University of California Press.

———. 1995. "Crying When She's Born, and Crying When She Goes Away": Marriage and the Idiom of the Gift in Pahansu Song Performance. In *From the Margins of Hindu Marriage: Essays on Gender, Religion and Culture* (eds.) L. Harlan and P. B. Courtright. Oxford: Oxford University Press.

Raheja, G. G., and A. G. Gold 1994. *Listen to the Heron's Words: Reimagining Gender and Kinship in North India*. Berkeley: University of California Press.

Ramanujan, A. K. 1989a. Is There an Indian Way of Thinking? An Informal Essay. *Contributions to Indian Sociology* (N.S.) 23, no. 1, 1–39.

——. 1989b. Where Mirrors Are Windows: Towards an Anthology of Reflections. *History of Religions* 28, no. 3, 187–216.

——. 1992. Food for Thought: Towards an Anthology of Hindu Food Images. In *The Eternal Food: Gastronomic Ideas and Experience of Hindus and Buddhists* (ed.) R. S. Khare. Albany: State University of New York Press.

Ramirez, P. 2000a. *De la Diaparition des Chefs: Une Anthropologie Politique Népalaise*. Paris: CNRS.

——. 2000b. Subjects and Citizens: Rural Headmen in Argha-Khanhci. In *Resunga, the Mountain of the Horned Sage: Two Districts in Central Nepal* (ed.) P. Ramirez. Lalitpur: Himal Books.

Ranjan, P. 1999. *Swasthāni*. Kathmandu: Spiny Babbler.

Reinhard, J. 1994 [1976]. Shamanism Among the Raji of Southwest Nepal. In *Spirit Possession in the Nepal Himalaya* (eds.) J. T. Hitchcock, and R. L. Jones. New Delhi: Vikas Publishing House.

Renou. L. 1953. *Religions of Ancient India*. London: The Athlone Press.

——. 1957 [1947]. *Vedic India*. Calcutta: Susil Gupta Private, Ltd.

——. 1961. *Hinduism*. New York: George Braziller.

Rivière, P. 1995. Houses, Places, and People: Community and Continuity in Guiana. In *About the House: Levi-Strauss and Beyond* (eds.) J. Carsten and S. Hugh-Jones. Cambridge: Cambridge University Press.

Roulet, M. 1996. Dowry and Prestige in North India. *Contributions to Nepalese Studies* 30, no. 1, 89–107.

Rutter, D. E. 1993. *Eating the Seed: The Use of Foods in the Structuring and Reproduction of Social Relations in a Nepali Chetri Community*. Thesis. London School of Economics.

Sax, W. S. 1990. Village Daughter, Village Goddess: Residence, Gender, and Politics in a Himalayan Pilgrimage. *American Ethnologist* 17, no. 3, 491–512.

——. 1991. *Mountain Goddess: Gender and Politics in a Himalayan Pilgrimage*. New York and Oxford: Oxford University Press.

Selwyn, T. 1980. The Order of Men and the Order of Things: An Examination of Food Transactions in an Indian Village. *International Journal of the Sociology of Law* 8, no. 3, 297–317.

Sharma, U. 2001. Trust, Privacy, Deceit, and the Quality of Interpersonal Relationships: "Peasant" Society Revisited. In *An Anthropology of Indirect Communication* (eds.) J. Hendry and C. W. Watson. London: Routledge.

Shulman, D. 1980. *Tamil Temple Myths: Sacrifice and Divine Marriage in the South Indian Śaiva Tradition*. Princeton: Princeton University Press.

——. 1984. The Enemy Within: Idealism and Dissent in South Indian Hinduism. In *Orthodoxy, Heterodoxy and Descent in India* (eds.) S. N. Eisenstadt, R. Kahane, and D. Shulman. Berlin, New York, and Amsterdam: Mouton Publishers.

——. 1985. Kingship and Prestation in South Indian Myth and Epic. *Asian and African Studies* 19, 93–117.

——. 1993. *The Hungry God*. Chicago: The University of Chicago Press.

Skinner, D. G. 1990. *Nepalese Children's Understanding of Self and the Social World: A Study of a Hindu Mixed Caste Community*. Thesis. The University of North Carolina–Chapel Hill.

Skinner, D., and D. Holland 1998. Contested Selves, Contested Femininities: Selves and Society in Process. In *Selves in Time and Place: Identities, Experience and History in Nepal* (eds.) D. Skinner, A. Pach III, and D. Holland. Lanham, MD: Rowman & Littlefield Publishers, Inc.

Slusser, M. S. 1982. *Nepal Mandala: A Cultural Study of the Kathmandu Valley*. Princeton: Princeton University Press.

Smith, B. K. 1989. *Reflections on Resemblance, Ritual, and Religion*. Oxford: Oxford University Press.

——. 1990. Eaters, Food, and Social Hierarchy in Ancient India: A Dietary Guide to a Revolution of Values. *Journal of the American Academy of Religion* 58, no. 2, 177–205.

Smith, B. K., and W. Doniger 1989. Sacrifice and Substitution: Ritual Mystification and Mythical Demystification. In *NVMEN* 36, 189–224.

Snodgrass, A. 1992 [1985]. *The Symbolism of the Stupa*. Delhi: Motilal Banarsidass.

——. 1994 [1990]. *Architecture, Time and Eternity: Studies in the Stellar and Temporal Symbolism of Traditional Buildings*. Vol. 1 (Śata-Pitaka Series, Indo-Asian Literatures, 356). New Delhi: International Academy of Indian Culture and Aditya Prakashan.

Standing, H. 1981. Envy and Equality: Some Aspects of Munda Values. In *Culture and Morality: Essays in Honour of Christoph Von Furer-Haimendorf* (ed.) C. Mayer. Delhi: Oxford University Press.

Stevenson, S. 1971 [1920]. *The Rites of Twice Born*. New Delhi: Oriental Books.

Stone, L. 1977. *Illness, Hierarchy, and Food Symbolism in Hindu Nepal*. Thesis. Brown University.

——. 1978. Food Symbolism in Hindu Nepal. *Contributions to Nepalese Studies* 4, no. 1, 47–65.

——. 1983. Hierarchy and Food in Nepalese Healing Rituals. *Social Science & Medicine* 17, no. 14, 971–78.

——. 1988. *Illness Beliefs and Feeding the Dead in Hindu Nepal: An Ethnographic Analysis*. Studies in Asian Thought and Religion, 10. Lewiston, NY: The Edwin Mellen Press.

Strathern, M. 1980. No Nature, No Culture: The Hagen Case. In *Nature, Culture and Gender* (eds.) C. P. MacCormack, and M. Strathern. Cambridge: Cambridge University Press.

——. 1988. *The Gender of the Gift*. Berkeley: University of California Press.

——. 1999. *Property, Substance, and Effect, Anthropological Essays on Persons and Things*. London: The Athlone Press.

——. 2000. Environment Within: An Ethnographic Commentary on Scale. In *Culture, Landscape, and the Environment* (eds.) K. Flint and H. Morphy (the Linacre Lectures 1997). Oxford: Oxford University Press.

Tambiah, S. J. 1973. From Varna to Caste Through Mixed Unions. In *The Character of Kinship*. (ed.) J. Goody. Cambridge: Cambridge: University Press.

——. 1984. *The Buddhist Saints of the Forest and the Cult of Amulets: A Study in Charisma, Hagiography, Sectarianism, and Millennial Buddhism*. Cambridge: Cambridge University Press.

Tingey, C. 1990. *Heartbeat of Nepal: The Pañcai Bājā*. Kathmandu: Royal Nepal Academy.

Toffin, G. 1977. *Pyangaun, Communaute Newār de la Vallee de Kathmandu: la Vie Materielle.* Paris: CNRS.

Trautmann, T. 1981. *Dravidian Kinship*. Cambridge: Cambridge University Press.

Trawick-Egnor, M. 1978. *The Sacred Spell and Other Conceptions of Life in Tamil Culture*. Thesis. The University of Chicago.

Trawick, M. 1990. *Notes on Love in a Tamil Family*. Berkeley: University of California Press.

Tridevi, K. 1989. Hindu Temples: Models of a Fractal Universe. *The Visual Computer* 5, no. 4, 243–58.

Turner, R. L. 1996 [1931]. *A Comparative and Etymological Dictionary of the Nepali Language*. New Delhi: Allied Publishers, Ltd.

Turner. V. 1964. Betwixt and Between: The Liminal Period in Rites of Passage. In *Symposium on New Approaches to the Study of Religion* (ed.) J. Helm. Proceedings of the 1964 Annual Spring Meeting of the American Ethnological Society. Washington, DC: American Ethnological Society.

——. 1969. *The Ritual Process: Structure and Anti-Structure*. London: Routledge & Kegan Paul.

——. 1977. Sacrifice as Quintessential Process Prophylaxis or Abandonment? *History of Religions* 16, no. 3, 189–215.

——. 1978. Comments and Conclusions. In *The Reversible World: Symbolic Inversion in Art and Society* (ed.) B. A. Babcock. Ithaca, NY, and London: Cornell University Press.

Upadhyaya, H. K. 1996. Rice Research in Nepal: Current State and Future Priorities. In *Rice Research in Asia: Progress and Priorities* (eds.) R. E. Evenson, R. W. Herdt, and M. Hussain. Wallingford: Cab International and the IRRI.

Vajracharya, G. V. 1997. The Adaptation of Monsoonal Culture by Rgvedic Aryans: A Further Study of the Frog Hymn. *Electronic Journal of Vedic Studies* 3, no. 2, 1–13, http://www1.shore.net/~india/ejvs/ejvs0302/ejvs0302.txt.

Vale de Almeida, M. 1996. *The Hegemonic Male: Masculinity in a Portuguese Town*. Oxford: Berghan Books.

Volwahsen, A. 1969. *Living Architecture: Indian*. London: Macdonald.

Wagner, R. 1986. *Symbols That Stand for Themselves*. Chicago: The University of Chicago Press.

——. 1991. The Fractal Person. In *Big Man and Great Man: Personifications of Power in Melanesia* (eds.) M. Godlier and M. Strathern. Cambridge: Cambridge University Press.

Waterson, R. 1997 [1990]. *The Living House: An Anthropology of Architecture in South-East Asia*. London: Thames and Hudson.

Wayman, A. 1982. The Human Body as Microcosm in India, Greek Cosmology, and Sixteenth-Century Europe. *History of Religions* 22, no. 2, 172– 190.

White, D. G. 2000. Introduction, Tantra in Practice: Mapping a Tradition. In *Tantra in Practice* (ed.) D. G. White. Princeton: Princeton University Press.

Witzel, M. 1987. The Coronation Rituals of Nepal, with Special Reference to the Coronation of King Birendra (1975). In *Heritage of the Kathmandu Valley: Proceedings of an International Conference in Lübeck, June 1985* (eds.) N. Gutschow and A. Michaels (Nepalica 4/20). Sankt Augustin: VGH Wiessenschaftsverlag.

———. 1992. Meaningful Ritual: Vedic Medieval and Contemporary Concepts in the Nepalese Agnihotra Ritual. In *Ritual, State and History in South Asia: Essays in Honour of J. C. Heesterman* (eds.) A. W. van den Hoek, D. H. A. Kolff, and M. S. Oort. Leiden: E. J. Brill.

Wojtilla, G. 1985. Kasyapiyakrsisukti: A Sanskrit Work on Agriculture II, English Translation. *Acta Orientalia Academiae Scientiarum Hung* 39, no. 1, 85–136.

Zemon-Davis, N. 1978. Women on Top: Symbolic Sexual Inversion and Political Disorder in Early Modern Europe. In *The Reversible World: Symbolic Inversion in Art and Society* (ed.) B. A. Babcock. Ithaca, NY, and London: Cornell University Press.

# Index


*abir,* 84n24, 122, 124–25, 127, 168, 211–13
*achetā,* 122, 125, 127–28, 132, 134–35, 137nn32–33, 137n45, 185; as embodied representation of goddess, 169, 212; meaning of, 168–69, 211–13; warding off evil, 169
action-centered approach to social life, 241–42. *See also* Gell, A.
actor network theory, 182n66
actors, actions and, 66, 153–54, 207–8; in Hinduism, 66, 102
Aditi, 99
agency, 11, 151–52, 173–75, 182n65, 241–42; sensitive perspective, 215. *See also* embodied representation; Gell, A.; gender, sensitivity to; sexuality, work and
Agni, 93; as cosmic mouth, 145; identified with ritual actor, 100, 130–31, 149, 164–65, 170; as master of *jagya,* 132; *saṁskāras* of, 130–31, 149, 164–65; and *tapas,* 101–2, 165; in Vedic sacrifice, 101, 144, 170; as womb, 101, 124, 130–31. *See also* fire; heat; *tapas*
*agnicayana,* 98, 109n50, 118, 147; and ritual womb, 99
agriculture: annual cycle of crops, *187*; imagining of, 192; sexual labor and, 192–94
alignment with time and space: in Hindu life, 11, 172–73, 177, 240; in *jagya,* 118–22, 125, 141, 149–50, 177; in marriage, 71–72, 74–75. *See also* fractal; ritual efficacy
ambivalence: of childhood, 56–64, 196; of goddess, 92–93, 95; of human marriage, 92–93; of *moksa* and renunciation, 104; of power (*śakti*), 29, 56, 68, 71, 92; of rice seedlings, 196; of sexuality, procreation and divine marriage, 92–96, 104; of wife, 9, 65, 68, 71, 75, 78–79, 81; of women in general, 49, 71
Annapurna, 206
annual oscillation of trust and cooperation, 221, 231–32
anxiety, 27; regarding adolescent girls, 61–63; regarding evil, 29–33; regarding nonhuman forces, 30–33, 47; within social milieu, 27–28, 33–40, 45–47, 77–80; regarding the wife, 65–71, 79. *See also* ambivalence; evil; mistrust; *śakti*

Appadurai, A., 33–34, 50n12, 86n68
*ardhanārīśvara,* 82
*ardhāngini*, 81–82
asceticism, 92–96, 100–106; divine sex and, 93; sexuality, procreation and, 95; Upanayana and, 100, 104–5. *See also* divine marriage
*ātman*, 42–43, 51n33; householder and family as, 42, 78, 165; as part of cosmic fractal, 5; as village fractal dimension, 42–43, 78–79; wife as, 78
Aubriot, O., 186, 192, 217n2, 217n11, 219n53
*axis mundi*, 158, 171

Babb, L. A., 31, 35, 52n51, 83n3, 85n39, 86n62, 94, 105, 107n3, 107n9, 143–44, 166–67, 180n30
background potentiality, 235–36, 241
Bahun, 13, 23n38, 44. *See also* Brāhmans
*bali*, 142–44, 179n5
*baran* (*garnu*), 128–29, 200
*bartoon*, 16, 51n 32, 56, 61, 104–5, 109n65, 113, 118, 145, 180n37. *See also* Upanayana
Bāstu (Purusa), 41, 46, 51n34, 112, 121–23, 128, 136n16, 137n35, 137n39, 141, 154–65, 170, 177, 180n38, 199–201, 207, 214. *See also* house
Bāstu *pūjā,* 113, 121–23, 131, 136n29, 141, 144, 155, 158–65, 180n38, 201; fractal of matrimonial encompassment and, 158–65. *See also* house
*bāsudhārā*, 118, 147–48
Beck, B. E. F, 22n20, 22n23, 86n62, 94, 107n3, 136n28, 137n37, 149, 151, 165–66, 171, 173, 179n6, 179n16, 180n27, 182n60, 216, 218n50
*bedi*, 115, 119, 123–28, 130–31, 135n1, 137n35, 142, 147, 154–57, 166, 192, 207; as feminine locus of *śakti,* ritual womb, 124, 130, 145, 154–57, 166, 214; matrimonial encompassment and, 155, 157. *See also vedi*
Bennett, L., 23n38, 35, 50n18, 53–54, 61, 63, 67–68, 84n23, 84nn35–36, 104, 112, 142, 144, 147, 168, 179n10, 180n30, 212, 217n6
*bhakari*, 196, 198–201, 204, 207, 227–28
*bhat*, 34–36, 145; vulnerability and, 35. *See* food
*bhāters*, 35–37, 132, 143, 200; danger in, 35–36; as social glue, 36
Bhūme, 32, 58, 95, 154–59, 224; Annapurna and, 206; in Bāstu *pūjā,* 123, 158–59, 163–64, 177; house and, 42, 137n39, 164; *jagya* and, 112, 121, 124, 128, 132, 145, 154–58, 164–65, 194; origin of rice agriculture and, 191; pregnancy of, 153, 195; rice and, 11, 107n14, 153, 155, 190–92, 194–95, 199–206; temples of, 190–91. *See also See also bedi*; *kunya*; *vedi*
*bhūt*, 32, 78
Biardeau, M., 109n56, 143, 146
*bir*, 45–47, 107n13, 226, 229, 232–34, 236n11, 248; collective "exorcism" of, 233; hierarchy of wealth and prestige and, 46; as negative embodied emotions, 45–47, 232; sent with bride as part of her dowry, 66, 68, 78–79
birth, 99; death and, 164; of house, 42. *See also* pregnancy; procreation
*bitulo*. *See* pollution, impurity
*biu rākhnu*. *See* rice agriculture
body, 51n33, 81; cosmos as, 157–58; of husband, 79; men vis-á-vis women, 62; mind and, 42–43; as temple, 5, 42, 157–58, 204, 215
*boksi,* 32, 45, 67, 248
Bouillier, V., 51n39, 85n51
Brahma, 168, 171
Brāhmaṇ, 5, 43

Brāhmaṇs, 1–2, 12–13, 35
brothers, 68–69
Brown, P., 23n30, 219n61
*buhāri*, 36, 64–68, 70, 75, 78–79
Burckhardt, T., 181n54
*buṭi biṭo*. *See* rice agriculture
*byar*. *See* rice agriculture

calendar, 222
*cāmal*, 35, 127, 152; as *dān*, 35, 58–59, 152; inner *śakti* of, 210–13. *See also* rice
*canduwā*, 124
canonical categories, 142
captivation, 97, 165–66, 209
caste system, 6, 16, 23n43, 40, 73–74, 224. *See also* Bahun; Chetri; Damāi; status
*cāturmāsa*, 188–89
*caurāsī godān,* 113, 134
cause, effect and, 150–51, 164, 174–77, 201
chaos, organization of, 1, 243. *See also* entropy
chastity. *See* virginity
Chetri, 23n38, 23n40, 37, 51n39, 53–54, 85n46, 96, 130n70, 143–44, 146, 168, 180n30, 186, 188, 192, 196, 204, 217n10, 217n13, 219n54
child marriages, 61–64
children, 33, 38, 56–57, 60–64, 222, 224
*chokā lagne*. *See* evil eye
clan. *See kul*
clan villages, 13–14
coercive subordination, 33
commensality, 36–37
Community Party of Nepal (CPN), 21
community, perspectives of, 232–33
competition, cooperation v., 221–22
completeness, 77, 101, 127–28, 151, 154, 158, 164, 173, 178, 183, 204
complexity, 238, 243
concealment: of answers, 248; of emotion, 37; of prestige, 28, 40–41, 43, 45, 230–323; of wealth, 28, 40–41, 43, 45, 230–32
conceptual overflow, 178
conjugal bond, 9, 55. *See also* marriage
connections, displaying of, 149–51
containment, 75. *See also* control
control: of change, 231; evil forces and, 33; of house, 78–79; of *śakti,* 92, 163; of sexuality, 94; of women, 67, 71, 77, 237. *See also* containment
cooking, 34–35, 144–46, 238; *dai halnu* and, 230; procreation and, 207; rice, 70. *See also kunya*
cooperation, 47, 189; competition v., 221–22; contractual, 39; lack of, 27
cosmogony. *See* creation myths
cosmology, 89, 96–102, 191, 198, 202, 210; agrarian, 235; gendered, 141, 177. *See also* fractal; Hinduism, Hindu world
cosmos, 101–2, 184, 198, 202, 210, 216; emotionality of, 30–33; food and, 30–33; fractality of, 2–8, 5, 11, 141, 177, 240; gender of, 151, 240; hunger of, 30, 102; male body, 157–58; merger, identification with, 172–77; outline regarding, 9–10; twin aspects of reality in, 46–47, 232, 239. *See also* alignment with time and space
CPN. *See* Community Party of Nepal
creation myths: Thamgarian, 149, 191; Vedic, 96–98, 163, 175, 191, 194. *See also* mythology; procreation
crop cycle, *187,* 221–22, 235
*culo*, 36, 42–43, 144–45
culture, nature v., 48

*dai halnu*, 79, 184, 204–6, 216, 221–36; analysis of, 228–33; conclusions regarding, 235–36; humor during, 224–25, 229, 234; operation on perceptions of self and community of, 232–33; revelation of scale of wealth and prestige in, 226–27, 230–33; sexuality and, 224, 229;

status and, 224, 241; women and, 233. *See also bir*; hierarchy; inversion; trust
*dāijo*, 40, 66. *See also bir*
*dakṣinā*, 58–60
Damāi, 196
*dān*, 55, 57–62, 129, 135; collectors of, 60; gender of, 59; intrinsic worth of, 60; logic of transference, 58–61; as paid service, 60; transference of, 72, 76, 135, 151–53, 211–12. *See also* danger
danger, 32, 53–54, 68–69, 71, 233; *dān* and, 58–59; pubescent girls and, 61; twilight and, 32
Daniélou, A., 5, 30, 94–96, 107n10, 107n17, 136n28, 175, 179n12, 179n14
*darśan*, 169
Dasaī, 14, 57, 142–43, 169, 188, 213, 221–22
Das, V., 40, 75, 86n68, 179n5, 179n11, 182n60
daughter-in-law. *See buhāri*
daughters, 54, 69. *See also kanyā*
death: annihilation of, 164–65, 178; birth and, 164; fire and, 144; hunger and, 97; *jagya* and, 117, 120, 144, 178; life from, 102, 106, 240; orgasm and, 97; *śakti* and, 210; time and, 149
deities, relationships with, 31. *See also individual gods and goddesses*
demons, 233–34
Devī, 53–54, 127, 169, 188
*dhān*, 127, 152, 210–11, 229. *See also* rice
*dhān khuṭnu*. *See* rice agriculture
*dhān ropnu*, 195–97. *See* rice agriculture
*dharma*, 69
*dhiki*, 210–12
*dhognu*, 57–59, 128, 135, 169, 193
*dhup halnu*, 203
*dikagadges*, 121, 128
*dīkṣā*, 100–1
distributed personality, 173–77, 242
divine marriage, 91–106, 158–65
divine sexuality, 91–92, 150, 163, 239; ascetic, 92–96; procreation, 96–102, 163; renunciation in, 100–1; shifting moods in, 102; *tapas* and, 101–2
domination, 72–77. *See also* control
Doniger, W., 40, 76, 83n9, 83n18, 87n93, 95, 105, 107n14, 108n37, 109n51, 109n56, 109n63, 143, 152, 217n21
Dumont, L., 5–6, 22n26, 50n15, 70, 73–74, 86n68

Eck, D., 94, 107n9, 137n37, 138n52, 166, 179n16, 179n19
effect, cause and, 150–51, 164, 174–77, 201
efficacy, 11, 156. *See also* ritual efficacy
Eglash, R., 21n3, 22n10, 23n37
Eliade, M., 24n46, 95, 98–99, 101, 108n34, 108n37, 108n47, 108n49, 109n51, 109n56, 136n28, 165
embodied representation, 10, 112–13, 115–17, 154, 242. *See also achetā*; *jagya*; rice
emotions: concealment of, 37; of cosmic, semidivine forces, 30–33; management of, 38. *See also* cosmos
encompassment: in Melanesia, 76; nonmarital. *See* rice. *See also* Dumont, L.; hierarchy; matrimonial encompassment
entropy, 150, 173
evil, 32–37, 44, 48, 52n54, 57–60, 62, 75–77, 84n38, 85n46, 108n41, 122, 125, 129, 132, 135, 188, 190, 194, 199, 205, 218n44, 225–26, 229, 231, 233; control over, 33; in Hinduism, 28–30; in social life, 28, 39, 44–48, 51n30, 57, 77–79, 139n74, 231–33. *See also achetā*; *bhut*
evil eye, 35, 44–45, 56, 223
exogamy, 64, 66

face, loss of, 211
fear, 33, 234. *See also* anxiety
females. *See* women
fertility: procreation v., 95; sacrifice and, 98–99; and *tapas,* 103. *See also* *saha*
fidelity, power of, 69
fire, 130; death and, 144; sexuality of, 145; *tapas and,* 101–2. *See also* Agni; heat; *tapas*
fire ceremony. *See hom*
food, 42, 167–69, 238; children and, 56; emotion and, 30–33; in *jagya,* 131; meaning of, 34–36; mistrust and, 34–37; power of, 35; shortages of, 48; status of, 50n15; trust and, 229
fractal: cosmos and, 2–8, 11, 30, 141, 149, 156–58, 177–78, 240; definition of, 2–4, 22n7; dimensions of, 4, 156–57, 178, 180nn31–32; divinity and, 6; effective dimensions of, 4, 157; Euclidean dimensions v., 4; in Hindu world, 1–2, 5–8, 10–12, 17, 22n12, 27, 184, 241–43; *hom* and, 243; humanities and, 1; introduction to, 1–2; *jagya* as, 113, 243; Mandelbrot, 1–4; as metaphor, 2, 10; micro and macrocosms and, 4; relative aspect of, 4; ritual, 167, 180n32; scale and, 2–4, 157, 177, 240; self-similarity in, 2–3; traditional societies and, 4–5, 10; in Vedic sacrifice, 98. *See also* hologram, fractal v.; matrimonial encompassment; ritual efficacy; self-similarity
fractal, gendered dimensions. *See* matrimonial encompassment
Fuller, C. J., 22n17, 49n7, 52n51, 83n3, 86n62, 86n67, 93–95, 104, 106n1, 107n3, 107n5, 107nn9–12, 109n56, 143–44, 146, 167, 179n16, 180n25, 217n13, 226
von Fürer-Haimendorf, C., 23n39, 36–37, 67, 83n5, 83n19
Gaṇeś, 93, 127, 130
Ganges, 123, 127, 148, 151
*Garbhagṛha,* 87n80, 94, 104, 179n17, 214–16, 219n62. *See also* Hindu temple; ritual: womb
gastro-cosmology, 30, 34
Gell, A., 107n9, 108n32, 109n55, 173–76, 241–42. *See also* actor network theory
gender, 11, 53; of action, 153; agriculture and, 193–94; cosmos and, 151, 240; hierarchy and, 74; life and, 152; attributed to nonhuman entities, 151–56; *rekhi* and, 127; rice and, 11, 153–54, 185, 209–13; sensitivity to, 150–56; Tamil perceptions of, 152; temples and, 215; Vedic sacrifice and, 98–100. *See also dān*; *dhiki*; *jagya*; rice; rice agriculture
*ghar*, 43, 64–65, 67, 70, 79
*ghee. See ghyu*
*ghyu,* 35, 117–18, 131–32, 142, 144–45
gift. *See dān*
god. *See* Īśwar
*godān. See jagya*
goddess. *See* Devī
Gold, A. G., 53, 63, 71–72, 83n6
Gonda, J., 22n19, 35, 52n46, 107n21, 107nn9–10, 108n49, 109n54, 138n58
Gray, J., 52n49, 83n5, 83n19, 84n23, 84n33, 85n46, 87n88, 109nn60–62, 138n58, 143–44, 146, 179n10, 181nn49–50, 217n6, 219n54
groom, ridiculing of, 66

*hakuwā*, 202–3, 208
harmony. *See* completeness
heat, 101, 166–67, 216. *See also* Agni; fire; *tapas*
Heesterman, J. C., 32, 58, 83n20, 84n21, 85n45, 86n65, 96, 99, 101, 108n38, 108n41, 108nn46–49, 139n69, 156–58, 164, 167, 170, 172
hierarchy, 46, 66, 150, 168; cyclical annual process of, 231–32; divine,

230–32; encompassment and, 73–74; gender and, 74; of wealth, 46, 241
Hinduism, Hindu world: anthropology and, 2, 6, 237; architecture of, 6; categories within, 102, 146, 178; complexity in, 238; contextual sensitivity of, 146; cosmology of, 2–8; divine sexuality in, 91–106; entropy and, 150, 173; ethno-sociological approach to, 6; evil and, 28–30; fluid perceptions of reality, 146; gastro-cosmological vision, 30, 34; ideological transition in, 102–4; mythology, 1, 5–6, 9, 91, 96–97, 100, 113–14, 163, 175, 216, 239–40; rebirth in, 74, 92, 100–1, 104–5; symbols and, 10, 115–17. *See also* actors, actions and; alignment with time and space; cosmology; creation myths; fractal; gender; Hindu temple
Hindu temple: and fractal, 6; Newari pagoda, 214, 218n48; Śikara, 6–7, 11, 184, 213–16; development of the Śikara, 11, 184, 213–16; village temple, 214, 218n48. *See also* *Garbhagṛha*
Holland, D., 63, 84n35, 85n39, 85nn51–52
hologram, fractal v., 7–8
*hom*, 10, 71, 111, 123–24, 128–32, 144–47, 159–60, 183, 215; agriculture and, 194; fractality of, 243; *homa* v., 144; *kunyā* as, 207–8
*homa*, *hom* v., 144
*Homo Hierarchicus: Essai sur le système des castes* (Dumont), 73
honor, 38
hostility, 9, 47–48, 55, 228, 239. *See also* mistrust
house: birth of, 42; construction of, 145; control of, 78–79; duality of, 232; ideal of, 206; identity with, 41–46, 78, 159, 165, 230, 232–33; immortality of, 165; layout of, 43–44; marriage of, 42, 113, 158, 160, 165; prestige and, 40–41; purity axis of, 43–44; rice and, 28; as temple, 41, 77; trust axis of, 44–45. *See also* Bāstu; Bāstu *pūjā*
humor, 224–25, 229, 234
hunger, 47; of cosmos, 102; death and, 97; metaphysics of, 30
husbands, 68–69, 75–76, 79, 238
husking. *See* rice agriculture
hypergamy, 61

*ijjat*: concealment of, 28, 40–41, 43, 45, 230–33; criteria for, 40; display of, 230; hierarchy, annual cycle of, 231–32. *See also* status; wealth
illness, 20
image/s: dialectics of, 77–82, 183, 239; meaning of, 80, 237; sorcery, 10, 173–74, 177, 182n64
immortality, 103, 165; conditional, 105, 215, 238; death and, 106; procreation and, 96–102
Inden, R. B., 23n32, 29, 83n6, 83nn18–19, 85n50, 87n73, 87n87, 107n21, 109n56, 109n59, 109n61, 116, 135n11, 151, 157, 167, 171–73, 175, 179n16, 181n43, 214
India: northern, 186, 214; southern, 55, 80–81, 92, 152, 192, 208, 239
information: lack of, 47; management of, 38
initiation. *See bartoon*; Upanayana
inner self, social persona v., 47, 232–33, 239
intercourse, sexual, 163–64, 193
interdependence, 221
interpersonal relationships, 1
inversion: of hierarchy, 221–36; introduction to, 221–22; of social structure, 221–36; of trust, 221–36
irrigation, 189, 201, 222
Ishii, H., 179n10, 188, 192, 217n3, 236n3
Īśwar, omnipresence of, 6, 29

Jackson, W. J., 4, 21n3, 22n13, 22n27, 23n31, 23n37
*jagya*, 10–11, 91, 183; analysis of, 141–78; Bāstu in, 121–23; conclusions regarding, 177–78; construction of, 117–28, 141, 147; core of, 111; cost of, 113; death and, 117, 120, 144, 178; efficacy of, 141, 150, 156, 170–77, 240; elaborated, 120–22, 124, 157; embodied representations in, 115–17; encapsulating infinity, 178; ethnography of, 111–35; final rites of, 132–35; food in, 131, 142–45; fractality of, 113, 243; gender in, 151–66; *godān*, 59, 116; *hom* in, 111, 123–24, 128–32, 142, 144–47, 159–60; infinity and, 178; introduction to, 111–17, 141–47; *jagya lāunu* in, 117–28; *kartā* of, 10, 114–15, 118–19, 125, 128–32, 135, 149–50, 155–56, 164–66, 170, 176, 215; as *kartā*'s distributed personality, 173–77; *khāldo*, 124, 127, 130, 145, 155, 159–60, 163; *lingas* in, 119–23, 127–28, 136n28, 154–55, 157, 160–62; matrimonial encompassment and, 119–28, 156, 178; as microcosm, 112–15; *mul kalas*, 127–28, 166–67; *naibedde*, 117–18, 124; *phal* of, 121, 177; *prāyaścitta godān*, 128–29; as prototype of north-Indian Śikara Hindu temple, 213–16; *purbānga* in, 112, 117–18, 147–49; rice cultivation as, 185–216; as ritual enclosure, 112–15, 146–47; as ritual performance, 112, 118, 146–47; sexual labor in, 151–56; simple, 119–20, 157; *snān*, 112, 132–33, 166–67; space, alignment in, 118–22, 147–49; terminology of, 142–46; time, alignment in, 118–22, 147–49; village v. agricultural, 241; *yajña* v., 142–44, 146. *See also* Agni; Bhūme; *dai halnu*; death; food; fractal; rice; rice agriculture; time
*jajamāna*, 98–101, 170, 172, 176
*jajmān*, 142
*janaï*, 16, 105, 123–24
*jānne mānche*, 17
*jaro*, 205–9. *See also* heat; *kunya*; rice agriculture
*jat*, 23n43, 64, 125, 188
*juro*. *See kunya*; rice agriculture
*jutho*. *See* pollution, impurity

*kalas*, 127–28, 167
Kāli, 53–54, 64
Kāli temple, 14, 32, 57, 143, 214, 247
Kāli *yug*, 29, 49, 114
*kanikā*. *See* rice
*kanyā*, 65–66, 70, 76; married, 62–63; pubescent, 62–63; as virgin goddess, 56–61, 123, 125, 153. *See also kanyādān*
*kanyādān*, 55, 58–61, 65–66, 72, 212; *bitulo,* 62, 65; married, 61–64, 213. *See also dān*; *kanyā*
*karma*, 175; written on forehead, 133–34
*kartā*, 10, 114–15, 118–19, 125, 128–32, 135, 149–50, 155–56, 176, 215; distributed personality of, 173–77; rebirth of, 135, 164–66, 170
*khājā*, 198, 227
*khāldo*. *See jagya*
*khalo garo*. *See* rice agriculture
*khānā*, 36, 42
Khanal, Prakash Ram, 134
Khare, R. S., 34, 50nn14–15, 179n6, 212
*khetālo*. *See* rice agriculture
*khuṭṭa pāni*, 18, 66
kings, 95–96, 170–72, 176
kinship, 1, 9, 28, 49, 55, 64–65
Knipe, D. M., 99–101, 108n39, 109n53, 109n59, 179n16, 215
Kondos, V., 63, 67, 83n5, 84n37, 87n88, 181n50, 219n54

Kramrisch, S., 22n18, 22nn22–23, 23n30, 71, 77, 87n93, 93–94, 99, 107n8, 107n18, 108n49, 109n58, 122, 136n28, 137n42, 137nn36–38, 149, 151, 180n34, 181n54, 181n59, 213–15, 219n59, 219n61
Kuiper, F. B. J., 77, 179n13, 191
*kul*, 13–14, 36–37, 76, 129
*kul deutā pūjā*, 36–37
*kunya*: as Bhūme's temple, 201–9; cooking in, 205–13; as cosmos, 202, 204; *hom* in, 207–9; as householder, 204; *jaro* of, 205, 207–9; *juro* (pinnacle) of, 204, 214, 222, 224; location of, 207; as prototype of north-Indian Śikara Hindu temple, 213–16; status and, 214. *See also* rice agriculture

labor. *See* work
Lakṣmī, 116
life: death and, 102, 106, 240; gender and, 152; of rice, 210
life stages: sacrifice and, 100–2; of women, 49, 55–67, 83n10
liminality, 228–30
*linga*, 119–23, 127–28, 136n28, 154–55, 157, 160–62. *See also* Śiva-*liṅgam*
low castes ("Untouchables"), 13–15, 23n39, 24n53, 40, 56, 67, 76, 152, 191, 193, 196, 200, 224

macrocosm, 4, 10, 42, 75, 99, 141, 150–66, 172–73. *See also* microcosm
Mahaboudha of Patan, shrine of, 6–7
*māiti*, 61, 63–65, 70, 75, 155
Malamoud, C., 30–31, 35, 86n65, 96–99, 108n34, 108n36, 108n47, 109n54, 136n20, 175, 179n6, 181n54, 182n69, 191
male: attitude towards menstruation, 62–67, 75–76, 212; body, skin, 157–58; cosmos as body of, 157–58; perspective on women, 54, 67–71, 212, 239; pregnancy, 97, 213
Mallaya, N. V., 107n18, 215
man: dual personhood of, 232, 239; *śakti* of, 143, 210; woman v., 55, 68, 74, 242. *See also* distributed personality; house; personhood
*maṇḍala*, 171
Mandelbrot, Benoit, 1, 3–4
*mandir*. *See* temples
marriage, 71–77, 113, 146, 156; ambivalence of, 92; child, 61–64; divine, 91–106, 158–65, 239; as euphemism for sex, 163; of house, 42, 113, 158, 160, 165; imagery of, 9, 55, 154–55; intercourse and, 163–64; prescriptions for, 64; requirement of, 77; status and, 64–65; Tamil perceptions of, 80–82; transcendence through, 76. *See also* *ardhanārīśvara*; *ardhāngini*; matrimonial encompassment
Marriott, M., 6, 23n32, 35, 43, 50n14, 54, 62, 66, 82n2, 146, 166–67, 175, 179n6, 182n71, 236n14
*masān ghāt*, 32
matrilinearity, 80
matrimonial encompassment, 1, 9–10, 28, 49, 55, 71–77; agriculture and, 183, 185, 198–201, 203–9, 216; conclusions concerning, 237–41; divine, 65, 91, 106, 150, 176, 178, 183, 206, 235–36; embodied symbolism of, 72; fractal of, 82, 98, 106, 141–42, 149, 156–58, 167, 177–78, 180nn31–32, 186, 210, 216, 237–38, 240–43; Genesis and, 73–74; Hindu temple and, 213–16; ideal of, 82, 150, 183, 241; inversion of, 69, 233; meaning of, 72–77; rice and, 185–216; Vedic cosmogony and, 96–98; vicious circle of mistrust and, 55, 77–80, 235. *See also agnicayana*; *ardhānarīśvara*; *ardhāngini*; Dumont, L.; fractal; marriage; rice; rice agriculture; sexuality

matrimonial encompassment: in ritual. *See* Bāstu *pūjā*; *bedi*; *jagya*
Meister, W. M., 23n30, 137nn36–37, 219n61
menstruation, 62–65, 68, 212–13
Michell, G., 23n30, 151, 181n59, 215, 219n61
microcosm, 4–5, 10, 41–42, 75, 99, 150–66, 172–73. *See also* macrocosm
mind, body and, 42–43
misfortune, 31
mistrust, 9, 11–12, 27–28, 34–37, 45–48, 55, 70, 232; fluctuation of, 48; reemergence of, 228; self-fulfillment of, 47; vicious cycle of, 9, 239. *See also* trust
*mit*, 38
*mithuna*, 71, 94, 98, 123, 155, 165–66, 215. *See also* divine sexuality
*moksa*, 91–92, 94, 102, 104–6
monsoon, 188–89, 236
moral substances, transference of, 58. *See also dān*
*mul kalas*. *See jagya*
*murti*, 37
mutual aid. *See parma*
mythology, 1, 5–6, 9, 91, 96–97, 100, 113–14, 163, 175, 216, 239–40

*nāghnu*, 193
*nag, nagini*, 205, *208*
*naibedde*. *See jagya*
Newārs, 144, 202, 208, 214, 240
*nivek*, 50n24, 198, 217n15
nonhuman entities, 30, 32–34, 47, 151–53, 242
*nuwaran*, 74

observational distance, 157. *See also* perceptual resolution
observer's point of view, 34
O'Flaherty, W. D. *See* Doniger, W.
Ohnuki-Tierney, E., 41, 186, 210, 217n5, 217n8, 235
orgasm, 97

*paīco*, 39. *See also* reciprocity
*pāp*, 39, 58, 65, 132, 173
Parikchit, 113–14, 121
*parma*, 39, 189, 192, 198, 217n15, 222
Parry, J. P., 6, 35, 58–59, 66, 82, 83nn19–20, 84n21, 85n39, 85n46, 86n68, 87n87, 109n59, 143, 167, 179n5, 186, 217n7, 218n28
patriline, ideology of, 69–70, 239
People's War, 21
perceptual resolution, in social life, 70, 169
personhood, 41–47, 67–71, 77–80, 232–33, 235, 238–39, 242
*phal*, 121, 177
*phuknu*, 249
pollution, impurity, 35, 62, 65, 67, 136, 143, 145, 188, 203, 205, 210, 219
power, problem of, 29. *See also śakti*
*pradakṣinā*, 60, 118, 124–25, 132, 157
Prajapati, 96–101, 163, 170–71, 175–76
*prasad*, 67, 124, 132, 134, 141–42, 167–69, 190, 200, 211–13
*pratimā*, 112. *See also* embodied representation
*prāyaścitta godān*. *See jagya*
pregnancy, 97, 145, 163–64, 195, 207, 213
premarital sex, 154
prestige. *See ijjat*
priests, 14–16, 40, 59–60, 192, 205, 248
procreation, 91–93, 155; cooking and, 207; decline of, 102–5, 239–40; fertility v., 95; immortality and, 96–102; sacrifice and, 96–98, 144, 240; sexuality and, 96–102, 163
puberty, 56–57
public male discourse, 69–70, 239
*pūjā*, 10–11, 31, 34–35, 58, 102, 111–13, 128, 142–45, 166; Bāstu, 113, 122–23, 144, 158–65, 201; oral nature of, 143; rice cultivation as, 200. *See also hom*; *jagya*; *ṭikā*
*purān*, 113–15, 121, 128–30, 142. *See also* mythology

Purāṇic mythology, 1, 9, 91, 100, 163
*purbāṅga*, 112, 117–18, 147–49
purity, 61, 224; of children, 56–57; of cook, 34; of house, 43–44; of rice, 35

Raheja, G. G., 35, 53, 57, 62, 66, 70–71, 83n6, 83nn19–20, 87n87, 217n13
*rājasūya*, 164, 167, 170–72
*rakṣā-bandhan*, 72
Ramanujan, A. K., 82n2, 146, 212
Ramirez, P., 23nn39–40, 36, 49n2, 50n14, 84n23, 169, 181n49, 186, 212
rebirth, 74, 94, 96–97, 100–101, 104–5, 144, 181n57, 215, 238; of *kartā*, 131, 164–66, 170; of king, 171–72; of rice, 207–8; ritual, 94, 96–97, 100–106, 108n46, 131, 144, 149, 155, 164–67, 170–73, 215, 240; status and, 105; unlimited, 164–65; of village, 235. *See also bartoon*; Upanayana
reciprocity, 27, 39
*rekhi*, 125–27, 138nn47–48, 141, 157, 160, 165–66
relationships: contractual nature of, 39, 234; with deities, 31; interpersonal, 1. *See also* social fabric, world
Renou, L., 51n35, 75, 83n13, 87n74, 96, 99, 103, 107n24, 108n27, 108n36, 108n49, 109n54, 109n56, 135n6, 137n44, 172, 175, 180n33
renunciation. *See* asceticism
reproduction. *See* procreation
reputation, 38. *See also* status
rice, 11–12, 84n26; ambivalence of seedlings, 196; cooked, in fields, 203; cooked, in *kunya*, 205–9; cooking of, 145; culture of, 185–89; *dai halnu* of, 79, 184, 204–6, 216, 221–36; as conduit for *dān*, 59–61, 72, 76, 135; as embodiment of *dān*, 59–60, 72, 76, 135, 185; as feminine background potentiality, 241; fields, 32, 155, 189–91, 241; gender of, 11, 153–54, 183, 185, 195–96, 209–13; house and, 28; introduction to, 185–94; in Japan, 186, 210; *kanikā*, 210–12; measurement of, 226, 230; nonmatrimonial encompassment of, 76, 87n79, 209–13; rebirth of, 207–8; *saha* of, 45, 191–92, 206–8, 226, 235; *sāli* of, 188; trust and, 34–37, 229; as wealth and scale of prestige, 40–41, 230–32. *See also achetā*; ambivalence; *bhat*; Bhūme; *cāmal*; cooking; embodied representation; encompassment; gender; *hakuwā*; house; *jagya*; life; matrimonial encompassment; *pūjā*; purity; rebirth
rice agriculture, 1; *bhakāri bādnu* of, 198–201; *biu rākhnu*, 185, 194–95, 210; *buṭi biṭo*, 203–4, 207, 214, 224; *byar*, 195–96; *dhān kutna*, as sacrifice, 210; *dhān ropnu*, 185, 195–97; gender, sexuality and, 192–96, 199, 203–5, 213–16, 224; imagining of, 192–94; as *jagya*, 194–209, 241; *khalo garo*, 189–91, 201–3, 206, 224–26; *khetālo*, 198–200, 226, 229; matrimonial encompassment in, 198–209; origin of, 191–92; as pregnancy, 153–54, 195, 199, 201, 207–8; seedlings, 196, 199; sexual labor in, 153–54, 193–97; status and, 192; trust and, 183; violence and, 192; Viṣṇu and, 96. *See also dai halnu*; *kunya*
ritual: barrenness of, 104; enclosures, 112–15, 147, 151, 157, 214; purity, 34, 43, 46, 56–57; womb, 91, 93–95, 99–106, 104, 130–31, 163–67, 172–73, 178, 179n17, 179n20, 181n57, 207–8, 212. *See also baran*; *bartoon*; Bāstu *pūjā*; *bāsudhārā*; *bedi*; *caurāsī godān*; *dakṣinā*; *dān*; *darśan*; *dhup halnu*; *dīkṣā*; *hom*; *homa*; *jagya*; *kalas*; *kartā*; *khāldo*; *kul deutā pūjā*; *linga*;

*mit*; *mul kalas*; *pūjā*; *purān*; *purbāṅga*; *rekhi*; *samidhā*; *saṁskāra*; *saṅkalpa*; *snān*; *tapas*; temples; *ṭikā*; *toran*; *yoni*
ritual efficacy, 141, 150, 170–77, 240
Rutter, D. E., 45, 52n49, 83n5, 84n36, 87n88, 96, 109n59, 128, 138n58, 139n70, 168, 179n10, 180nn29–30, 181n50, 186, 192, 196, 201, 208, 217n2, 217n6, 217nn9–10, 217n13, 219n54

sacred thread. *See janaï*
sacrifice, 91–92, 122, 159, 170, 238; cooking and, 144–46; fertility and, 98–99; fractality of, 98; gendered, 98–100; *hom* and, 142–46; initiation and, 100–1; internalization of, 102; life stages and, 100–2; procreation and, 96–98, 240; Vedic, 1, 216
*saha*, 45, 52n46, 135, 191–92, 206, 226, 235. *See also* fertility
*sāit*, 137n34, 173
*śakti*, 29, 55–57, 59, 141, 148, 151–52, 215; of Bhūme, 201, 205; control of, 92, 163; death and, 210; divine, 166–69; flow of, 75; fluidity of, 78; of men, 143, 210; in *prasad*, 167–69; of rice, 96, 212; of Śiva, 127; of *snān*, 132; of women, 62–64, 68–69, 71–72, 75; of women, protects husbands, 68–69. *See also* sexuality
*sāli*. *See* rice
*samidhā*, as ritual womb, 130–31
*saṁsāra*, 29, 94, 106
*saṁskāras*, 10, 42, 56, 71, 92, 100, 113, 128, 130–31, 146, 149, 194
*saṅkalpa*, 128–29, 149
*satī*, 82
scale, 43, 50n25, 157, 177, 216; absence of absolute, 53; fractals and, 2–4, 240; revelation of, 230–32
seclusion, of young women, 62–63
second birth. *See* rebirth
seed grain, 212
seedlings, of rice. *See* rice agriculture
self-fulfillment, of mistrust, 47
self-similarity, 2–3, 6, 156, 167, 177. *See also* fractal
sexuality: agriculture and, 193–94; ambivalence of, 92, 96; asceticism and, 95; control of, 94; *dai halnu* and, 224, 229; divine, 91–106, 163, 239; fire and, 145; marriage as euphemism for, 163; matrimonial encompassment and, 163–65; premarital, 154; reference to, 224, 229; taboo of, 154–55; Tantric, 103; temples and, 215; of women, 212; work and, 151–56, 193. *See also śakti*
sexual labor. *See* agriculture; *jagya*; rice agriculture
Shulman, D., 7, 22n15, 22n25, 23n36, 29–30, 56, 60, 62–63, 75, 84n21, 86n58, 86n62, 87n74, 87n93, 92–93, 95, 97, 99, 102, 104, 106n1, 107n3, 108n48, 137n37, 143, 146, 158, 172, 175, 178, 180n25, 194, 215–16
Śikara. *See* Hindu temple
sisters, 9, 54, 57, 63, 68–69, 75, 83n14, 147
Śiva, 6, 93–95, 121, 127, 130, 148, 163, 171, 192, 224
Śiva-*liṅgam*, 163; as fractal, 6; as symbol of asceticism/procreation, 94
Skinner, D. G., 35n37, 63, 83n5, 84n35, 85n39, 85nn51–52
Smith, B. K., 40, 76, 83n18, 98–99, 105, 107n14, 108n34, 108nn36–37, 108n40, 108n46, 108nn48–49, 109n51, 109n56, 109n63, 112, 142–43, 170, 172, 182n67, 194, 217n21, 217n24
*snān*, 112, 132–33, 166–67
Snodgrass, A., 22n22, 109n58, 136n20, 136n28, 179n14, 181n54, 181n59, 219n61
social fabric, world, 27–49; annual oscillation within, 135, 188–89, 221,

231–32; commensality and, 33–37; concealment of feelings and, 37–40; evil and, 28–30; governing elements of, 30–40; hierarchy of, 46, 66; introduction to, 27–30; personhood; resilience of, 221; sociality, sociability within, 27–28, 36, 232; uniformity, facade of, 37–46. *See also* annual oscillation of trust and cooperation; anxiety; concealment; cooperation; emotions; *dai halnu*; evil; food; gastro-cosmology; hierarchy; hostility; mistrust; *parma*; relationships; trust; wealth
social fabric, world: inversion of. *See dai halnu*
social fabric, world: prestige within. *See ijjat*
social persona, inner self v., 47, 232–33, 239. *See also* personhood
solidarity, 222; lack of, 34, 36
sorcery, 10, 20, 45–46, 48, 67; DIY, 176; image, 10, 173–74, 177, 182n64
sowing. *See* rice agriculture
*sṛṣṭi*, 77, 120, 154–55, 158, 164
status: agriculture and, 192; *dai halnu* and, 224, 241; of food, 50n15; *kunyā* and, 214; marriage and, 64–65; purity, 34, 56–57, 61, 224; rebirth and, 105; ritual, 34, 57; wealth and, 28. *See also* caste system; *ijjat*
Stone, L., 34, 48, 50n14, 52n45, 52n49, 83n5, 212, 217n2, 217n6, 217n9
Strathern, M., 17, 22n10, 27, 48, 50n25, 76, 78, 179n15, 182n72
suspicion. *See* mistrust
*Swasthāni*, 93
symbols, 10, 112, 115–17, 136n14, 147, 168, 188, 194, 242. *See also* embodied representation; Hinduism, Hindu world

*talo,* 41, 43–44, 46, 227, 230, 232–34. *See also* house
Tambiah, S. J., 57, 86n69
Tamils, 55, 80–81, 92, 152, 208
Tantra, 103, 171, 215, 240
*tapas*, 91, 101–3, 105, 166, 207–9, 216. *See also* Agni; fertility; fire; heat
temples, 112, 149, 151, 166, 173; Bhūme tree, 189–91; body as, 5, 42, 157–58, 204, 215; construction of, 171; gender and, 215; house as, 41, 77; *kunyā* as, 201–9, 213–16; living, 77–78; sexuality and, 215; as womb, 94, 104. *See also* Hindu temple
Tihar, 68–69, 75, 147, 221–22
Tij, 68–69, 233
*ṭikā*, 11, 122, 128–29, 132–35, 142–43, 167–69, 199–200, 211–12
time: death and, 149; *jagya* and, 118–22, 147–49
Toffin, G., 203, 217n4
*toran,* 124–25
traditional societies, fractals and, 10
transmigration, 5
transplantation. *See dhān ropnu* (rice agriculture)
Trawick-Egnor, M., 7, 22n24, 23n34, 53, 80–81, 87n72, 87n80, 87nn84–85, 87n90, 116, 151–52, 180n40, 217n24
Trawick, M., 87n85. *See also* Trawick-Egnor, M.
trust, 232; annual cycle of, 228–322; as axis in house, 44; food and, 229; illusion of, 233–34; mistrust v., 221–22; rice and, 183. *See also* mistrust

uncertainty. *See* ambivalence; evil; *śakti*
Upanayana, 92, 100–101, 104–5, 146. *See also bartoon*
*upaniṣads*, 172

Vastu, Bāstu v., 122–23
*Vāstupuruṣamaṇḍala*, 122
*vedi*, 99, 108n38, 124. *See also bedi*; ritual womb
Vedic mythology, 1, 9, 91

vegetal ritual enclosures, 214. *See also* Hindu temple
violence: agriculture and, 192; women and, 85n47
virginity, 211–12. *See also kanyā*; rice; women
Viṣṇu, 95–96, 121, 127–28, 130, 148, 152–53, 158, 170, 208
Volwahsen, A., 22n22, 22nn28–29, 23n30, 137n37, 151, 181n59, 219n61

wealth: annual oscillation of, 231–32; change in, 46, 231, 241; concealment of, 28, 40–41, 43, 45, 230–32; display of, 230; rice and, 185, 230. *See also ijjat*; status
Witzel, M., 178n2, 181n56, 181n58, 189
womb. *See* Agni; *agnicayana*; *bedi*; *Garbhagṛha*; Hindu temple; ritual; *samidhā*; temples
women: adolescent, 62–63; ambivalence of, 9, 78; autonomy of, 76; blaming of, 67–68, 70, 79; as *boksis*, 67; as *buhāri*, 36, 64–68, 70, 75; control of, 67, 71, 75, 77, 237; *dai halnu* and, 233; danger and, 68–69, 71, 233; as daughters, 54, 69; fidelity of, 69; as guests in own houses, 67; hybridity of, 54, 239; imagined domination of, 72–77; introduction to, 53–55; as *kanyā*, 56–58, 60–66; as *kanyādān*, 58–61, 65–66, 72; life stages of, 49, 55–67, 83n10; male perspective of, 21, 54, 62–71, 212, 233, 239; as married *kanyā*, 63–64; menstruation of, 62–65, 68, 212–13; men v., 55, 68, 74, 242; multiple perspectives of, 67–71, 103, 239; perceptual resolution of, 70; permeability of, 62, 152, 233; rebirth of, 74; *śakti* of, 62–64, 68–69, 71–72, 75; seclusion of, 62–63; sexuality of, 212; as sisters, 9, 54, 57, 63, 75, 83n14, 147; unmarried, 62–63; various social perspectives on, 54, 67–71; violence against, 85n47; virginity of, 63–65, 65; as wives, 9, 54, 66, 68–69, 75–76
work: attitude towards, 223; auspiciousness of, 222; exchange. *See nivek*; *parma*; sexuality and, 151–56, 193

*yajña*, 142–44, 146
*yantra*, 165
*yogi*, 105
*yoni*, 99, 124, 155, 159, 163–64. *See also* ritual womb; women

## About the Author

**Gil Daryn** earned his Ph.D. in Anthropology from the University of Cambridge, UK, and is currently a British Academy postdoctoral Fellow at the School of Oriental and African Studies, University of London.